WITHDRAWN
FROM THE VPI & SU
LIBRARY COLLECTION

IUTAM SYMPOSIUM ON ADVANCES IN MATHEMATICAL MODELLING
OF ATMOSPHERE AND OCEAN DYNAMICS

FLUID MECHANICS AND ITS APPLICATIONS
Volume 61

Series Editor: R. MOREAU
MADYLAM
Ecole Nationale Supérieure d'Hydraulique de Grenoble
Boîte Postale 95
38402 Saint Martin d'Hères Cedex, France

Aims and Scope of the Series

The purpose of this series is to focus on subjects in which fluid mechanics plays a fundamental role.

As well as the more traditional applications of aeronautics, hydraulics, heat and mass transfer etc., books will be published dealing with topics which are currently in a state of rapid development, such as turbulence, suspensions and multiphase fluids, super and hypersonic flows and numerical modelling techniques.

It is a widely held view that it is the interdisciplinary subjects that will receive intense scientific attention, bringing them to the forefront of technological advancement. Fluids have the ability to transport matter and its properties as well as transmit force, therefore fluid mechanics is a subject that is particulary open to cross fertilisation with other sciences and disciplines of engineering. The subject of fluid mechanics will be highly relevant in domains such as chemical, metallurgical, biological and ecological engineering. This series is particularly open to such new multidisciplinary domains.

The median level of presentation is the first year graduate student. Some texts are monographs defining the current state of a field; others are accessible to final year undergraduates; but essentially the emphasis is on readability and clarity.

For a list of related mechanics titles, see final pages.

IUTAM Symposium on Advances in Mathematical Modelling of Atmosphere and Ocean Dynamics

Proceedings of the IUTAM Symposium
held in Limerick, Ireland,
2–7 July 2000

Edited by

P.F. HODNETT

Department of Mathematics and Statistics,
University of Limerick,
Limerick, Ireland

KLUWER ACADEMIC PUBLISHERS
DORDRECHT / BOSTON / LONDON

A C.I.P. Catalogue record for this book is available from the Library of Congress.

ISBN 0-7923-7075-9

Published by Kluwer Academic Publishers,
P.O. Box 17, 3300 AA Dordrecht, The Netherlands.

Sold and distributed in North, Central and South America
by Kluwer Academic Publishers,
101 Philip Drive, Norwell, MA 02061, U.S.A.

In all other countries, sold and distributed
by Kluwer Academic Publishers,
P.O. Box 322, 3300 AH Dordrecht, The Netherlands.

Printed on acid-free paper

All Rights Reserved
© 2001 Kluwer Academic Publishers
No part of the material protected by this copyright notice may be reproduced
or utilized in any form or by any means, electronic or mechanical,
including photocopying, recording or by any information storage and
retrieval system, without written permission from the copyright owner.

Printed in the Netherlands.

CONTENTS

Preface ix

Keynote Lectures

D. Durran: Open boundary conditions: fact and fiction 1

P. Gent: Paramaterising Eddies in Ocean Climate Models 19

P. Lynch and R. McGrath: Boundary Filter Initialization of the HIRLAM Model ... 31

M. McIntyre: Balance, potential-vorticity inversion, Lighthill radiation and the slow quasi-manifold. 45

J. Pedlosky: Kelvin's theorem and the oceanic circulation in the presence of islands and broken ridges 69

Contributed Papers

F. Baer, J.J. Tribbia and M. Taylor: Global and regional atmospheric modelling using spectral elements 81

J.M. Baey and X. Carton: Piecewise-constant vortices in a two-layer shallow-water Flow 87

J.R. Bates and V.A. Alexeev: A dynamical stabilizer in the climate system: a mechanism suggested by a simple model and supported by GCM experiments and an observational data study 93

E.S. Benilov: Waves on the beta-plane over topography 99

M. Benjelloun, X. Carton: Asymptotic models and application to vortex dynamics 105

Stephen D. Burk, Tracy Haack and Richard M. Hodur: Orographically forced variability in the coastal marine atmospheric boundary layer 111

Antonietta Capotondi and Mike Alexander: The influence of thermocline Topography on the oceanic response to fluctuating winds: a case study in the tropical North Pacific 119

P.F. Choboter and G.E. Swaters: Modelling the dynamics of abyssal equator-crossing currents 125

Peter C. Chu: Toward accurate coastal ocean prediction 131

P.C. Chu, Shihua Lu and Chenwu Fan: An air-ocean coupled nowcast/forecast system for the East Asian marginal seas 137

Klara Finkele and Peter Lynch: Development and Simulation of Atlantic storms during FASTEX 143

A. Gluhovsky and C. Tong: Low-order models of atmospheric dynamics with physically sound behaviour 147

Roger Grimshaw and Georg Gottwald: Models for instability in geophysical flows .. 153

P.F. Hodnett and R. McNamara: Baroclinic structure of a modified Stommel – Arons model of the abyssal ocean circulation 161

Rui Xin Huang: The available potential energy in a compressible ocean 167

C. W. Hughes: The role of bottom pressure torques in the ocean circulation 173

N. Keeley and A.A. White: Quasi-geostrophic potential vorticity for a generalised vertical coordinate formulation and applications 179

P. Lionello and J. Pedlosky: On the effect of a surface density front on the interior structure of the ventilated thermocline 183

Mankin Mak: Non-hydrostatic barotropic instability 191

John L. McGregor and Martin R. Dix: The CSIRO conformal-cubic atmospheric GCM 197

N. Nawri: On the origin of helical vortex flow 203

M.K. Reszka and G.E. Swaters: Baroclinic instability of bottom-dwelling currents .. 209

G.M. Reznik and G.G. Sutyrin: General properties of baroclinic modons over topography 215

Rein Rööm and Aarne Männik: Acoustic filtration in pressure-coordinate models. Basic concepts and applications in non-hydrostatic modelling 221

I.A. Sazonov, S.D. Danilov, Yu.L. Chernousko and V.G. Kochina: Oscillatory regimes of forced zonal flow 227

A. Schiller: Improving climate simulations in the tropical oceans 233

D.M. Sonechkin: Synchroneity of the low-frequency planetary wave dynamics and its use to create a model for the numerical monthly weather forecasting 239

G.G. Sutyrin, I. Ginis and S.A. Frolov: Physical mechanisms of nonlinear equilibration of a baroclinically unstable jet over topographic slope 245

Gordon E. Swaters: Evolution of near-singular jet modes 251

Remi Tailleux and James C. McWilliams: Linear Resonance, WKB breakdown, and the coupling of Rossby waves over slowly-varying topography 259

John Thuburn and Thomas W.N. Haine: Nonoscillatory advection schemes with well-behaved adjoints 265

Bruce Turkington: A statistical equilibrium model of zonal shears and embedded vortices in a Jovian atmosphere 271

J. Vanneste: The impact of small-scale topography on large-scale ocean dynamics ... 279

S. Yoden and M. Taguchi: Numerical experiments on intraseasonal and interannual variations of the troposphere-stratosphere coupled system 285

List of Participants 291

PREFACE

The goals of the Symposium were to highlight advances in modelling of atmosphere and ocean dynamics, to provide a forum where atmosphere and ocean scientists could present their latest research results and learn of progress and promising ideas in these allied disciplines; to facilitate interaction between theory and applications in atmosphere/ocean dynamics. These goals were seen to be especially important in view of current efforts to model climate requiring models which include interaction between atmosphere, ocean and land influences.

Participants were delighted with the diversity of the scientific programme; the opportunity to meet fellow scientists from the other discipline (either atmosphere or ocean) with whom they do not normally interact through their own discipline; the opportunity to meet scientists from many countries other than their own; the opportunity to hear significant presentations (50 minutes) from the keynote speakers on a range of relevant topics. Certainly the goal of creating a forum for exchange between atmosphere and ocean scientists who need to input to create realistic models for climate prediction was achieved by the Symposium and this goal will hopefully be further advanced by the publication of these Proceedings.

The symposium programme contained 5 keynote lectures; 54 are oral presentations; 5 poster presentations. These proceedings consist of 5 keynote lectures listed in alphabetical order by name of first author and 34 contributed papers ordered in the same way. There was a total of 73 participants (listed later) in the Symposium with the following representations by country: China 1, Japan 3, U.S.A. 26, Canada 6, Switzerland 1, Australia 4, Denmark 3, France 3, Italy 2, India 1, U.K. 9, Ireland 10, Russia 2, Estonia 2.

The Scientific Committee for the Symposium was appointed by the Bureau of IUTAM and contained the following members: F. Baer (U.S.A.), J.R. Bates (Denmark), P. Gent (U.S.A.), R. Grimshaw (Australia), P.F. Hodnett (Ireland, Chairman), A. Hollingsworth (U.K.), Y. Kimura (Japan), K. Moffatt (U.K., ex officio), R. Salmon (U.S.A.). This committee selected the participants to be invited and the papers to be presented at the Symposium.

The success of the Symposium would not have been possible without the excellent work of the Local Organising Committee which contained the following members: E. Benilov, E. Gath, D. Hodnett, P.F. Hodnett (Chairman) – all Limerick, M. Hayes (Dublin), P. Lynch (Vice-chairman, Dublin), P. O'Donoghue (Galway).

The Local Organising Committee wishes to thank the following organisations for their sponsorship and financial support of the Symposium.

Aer Lingus
College of Informatics and Electronics, University of Limerick
Convention Bureau of Ireland
Department of Mathematics and Statistics, University of Limerick
Enterprise Ireland
International Union of Theoretical and Applied Mechanics
International Union of Geodesy and Geophysics
Kluwer Academic Publishers
Marine Institute (Ireland)
Met Eireann
Office of Naval Research (London)
President, University of Limerick
Royal Irish Academy

The editor wishes to thank the authors of papers for their contributions and for their cooperation in responding to reviewers' comments; the anonymous reviewers who gave generously of their time and whose valuable work has helped improve the quality of the material in these Proceedings.

Limerick, January 2001   Frank Hodnett

OPEN BOUNDARY CONDITIONS: FACT AND FICTION

D.R. DURRAN
University of Washington
Department of Atmospheric Sciences
Box 351640
Seattle, WA 98195, USA

1. Introduction

Many atmospheric and oceanic phenomena occur in a localized region. When numerically simulating such phenomena it is not practical to include all of the surrounding fluid in the numerical domain. As a case in point, one would not simulate an isolated thunderstorm with a global atmospheric model just to avoid possible problems at the lateral boundaries of a limited domain. Moreover, in a fluid such as the atmosphere there is no distinct upper boundary, and most numerical representations of the atmosphere's vertical structure terminate at some arbitrary level.

When the computational domain is terminated at an arbitrary location within a larger body of fluid, the conditions imposed at the edge of the domain are intended to mimic the presence of the surrounding fluid. These boundary conditions should allow outward-traveling disturbances to pass through the boundary without generating spurious reflections that propagate back toward the interior. Boundary conditions designed to minimize spurious backward reflection are known as nonreflecting, open, wave-permeable, or radiation boundary conditions. The terminology "radiation boundary condition" is due to Sommerfeld, who defined it as the condition that "the sources must be sources, not sinks, of energy. The energy which is radiated from the sources must scatter to infinity; no energy may be radiated from infinity into... the field."

As formulated by Sommerfeld, the radiation condition applies at infinity; however, in all practical computations a boundary condition must be imposed at some finite distance from the energy source, and this creates two problems. The first problem is that the radiation condition itself may not properly describe the physical behavior occurring at an arbitrarily designated location within the fluid when that location is only a finite distance from the energy source. The second problem is that it is often very difficult to express the radiation condition mathematically at a boundary that is only a finite distance from the energy source. It is so difficult to translate Sommerfeld's physical description into a mathematical formula, that in most practical problems it is not known how to express the radiation condition as a solvable algebraic or differential equation involving the prognostic variables in the neighborhood of the boundary. Instead the exact radiation condition must be replaced by an approximation, and this approximation introduces an error in the mathematical model of the physical system that is distinct from the errors subsequently incurred when constructing a discrete approximation to the mathematical model.

There are several factors that prevent numerical models of the atmosphere and ocean from actually converging toward the exact solution as the spatial and temporal resolution is repeatedly refined. The foremost impediment to convergence is surely the enormous range of scales that must be included in geophysical models of high-Reynolds-number flow (i.e., in a hypothetical model without any parameterization of subgrid-scale eddies). The numerical solution cannot be expected to begin to converge until the spatial resolution becomes sufficiently fine that the true Reynolds number is an order of magnitude smaller than the "numerical Reynolds" number computed using the effective numerical diffusivity in place of true molecular viscosity. Nevertheless, it should be emphasized that errors in the specification of open boundary conditions also pose a limitation on the ability to replicate the true solution in an otherwise highly refined model. Due to the approximations that are generally required to mathematically formulate open boundary conditions for the nondiscretized continuous equations, the numerical solutions to those equations will not converge to the correct solution to the underlying physical problem as the spatial and temporal mesh is refined, but rather to the correct solution to the approximate mathematical problem determined by the approximate open boundary conditions.

2. Is a radiation boundary condition appropriate?

The unmodified radiation condition is obviously not appropriate in situations where information must be transmitted inward through the boundary, as might be the case when the evolution of the large-scale structure of the solution on the boundary is determined by observations or by data from a coarse-resolution simulation conducted on a larger domain. Nevertheless, even if one wishes to force the boundary tendencies toward values determined by an external dataset, it is still necessary to allow waves originating within a limited-area domain to pass outward through the boundary without spurious reflection. In the remainder of this paper we will therefore focus on the basic problem of developing wave-permeable boundary conditions without making specific provision for the inclusion of externally specified boundary data.

Consider a problem in which the surrounding fluid is initially quiescent and all the initial disturbances are contained within the computational domain. One might suppose that the radiation condition, applied at the edge of the computational domain, would constitute a correct open boundary condition for such a problem—but in general it does not. Two nonlinear waves that propagate past an artificial boundary within a fluid may interact outside the artificial boundary and generate an inward-propagating disturbance that should reenter the domain. This point was emphasized by Hedstrom (1979), who provided a simple example from compressible gas dynamics demonstrating that a shock overtaking a contact discontinuity must generate an echo that propagates back toward the wave generator.

Numerical simulations of nonlinear vertically-propagating mountain waves provide another example where the documented sensitivity of the numerical solution to the loca-

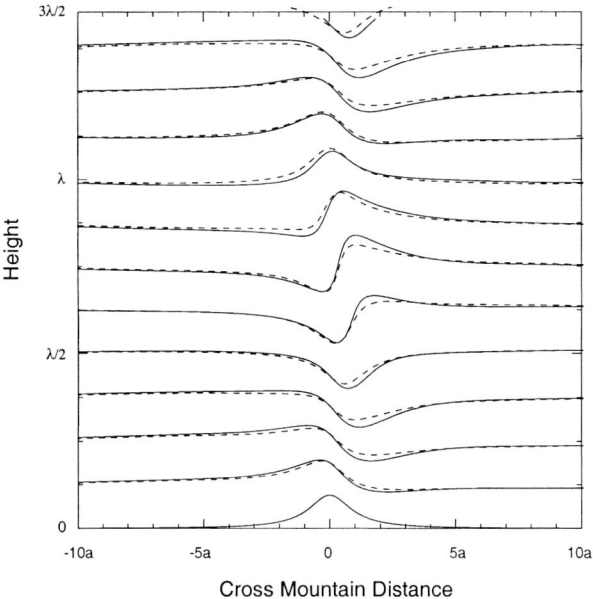

Figure 1. Streamlines in flow over a Witch of Agnesi mountain at nondimensional time $\tilde{t} = 44$ with the upper boundary at $z = 1.5\lambda$ (solid) or 2λ (dashed).

tion of the upper boundary (Durran and Klemp 1983, Klemp and Durran 1983, Scinocca and Peltier 1989, Epifanio and Durran 2000) may be less a consequence of inadequately approximating the radiation condition at the upper boundary than a failure to adequately capture important feedbacks on the low-level waves produced by downward propagating signals generated through wave-wave interactions occurring outside the numerical domain. A representative illustration of this sensitivity is shown in Fig. 1, which plots the streamlines in Boussinesq flow over a two-dimensional Witch of Agnesi mountain, $h(x) = h_0 a^2/(x^2 + a^2)$, from two numerical solutions computed using the model described in Epifanio and Durran (2000). The upstream stability and wind speed were constant with $N = 0.01$ s^{-1}, $U = 10$ ms^{-1}; the terrain parameters were $a = 10$ km, $h_0 = 600$ m. Thus $Nh_0/U = 0.6$, so the flow is moderately nonlinear, and $Na/U = 10$, so the flow is almost hydrostatic. Coriolis effects are ignored. In these simulations $\Delta x = 0.15a$ and $\Delta z = 0.4\lambda_z$, where $\lambda_z = 2\pi U/N$ is the vertical wavelength of a steady hydrostatic mountain wave. The lateral boundaries are placed extremely far upstream and downstream, at $x = -500a$ and $x = 700a$. The waves are initialized by gradually accelerating the flow from rest over $4\tilde{t}$, where $\tilde{t} = Ut/a$.

The solutions shown in Fig. 1 differ solely with respect to the location of the upper boundary. The solution shown by the solid lines was computed with an approximate radiation condition at $z = 1.5\lambda$; that shown by the dashed lines was computed with the upper boundary condition imposed at $z = 2\lambda$. One might suppose that this sensitivity to the location of the upper boundary is simply due to an inadequate approximation to the radiation condition, which was implemented using the local-in-time-and-space technique

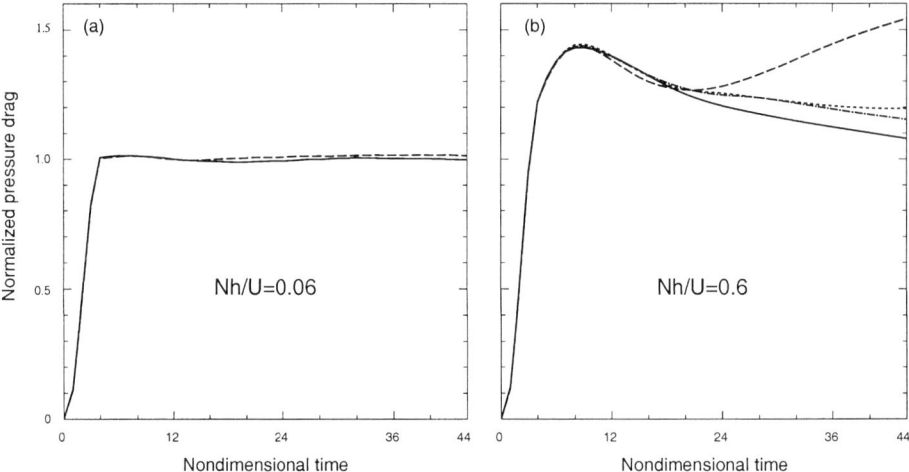

Figure 2. Cross mountain pressure drag, normalized by the drag associated with the corresponding linear solution for (a) $Nh_0/U = 0.06$ and (b) $Nh_0/U = 0.6$ when the upper boundary is at $z = 1.5\lambda$ (long dashed) or 2λ (solid). Also shown in (b) is the pressure-drag for cases with the upper boundary at 3λ (dot-dashed) and 3.5λ (short-dashed).

given in Durran (1999, p. 428), with $k_1 = 2\pi/a$, $k_2 = k_1/4$ and $k_3 = k_1/8$. However, a similar sensitivity is also obtained if the radiation condition is imposed by adding a very thick smoothly-varying sponge layer to the top of the model (Durran and Klemp 1983, Fig. 6). Moreover, the sensitivity to the location of the upper boundary is drastically reduced if the waves are made more linear by reducing h_0 by a factor of ten, as may be seen in the plots of the cross-mountain pressure drag versus time shown in Fig. 2 for the two cases: $Nh_0/U = 0.06$ and $Nh_0/U = 0.6$. The values plotted in Fig. 2 have been normalized by dividing the actual drag by the drag associated with the linear solution for flow over the same obstacle ($\pi \rho_0 NU h_0^2/4$). When $Nh_0/U = 0.06$, the solution generated by the nonlinear model is almost identical to the linear solution and there is almost no sensitivity to the height at which the radiation upper boundary condition is imposed.

When $Nh_0/U = 0.6$, finite-amplitude effects are no longer negligible, but the waves remain laminar and the subgrid-scale turbulence parameterization in the numerical model remains inactive. A semi-analytic steady-state solution for the nonlinear problem can be computed from Long's equation (Long 1953, Lilly and Klemp 1979). Since the waves are well-resolved, one might expect that the numerical solution could be made to converge to Long's solution by repeatedly refining the space-time mesh; however, difficulties with the boundary conditions prevent the numerical model from obtaining the correct result. Nevertheless, as suggested by those curves for the evolution of the drag in deeper domains plotted in Fig. 2b, increasing the depth of the domain decreases the sensitivity of the solution to the location of the upper boundary and delays the time at which the various solutions begin to diverge. At $\tilde{t} = 44$, the drags for the solutions for the two deepest domains (3λ and 3.5λ) bracket the correct value of 1.17 obtained from semi-analytic solutions to Long's equation.

Since similar sensitivities to the location of the upper boundary occur using very different numerical approximations to the radiation boundary condition (a local-in-space-and-time formulation and a deep "gentle" sponge layer), and since this sensitivity disappears in the small-amplitude limit, one can reasonably hypothesize that the basic problem arises from the use of the radiation condition itself. Nonsteady, nonlinear mountain waves appear to be generating downward propagating waves through wave-wave interactions. The truncation of the numerical domain at any finite height inevitably eliminates or distorts some of the internally generated downward propagating transients that should be present in the correct, nonsteady solution.

Errors resulting from a failure to incorporate inward-propagating signals generated by real physical processes occurring outside the boundaries of the computational domain cannot be avoided without enlarging the domain. Nevertheless, nonreflecting boundary conditions often become reasonable approximations to the true physical boundary condition as the size of the computational domain increases, provided that the local energy density of a disturbance arriving at the boundary is reduced as the result of wave dispersion or absorption in the large domain. When the disturbances arriving at the boundary are sufficiently weak, the governing equations in the region near the boundary can be approximated by their linearized equivalents, and it can be relatively easy to ensure that the radiation condition correctly describes the boundary conditions for the linearized system.

In summary, if the surrounding fluid is initially quiescent, the radiation boundary condition is usually the best statement of a physical condition that one might attempt to impose at an artificial boundary, but it is often only an approximation to the actual conditions that would hold at that boundary. As a consequence, even if a perfect implementation of the radiation boundary condition is present in some numerical model, one should not expect the model-generated solution to an arbitrary nonlinear problem to be approximately independent of the domain size unless the domain is sufficiently large.

3. Are One-Dimensional Test Problems Misleading?

One must learn to walk before learning to run, and in that spirit a great deal of effort has been devoted to the study of radiation boundary conditions for problems with two independent variables: time t and a single spatial coordinate x. While it is certainly true that open-boundary formulations that fail to work in x–t problems cannot be expected to perform well in higher-dimensional applications, it is not true that the most successful techniques for x–t problems can be trivially generalized to higher dimensions. Correct radiation boundary conditions for linear hyperbolic partial differential equations at each spatial boundary in an x–t domain can be obtained by setting the incoming characteristic variables to zero. Numerical details must still be considered to ensure that the equations for the out-going characteristic variables are treated in a stable, accurate manner, but prior to discretizing the governing equations, the radiation boundary conditions can be expressed mathematically without approximation.

On the other hand, in most time-dependent problems involving two or more spatial dimensions concepts related to characteristic variables are considerably less useful for the formulation of open boundary conditions. Two misconceptions that can easily develop after casual study of simple hyperbolic partial differential equations in an x–t problem are (1) that waves typically propagate along characteristics and (2) that setting the amplitude of an incoming wave to zero along the boundary is equivalent to specifying zero amplitude in the incoming characteristic variable in higher dimensional problems.

3.1. RAY PATHS AND CHARACTERISTIC CURVES

The ray paths along which geophysically significant waves such as gravity or Rossby waves propagate have essentially no relation to characteristic curves or their higher-dimensional generalizations (such as the Monge cones). Even in simple x–t problems, waves need not propagate along characteristics. As a first example, consider the linearized shallow-water equations for one-dimensional flow on an f-plane. Assuming the basic state is at rest, the linearized system reduces to

$$\frac{\partial u}{\partial t} + g\frac{\partial h}{\partial x} - f_0 v = 0, \tag{1}$$

$$\frac{\partial v}{\partial t} + f_0 u = 0, \tag{2}$$

$$\frac{\partial h}{\partial t} + H\frac{\partial u}{\partial x} = 0, \tag{3}$$

where u and v are the perturbation horizontal velocity components, h is the free-surface displacement about its mean depth H, and f_0 is the (constant) Coriolis parameter. Diagonalize the coefficient matrix of the spatial derivative by defining a vector of new unknown functions

$$\mathbf{v} = \begin{pmatrix} u - gh/c \\ v \\ u + gh/c \end{pmatrix},$$

with respect to which (1)–(3) transform to

$$\frac{\partial \mathbf{v}}{\partial t} + \begin{pmatrix} -c & 0 & 0 \\ 0 & 0 & 0 \\ 0 & 0 & c \end{pmatrix} \frac{\partial \mathbf{v}}{\partial x} + \begin{pmatrix} 0 & -f_0 & 0 \\ f_0/2 & 0 & f_0/2 \\ 0 & -f_0 & 0 \end{pmatrix} \mathbf{v} = \mathbf{0}.$$

The characteristics for this system are the curves satisfying $dx/dt = \pm c$ and $dx/dt = 0$.
Wave solutions to (1)–(3) have the form

$$(u', v', h') = \Re\left\{(u_0, v_0, h_0)e^{i(kx-\omega t)}\right\}, \tag{4}$$

provided that the frequency ω and wave number k satisfy the dispersion relation

$$\omega^2 = c^2 k^2 + f_0^2, \tag{5}$$

as may be demonstrated by substituting (4) into (1)–(3). Lines of constant phase, such as the locations of the troughs and crests, propagate at the *phase speed* ω/k, which from (5) is

$$\frac{\omega}{k} = \pm c\left(1 + \frac{f_0^2}{c^2 k^2}\right)^{1/2}$$

A compact group of waves travels at the group velocity $\partial\omega/\partial k$, which can also be computed from (5):
$$\frac{\partial\omega}{\partial k} = \pm c\left(1 + \frac{f_0^2}{c^2 k^2}\right)^{-1/2}$$

In the limit $|k| \gg f_0/c$, the phase speed and group velocity both approach the slope of a characteristic along which $|dx/dt| = c$. Nevertheless, for any finite value of k,
$$\left|\frac{\partial\omega}{\partial k}\right| < c < \left|\frac{\omega}{k}\right|,$$
and neither the lines of constant phase nor the wave groups follow trajectories that coincide with the characteristic curves. The magnitude of the group velocity, which is the rate at which energy propagates in a wave, is nevertheless bounded by c. Thus, if waves are generated by a compact source at some initial time, the disturbance will lie within the portion of the x–t plane bounded by characteristic curves emanating from each edge of the source region, i.e., the disturbance will lie within the conventionally defined "domain of dependence" for the hyperbolic system (Durran 1999, p. 45).

The loose connection between wave propagation and the characteristics in the preceding example can disappear altogether if the Coriolis parameter is a function of the spatial coordinate. Then a second type of wave, the Rossby wave, may appear as an additional solution. If f increases linearly in proportion to y, Rossby-wave solutions may exist with phase speeds in the negative-x direction. Neither the phase speeds nor the group velocities of these waves have any relation to the characteristic curves. It is not surprising that Rossby waves do not propagate along the characteristics, because the terms involving the undifferentiated functions of u and v play no role in the determination of the characteristics of (3), yet those same terms are essential for the maintenance of the Rossby waves.

The Euler equations governing inviscid isentropic motion in a density stratified fluid provide another example of a hyperbolic system that supports two types of waves, one of which propagates along ray paths that are completely unrelated to any characteristic curve. The essential properties of this system can be most simply examined in a two-dimensional context. Let x and z be the horizontal and vertical coordinates, and decompose the thermodynamic fields into a vertically varying basic state and a perturbation such that
$$\begin{aligned}\pi(x,z,t) &= \overline{\pi}(z) + \pi'(x,z,t),\\ \theta(x,z,t) &= \overline{\theta}(z) + \theta'(x,z,t),\\ c_p\overline{\theta}\frac{d\overline{\pi}}{dz} &= -g.\end{aligned} \qquad (6)$$

Here π is nondimensional Exner function defined as $(p/p_0)^{R/c_p}$ and $\theta = T/\pi$ is the potential temperature. The velocity components are decomposed as
$$u(x,z,t) = U + u'(x,z,t), \qquad w(x,z,t) = w'(x,z,t). \qquad (7)$$

The basic-state vertical velocity is zero to ensure that the basic state is a steady solution to the nonlinear equations. The linearized governing equations become
$$\left(\frac{\partial}{\partial t} + U\frac{\partial}{\partial x}\right)u' + c_p\overline{\theta}\frac{\partial\pi'}{\partial x} = 0, \qquad (8)$$

$$\left(\frac{\partial}{\partial t} + U\frac{\partial}{\partial x}\right)w' + c_p\bar{\theta}\frac{\partial \pi'}{\partial z} = g\frac{\theta'}{\bar{\theta}}, \qquad (9)$$

$$\left(\frac{\partial}{\partial t} + U\frac{\partial}{\partial x}\right)\theta' + \frac{\bar{\theta}}{g}N^2 w' = 0, \qquad (10)$$

$$\left(\frac{\partial}{\partial t} + U\frac{\partial}{\partial x}\right)\pi' + w'\frac{\partial \bar{\pi}}{\partial z} + \frac{R\bar{\pi}}{c_v}\left(\frac{\partial u'}{\partial x} + \frac{\partial w'}{\partial z}\right) = 0, \qquad (11)$$

where

$$N^2 = \frac{g}{\bar{\theta}}\frac{d\bar{\theta}}{dz}$$

is the square of the Brunt–Väisälä frequency.

Suppose that the reference state is isothermal. Then N^2 and the speed of sound $c_s = (c_p RT/c_v)^{1/2}$ are constant, and the preceding system can be simplified by removing the influence of the decrease in the mean density with height via the transformation

$$\tilde{u} = \left(\frac{\bar{\rho}}{\rho_0}\right)^{1/2} u', \qquad \tilde{\pi} = \left(\frac{\bar{\rho}}{\rho_0}\right)^{1/2}\frac{c_p\bar{\theta}}{c_s}\pi', \qquad (12)$$

$$\tilde{w} = \left(\frac{\bar{\rho}}{\rho_0}\right)^{1/2} w', \qquad \tilde{\theta} = \left(\frac{\bar{\rho}}{\rho_0}\right)^{1/2}\frac{g}{N\bar{\theta}}\theta'. \qquad (13)$$

Note that $\tilde{\theta}$ represents a scaled buoyancy and $\tilde{\pi}$ a scaled pressure. Let

$$\mathbf{v} = \begin{pmatrix} \tilde{u} & \tilde{w} & \tilde{\theta} & \tilde{\pi} \end{pmatrix}^T;$$

then the transformed equations have the form

$$\frac{\partial \mathbf{v}}{\partial t} + \mathbf{A}_1 \frac{\partial \mathbf{v}}{\partial x} + \mathbf{A}_2 \frac{\partial \mathbf{v}}{\partial z} + \mathbf{B}\mathbf{v} = 0, \qquad (14)$$

in which

$$A_1 = \begin{pmatrix} U & 0 & 0 & c_s \\ 0 & U & 0 & 0 \\ 0 & 0 & U & 0 \\ c_s & 0 & 0 & U \end{pmatrix}, \qquad A_2 = \begin{pmatrix} 0 & 0 & 0 & 0 \\ 0 & 0 & 0 & c_s \\ 0 & 0 & 0 & 0 \\ 0 & c_s & 0 & 0 \end{pmatrix},$$

$$B = \begin{pmatrix} 0 & 0 & 0 & 0 \\ 0 & 0 & -N & -S \\ 0 & N & 0 & 0 \\ 0 & S & 0 & 0 \end{pmatrix}, \qquad S = c_s\left[\frac{1}{2\bar{\rho}}\frac{\partial \bar{\rho}}{\partial z} + \frac{1}{\bar{\theta}}\frac{\partial \bar{\theta}}{\partial z}\right].$$

The eigenvalues of $\mathbf{A}_1$ are U, U, $U+c_s$, and $U-c_s$; those of $\mathbf{A}_2$ are 0, 0, c_s, and $-c_s$. The eigenvalues involving c_s give the speed at which sound waves in an unstratified fluid propagate parallel to the x and z coordinate axes. As will be demonstrated below, sound waves in an isothermally stratified atmosphere actually propagate at slightly different speeds due to the influence of the zero-order term in (14). The remaining eigenvalues relate to the speed at which fluid parcels are advected horizontally and vertically by the mean flow. None of these eigenvalues have any relation to the speed of propagation of gravity (or buoyancy) waves, which are the second type of fundamental wave motion supported by (14).

When the basic state is isothermal, S is constant, and wave solutions to (14) exist in the form

$$(\tilde{u}, \tilde{w}, \tilde{\theta}, \tilde{\pi}) = \Re\left\{(u_0, w_0, \theta_0, \pi_0)e^{i(kx+mz-\omega t)}\right\}, \tag{15}$$

provided that ω, k, and m satisfy the dispersion relation

$$(\omega - Uk)^2 = \frac{c_s^2}{2}\left(k^2 + m^2 + \frac{N^2 + S^2}{c_s^2}\right)$$
$$\pm \frac{c_s^2}{2}\left[\left(k^2 + m^2 + \frac{N^2 + S^2}{c_s^2}\right)^2 - \frac{4N^2 k^2}{c_s^2}\right]^{1/2}, \tag{16}$$

which is obtained by substituting (15) into (14). The second term inside the square root is much smaller than the first term in most applications (Durran 1999, p. 351), so (16) can be separated into a pair of approximate dispersion relations for the sound waves and the gravity waves. The dispersion relation for the sound waves,

$$(\omega - Uk)^2 = c_s^2\left(k^2 + m^2\right) + S^2 + N^2, \tag{17}$$

is obtained by taking the positive root in (16). In a manner analogous to the effect of the Coriolis force on gravity waves in the shallow-water system, the terms involving the product of N or S with the zero-order derivatives of the unknown variables introduce a slight discrepancy between the phase speeds and group velocities of the actual sound waves and those that might be suggested by the eigenvalues of $\mathbf{A}_1$ and $\mathbf{A}_2$.

The dispersion relation for the gravity waves is obtained by taking the negative root in (16), which to a good approximation yields

$$(\omega - Uk)^2 = \frac{N^2 k^2}{k^2 + m^2 + (S^2 + N^2)/c_s^2}. \tag{18}$$

Neither the phase speeds nor the group velocities of these waves have any relation to the eigenvalues of $\mathbf{A}_1$ and $\mathbf{A}_2$. Unlike sound waves, *gravity waves do not even approximately propagate along the characteristics*. There is no relation between the characteristics and the paths of the gravity-waves because some of the physical processes essential for gravity-wave propagation are mathematically represented by undifferentiated functions of the unknown variables, and as such exert no influence on the shape of the characteristics.

3.2. SPECIFICATION OF THE INCOMING "CHARACTERISTIC" VARIABLE

As suggested previously, exact and practically useful mathematical formulae for the specification of nonreflecting boundary conditions are seldom available in multidimensional problems. The difficulties are readily apparent in simple two-dimensional shallow-water flow. The two-dimensional shallow-water equations, linearized about a basic state at rest with depth H, may be written

$$\begin{pmatrix} u \\ v \\ h \end{pmatrix}_t + \begin{pmatrix} 0 & 0 & g \\ 0 & 0 & 0 \\ H & 0 & 0 \end{pmatrix} \begin{pmatrix} u \\ v \\ h \end{pmatrix}_x + \begin{pmatrix} 0 & 0 & 0 \\ 0 & 0 & g \\ 0 & H & 0 \end{pmatrix} \begin{pmatrix} u \\ v \\ h \end{pmatrix}_y = 0,$$

where the notation follows that used previously and Coriolis forces have been neglected. Suppose that a solution is sought in the limited domain $0 \le x \le L$, $0 \le y \le L$. In contrast to the one-dimensional case, the two-dimensional system cannot be reduced to a set of three scalar equations because no transformation of variables will simultaneously diagonalize both coefficient matrices. Nevertheless, the coefficient matrix multiplying the x-derivative can be diagonalized through the same change of variables used in the one-dimensional problem. Defining $d \equiv u - gh/c$ and $e \equiv u + gh/c$, the two-dimensional system becomes

$$\begin{pmatrix} d \\ v \\ e \end{pmatrix}_t + \begin{pmatrix} -c & 0 & 0 \\ 0 & 0 & 0 \\ 0 & 0 & c \end{pmatrix} \begin{pmatrix} d \\ v \\ e \end{pmatrix}_x + \begin{pmatrix} 0 & -c & 0 \\ -c/2 & 0 & c/2 \\ 0 & c & 0 \end{pmatrix} \begin{pmatrix} d \\ v \\ e \end{pmatrix}_y = 0. \quad (19)$$

Since only one coefficient matrix is a diagonal matrix, (19) is not as simple as the corresponding relation for one-dimensional flow. Nevertheless, (19) is useful for determining the number of boundary conditions that should be specified at a x-boundary. The signal in v and e is propagating inward through the boundary at $x = 0$, and the signal in d is propagating inward through the boundary at $x = L$. In order to obtain a well-posed problem, one might therefore attempt to specify two conditions at $x = 0$ of the form

$$e(0, t) = \alpha_1 d(0, t) + f_1(t), \qquad v(0, t) = \alpha_2 d(0, t) + f_2(t),$$

and one condition at $x = L$ of the form

$$d(L, t) = \alpha_3 e(L, t) + \alpha_4 v(L, t) + f_3(t). \quad (20)$$

This approach follows the guideline that the number of conditions specified at each boundary should be equal to the number of eigenvalues associated with outward propagation through each boundary. Even so, not all choices of $\alpha_1, \ldots, \alpha_4$ yield a well-posed problem. Oliger and Sundström (1978) and Sundström and Elvius (1979) provide details about various allowable values for $\alpha_1, \ldots, \alpha_4$. One way to obtain a well-posed problem is by choosing $\alpha_1 = \alpha_2 = \alpha_3 = \alpha_4 = 0$, which specifies e and v at inflow and d at outflow.

Unlike the one-dimensional case, it is not clear what values of e, v, and d should be prescribed to prevent waves impinging on the boundary from partially reflecting into an inward-propagating mode. Suppose that the incoming variable d is set to zero at $x = L$, and suppose that a wave of of the form $(u, v, h) = (\hat{u}, \hat{v}, \hat{h}) \exp[i(kx + \ell y - \omega t)]$ is approaching the boundary at $x = L$ as shown in Fig. 3. The amplitudes in u and h are related by the linearized x-momentum equation, which for this case with no mean flow yields

$$-i\omega \hat{u} + ikg\hat{h} = 0,$$

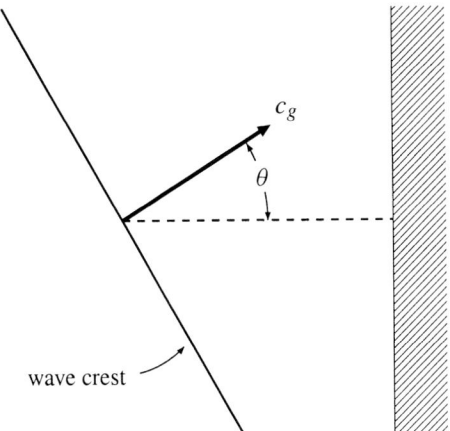

Figure 3. Wave crest approaching the "east" boundary at $x = L$.

or
$$\hat{u} = \frac{gk}{\omega}\hat{h}. \tag{21}$$

Using (21) and substituting for ω from the shallow-water dispersion relation for waves moving to the "east",
$$\omega = c(k^2 + \ell^2)^{1/2}, \tag{22}$$

the value of d at the right boundary becomes

$$d(L, y, t) = \left(\frac{gk}{\omega} - \frac{g}{c}\right)\hat{h}e^{i(kL+\ell y-\omega t)} = \frac{g}{c}\left(\frac{k}{(k^2+\ell^2)^{1/2}} - 1\right)\hat{h}e^{i(kL+\ell y-\omega t)}. \tag{23}$$

The condition $d(L, y, t) = 0$ cannot be satisfied without the simultaneous presence of a second reflected wave unless $\ell = 0$, or equivalently, unless the outward propagating wave is traveling at right angles to the boundary.

An expression for the amplitude of the spurious reflected wave generated by the $d(L, y, t) = 0$ boundary condition can be derived as follows. Let r be the amplitude of this reflected wave, and note that the condition $d(L, y, t) = 0$ cannot be satisfied at every point along $x = L$ unless the incident and reflected waves are linearly dependent functions of y and t. The frequency and the wave number parallel to the y-axis must therefore be the same in the incident and reflected waves, so according to the dispersion relation (22), the wave numbers parallel to the x-axis must have the same magnitude and opposite sign. The sum of the incident and reflected waves may therefore be expressed in the form
$$h(x, y, t) = \hat{h}e^{i(kx+\ell y-\omega t)} + r\hat{h}e^{i(-kx+\ell y-\omega t)}. \tag{24}$$

Here the first term represents the wave approaching the boundary, and the second term represents the reflected wave. Using (23) and (24) the condition $d(L, y, t) = 0$ becomes

$$\left(\frac{gk}{\omega} - \frac{g}{c}\right)\hat{h}e^{i((kL+\ell y-\omega t)} + r\left(\frac{-gk}{\omega} - \frac{g}{c}\right)\hat{h}e^{i((-kL+\ell y-\omega t)} = 0.$$

Solving for r yields

$$r = -\left(\frac{\omega - kc}{\omega + kc}\right) e^{2ikL}. \tag{25}$$

The strength of the reflected wave can be alternatively expressed as a function of the angle of incidence as follows. The lines of constant phase in Fig. 3 are parallel to the wave troughs and crests and satisfy $F(x, y) = 0$, where

$$F(x, y) = kx + \ell y - C;$$

C is a constant; and in the case shown in Fig. 3, k and ℓ are positive. The group velocity of the wave is both perpendicular to the lines of constant phase and parallel to the wave-number vector (k, ℓ), as may be confirmed from the relations

$$\left(\frac{\partial \omega}{\partial k}, \frac{\partial \omega}{\partial \ell}\right) = \frac{c^2}{\omega}(k, \ell) = \frac{c^2}{\omega}\nabla F.$$

As in Fig. 3, let θ be the angle by which the propagation of the wave deviates from the direction normal to the boundary, then $\tan\theta = \ell/k$. Using the dispersion relation (22) it follows that

$$k = \frac{\omega}{c}\cos\theta, \quad \ell = \frac{\omega}{c}\sin\theta. \tag{26}$$

Finally, substituting (26) into (25), one obtains

$$|r| = \left|\frac{\omega - kc}{\omega + kc}\right| = \left|\frac{1 - \cos\theta}{1 + \cos\theta}\right|,$$

again showing that this condition is perfectly nonreflecting only when the waves approach perpendicular to the boundary.

When d is set to zero at $x = L$ in two-dimensional shallow-water flow, at least one obtains an approximation to the correct radiation condition. If one attempts the same procedure with the continuous-stratified x–z problem governed by (14), the resulting boundary conditions will serve as an approximate radiation condition for sound waves but will not do anything to eliminate the spurious reflection of gravity waves impinging on the boundary. To be specific, the coefficient matrix of $\partial \mathbf{v}/\partial x$ in (14) will be diagonalized if $\mathbf{v}$ is transformed to $(\tilde{u} - \tilde{\pi}, \tilde{w}, \tilde{\theta}, \tilde{u} + \tilde{\pi})$. Assuming the flow is subsonic, the variable $\tilde{u} - \tilde{\pi}$ would be transmitted inward through an "east" boundary at $x = L$ at speed $U - c_s$. Will the relation $\tilde{u} - \tilde{\pi} = 0$ at $x = L$ therefore constitute a correct formulation of the radiation boundary condition?

As in the shallow-water case, consider a basic state with no mean flow for simplicity and suppose that a wave of the form $(\tilde{u}, \tilde{w}, \tilde{\theta}, \tilde{\pi}) = (\hat{u}, \hat{w}, \hat{\theta}, \hat{\pi}) \exp[i(kx + mz - \omega t)]$ is approaching the boundary at $x = L$. The x-momentum equation implies that

$$\hat{u} = \frac{c_s k}{\omega}\hat{\pi},$$

which, using the condition that $\tilde{u} - \tilde{\pi} = 0$ at $x = L$, may be expressed as

$$\left(\frac{c_s k}{\omega} - 1\right) \hat{\pi} e^{i(kL + mz - \omega t)} = 0. \tag{27}$$

Nontrivial solutions to (27) require $c_s = \omega/k$, or equivalently that *the trace speed[1] along the x-axis must be equal to the speed of sound*. If the outgoing wave is a sound wave propagating perpendicular to the boundary (so that $m = 0$), and if buoyancy effects are neglected ($N = S = 0$), then according to the sound wave dispersion relation (17), the wave will be transmitted without reflection. On the other hand, if the angle of incidence is not perpendicular to the boundary, or if buoyancy effects are not negligible, the condition $\tilde{u} - \tilde{\pi} = 0$ at $x = L$ is only an approximation.

By far the most serious problems associated with the strategy of setting the incoming variable $\tilde{u} - \tilde{\pi}$ to zero at $x = L$ involve the failure of this boundary condition to transmit gravity waves. The only gravity waves that will be transmitted with minimal reflection are those whose horizontal trace speeds are approximately equal to the speed of sound! These are certainly not the gravity waves of primary interest in most atmospheric and oceanic applications.

4. Are Dispersive Waves a Particular Challenge?

Previous researchers have often assumed that the Sommerfeld radiation condition at a lateral boundary may be expressed mathematically as

$$\frac{\partial \phi}{\partial t} + c^* \frac{\partial \phi}{\partial x} = 0, \qquad (28)$$

where ϕ is an arbitrary prognostic variable, c^* is some effective wave speed, and (28) is applied at an x boundary. In the case of one-dimensional subcritical linearized shallow-water flow, (28) is a correct statement of the radiation condition provided that at the "east" boundary, $c^* = U + c$, or at the "left" boundary, $c^* = U - c$.

Equation (28) is not, however, a correct formulation of the Sommerfeld radiation condition in more general physical systems. Suppose a wave of the form $\hat{\phi} \exp[i(kx + \ell y + mz - \omega t)]$ is approaching the boundary at $x = L$. If this wave is to be transmitted without reflection, it must satisfy (28) at $x = L$, in which case

$$(\omega - c^* k)\hat{\phi} e^{i(kL + \ell y + mz - \omega t)} = 0.$$

For a nontrivial solution, $\omega/k = c^*$; that is c^* *must be identical to the x-trace speed of the wave*. In general $\omega/k = c^*$ is a function of the vertical and/or horizontal wavenumbers and, in contrast to one-dimensional shallow-water flow, there is no single value of c^* that can transmit an arbitrary superposition of several waves through the boundary without spurious reflection. Note that *the difficulty in specifying c^* is not due to wave dispersion per se;* the same essential difficulty arises when formulating boundary conditions for nondispersive two-dimensional shallow-water waves.

5. Can c^* be effectively diagnosed from (28)?

One example, in which c^* is a function of wavelength, is gravity wave propagation in stratified flow. Pearson (1974) suggested that (28) may be used to radiate internal gravity

---

[1] The x trace speed, ω/k, is the apparent phase speed of the wave along the x-axis. Unless the wave is actually propagating parallel to the x-axis, the x trace speed exceeds the true phase speed $\omega(k^2 + m^2)^{-1/2}$.

waves in stratified flow by fixing c^* at the Doppler-shifted phase speed of the dominant vertical mode. As an alternative, Orlanski (1976) suggested calculating c^* at a point just inside the boundary from the relation

$$c^* = -\frac{\partial \phi/\partial t}{\partial \phi/\partial x}. \tag{29}$$

Variants of the Orlanski lateral boundary condition have become very widely used in limited-area geophysical modeling.

The Orlanski scheme appears to have the virtue of avoiding the specification of c^* as an arbitrary parameter, but in fact it seldom yields a meaningful value for c^*. In particular, the computed value of c^* often falls outside the range allowed by the CFL stability criteria for the numerical approximation to (28) and must be artificially reset to a stable value. For example, the value of c^* would need to be limited to the interval $0 \le c^* \le \Delta x/\Delta t$ to preserve the stability of an upstream approximation to (28) at the "east" boundary.

Some indication of the difficulties encountered in the calculation of c^* were documented by Durran, Yang, Slinn and Brown (1993), who investigated the accuracy with which (29) diagnosed the phase speed of numerically simulated shallow-water waves, for which the correct value of c^* is simply $\sqrt{gH}$. Durran et al. (1993) solved a a linearized shallow-water model governed by the following finite-difference equations

$$\frac{u_{i+1/2}^{n+1} - u_{i+1/2}^{n-1}}{2\Delta t} + U\left(\frac{u_{i+3/2}^n - u_{i-1/2}^n}{2\Delta x}\right) + g\left(\frac{h_{i+1}^{n-1} - h_i^{n-1}}{\Delta x}\right) = 0,$$

$$\frac{h_i^{n+1} - h_i^{n-1}}{2\Delta t} + U\left(\frac{h_{i+1}^n - h_{i-1}^n}{2\Delta x}\right) + H\left(\frac{u_{i+1/2}^{n+1} - u_{i-1/2}^{n+1}}{\Delta x}\right) = 0.$$

The velocity and height variables are spatially staggered such that the edge of the domain coincides with the outermost velocity points. At the initial time the velocity is zero and

$$\eta(x, 0) = g \sin^4(\pi x).$$

The solution was integrated forward 200 time steps on the domain $0 \le x \le 4$ km, with $\Delta x = 80$ m, $\Delta t = 0.48$ s, $c = 40$ ms^{-1}, $U = 10$ ms^{-1}. The boundary condition (28) was applied to each field at $x = 0$ and $x = 4$ km using upstream differencing to step between time levels $t - \Delta t$ and $t + \Delta t$. A reference solution was also computed on a large periodic domain $-8 \le x \le 12$ km. The large-domain solution may be considered free from boundary-induced errors, since the periodic domain is too large to allow any disturbance exiting the region $0 \le x \le 4$ km to reenter that region before the end of the integration.

The value of c^* at $x = 4$ km was calculated using using Miller and Thorpe's (1981) formula

$$c^* = -\frac{\Delta x}{\Delta t}\left(\frac{\phi_{b-1}^n - \phi_{b-1}^{n-1}}{\phi_{b-1}^{n-1} - \phi_{b-2}^{n-1}}\right), \tag{30}$$

where ϕ represents either u or η, b is the index of the gridpoint at the right boundary, and the final value of c^* is constrained to lie in the interval $0 < c^* < 0.95\Delta x/2\Delta t$. A similar

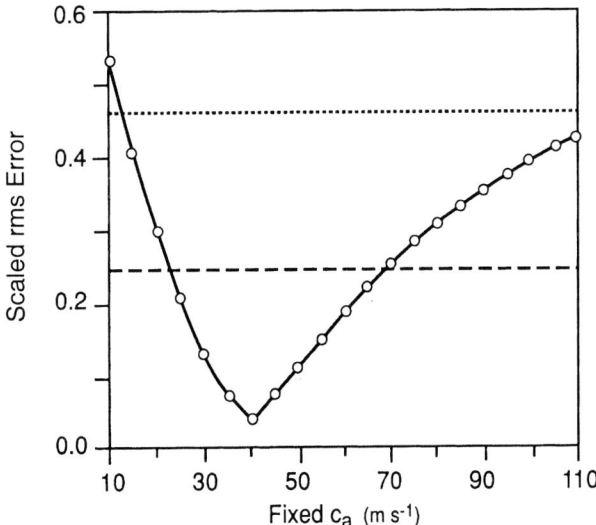

Figure 4. Cumulative rms boundary-induced error generated by the mis-specification of c^* as $U \pm c_a$, plotted as a function of c_a (solid curve). Also shown is the error for two variants of the Orlanski scheme which, being independent of c_a, appear as horizontal lines. The result using Orlanski's original formula for c^* is shown as the short-dashed line, the long-dashed line is obtained by computing c^* from (30). From Durran et al., (1993).

expression is evaluated at $x = 0$. This numerical approximation to (29) was found to give better results than the original formula proposed by Orlanski (1976).

Additional simulations were also conducted in which the gravity-wave phase speed was approximated by an arbitrary constant value c_a, and c^* was set to $U + c_a$ (or $U - c_a$) at the right (or left) boundary. Even when c_a was a relatively poor approximation to $\sqrt{gH}$, the constant-c_a approach produced better results than those obtained by continually estimating c^* from (30). This superiority is illustrated in Fig. 4, in which the scaled cumulative root-mean-square boundary-induced error is plotted as a function of c_a and compared with the error that develops when c^* is calculated using (30). The boundary-induced error is the difference between the limited-area and large-periodic-domain solutions for u and h. Details of the scaled rms error computation are given in Durran et al. (1993).

As evident in Fig. 4, the optimal choice of $c_a = \sqrt{gh} = 40$ ms^{-1} gives substantially less error than that obtained by calculating c^* at the boundary. It is certainly not surprising that fixing c_a at its theoretically correct value gives the best result. What is surprising is that fixed values of c_a are superior to those obtained through the evaluation of c^* via (30) over the wide range $25 \leq c_a \leq 65$ ms^{-1}. To understand why the diagnosis of c^* is inferior over such a wide range of c_a, it is instructive to examine the actual c^* values calculated at the boundaries. The values of c^* computed from (30) at the right-hand-boundary are plotted as a function of the time step in Fig. 5. At the initial time, a wave trough is present at the boundary; every 20 time steps thereafter another trough or crest arrives at the boundary. During the interval between the passage of the initial trough and the first crest (the first 20 steps), the c^* calculation yields a somewhat reasonable approximation to the correct value of $U + \sqrt{gH} = 50$ ms^{-1}. During the interval between the passage of

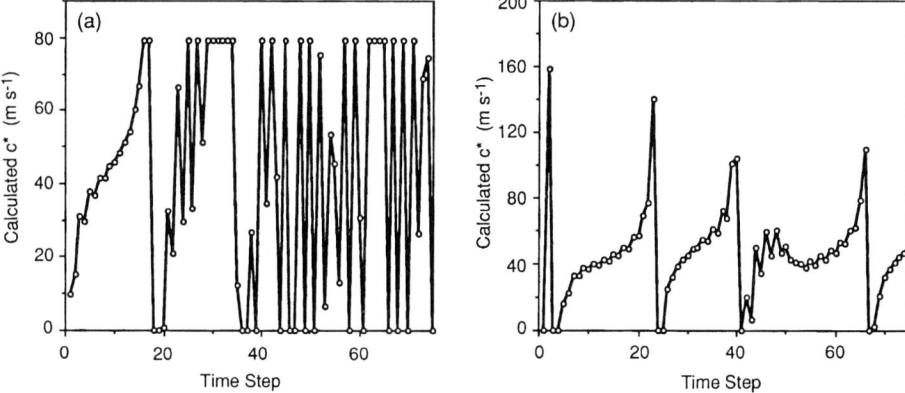

Figure 5. Plot of c^* as a function of time step as calculated from (30) using the solution on: (a) the limited domain, and (b) the wide periodic domain. After Durran et al., (1993)

the first crest and the second trough (the second 20 steps), the computed c^* is quite noisy. After the first 40 steps, c^* is drastically over- or under-predicted and set to its limiting values of 0 or 80 ms^{-1} in an almost $2\Delta t$ oscillation.

The most serious problem with the c^* calculation is not the fidelity with which the finite-difference formula (30) approximates the continuous expression (29), but the susceptibility of the calculation to a positive feedback between the generation of error at the lateral boundary and further degradations in the computed values of c^*. The effect of the cumulative boundary-induced error on the calculation of c^* may be appreciated by comparing Fig. 5a with Fig. 5b. Fig. 5b shows the right-boundary c^* obtained when the large-periodic-domain solution (instead of the limited-domain solution) is used in the evaluation of (30). When c^* is diagnosed from the large-periodic-domain solution, the estimation of c^* beyond the 20th time step is much improved and there is no systematic degradation in accuracy as the number of time steps increases. The largest errors are associated with the passage of troughs or crests in which case the denominator in (29) becomes small and the c^* calculation becomes very sensitive to numerical error.

The positive feedback between errors in c^* and errors in u and h is not surprising. Since the full equations support waves moving at both $U + c$ and $U - c$, (28) does not correctly describe the dynamics of the flow at the point point just inside the right boundary at which c^* is evaluated via (30). Unless only one of the two admissible waves is present, the c^* calculation cannot be expected to yield a meaningful result. Very early in the simulation there are no inward propagating waves adjacent to the lateral boundary, and a reasonable calculation of c^* is possible (compare the first 20 steps in Figs. 5a and 5b; note the difference in the vertical scales). As the simulation proceeds, initially small errors develop at the lateral boundary and induce inward propagating waves. These erroneously generated waves grow until (28) is no longer applicable, and the local calculation of the phase speed via (29) becomes essentially meaningless.

The difficulties encountered using an Orlanski-style open boundary condition in the one-dimensional shallow water system are discouraging since this is one of the few situations where there is a unique wave speed that might be calculable from (29). On the other hand, the one-dimensional shallow-water system is also one of the few systems in which there is no need for the Orlanski approach, because it is easy to specify c^* correctly. In systems that support waves moving at different speeds, it is not possible to specify a single value of c^* that will transmit both waves without reflection and one might suppose that it could be more effective to calculate some (presumably averaged) c^* via (29) than to fix c_a at a constant value. Nevertheless, as shown in Durran et al. (1993), distinctly better results are still obtained using a fixed value of c_a [instead of (29)] in a two-layer shallow-water system in which two waves with different speeds impinge on the boundary.

6. Better Formulations

The preceding sections have emphasized the shortcomings and limitations of some popular methods and their underlying assumptions. What approach might yield an improvement? Perhaps the most promising techniques are those that build of the pioneering work by Engquist and Majda (1977) who suggested replacing the true governing equation at an open boundary by a locally solvable partial differential equation whose dispersion relation is approximately satisfied by the outward directed waves arriving at the boundary. These partial differential equations are "one-way wave equations" in the sense that they are not even approximately satisfied by any inward directed wave, thereby minimizing the development of spurious reflections at the boundary.

The original formulations of Engquist and Majda (1977), which were proposed for two-dimensional shallow-water flow, have been generalized in a series of papers by Higdon. In particular, Higdon (1994) has suggested that an improved approximation to the radiation boundary condition for waves with dispersive x-trace speeds can be obtained by replacing the basic one-way wave equation (28) with the product of a series of one-way operators of the form

$$\left[\prod_{j=1}^{p}\left(\frac{\partial}{\partial t}+c_j\frac{\partial}{\partial x}\right)\right]\phi = 0, \tag{31}$$

where the set of c_j is chosen to span the range of x trace speeds associated with the waves appearing at the boundary. The ratio of the amplitude of the reflected wave generated by this boundary condition to that of an incident wave traveling parallel to the x-axis at trace speed c is

$$|r| = \prod_{j=1}^{p}\left|\frac{c-c_j}{c+c_j}\right|.$$

Although this scheme does not seem to have been used as a boundary condition for limited-area models of stably stratified flow, Higdon (1994) has successfully used to it simulate dispersive shallow-water waves on an f-plane. Experiments using (31) as a lateral boundary condition in simulations of stratified flow conducted by this author have suggested that a substantial improvement over the simplest one-way wave equation (28) can be obtained using only two phase speeds, i.e., by increasing p from one to two. Further in-depth analysis of the formulation of one-way wave equations for the approximation of open boundary conditions in both shallow-water and stratified flow is provided in Durran (1999, Chpt. 8).

Acknowledgments. This research was sponsored by NSF grant ATM-9817728.

References

Durran, D. R.: 1999, *Numerical Methods for Wave Equations in Geophysical Fluid Dynamics*, Springer-Verlag, New York. 465 p.

Durran, D. R. and Klemp, J. B.: 1983, A compressible model for the simulation of moist mountain waves, *Mon. Wea. Rev.* **111**, 2341–2361.

Durran, D. R., Yang, M.-J., Slinn, D. N. and Brown, R. G.: 1993, Toward more accurate wave-permeable boundary conditions, *Mon. Wea. Rev.* **121**, 604–620.

Engquist, B. and Majda, A.: 1977, Absorbing boundary conditions for the numerical simulation of waves, *Math. Comp.* **31**, 629–651.

Epifanio, C. C. and Durran, D. R.: 2000, Three-dimensional effects in high-drag-state flows over long ridges, *J. Atmos. Sci.* in press.

Hedstrom, G.: 1979, Nonreflecting boundary conditions for nonlinear hyperbolic systems, *J. Comp. Phys.* **30**, 222–237.

Higdon, R.: 1994, Radiation boundary conditions for dispersive waves, *SIAM J. Numer. Anal.* **31**, 64–100.

Klemp, J. B. and Durran, D.: 1983, An upper boundary condition permitting internal gravity wave radiation in numerical mesoscale models, *Mon. Wea. Rev.* **111**, 430–444.

Lilly, D. K. and Klemp, J. B.: 1979, The effects of terrain shape on nonlinear hydrostatic mountain waves, *J. Fluid Mech.* **95**, 241–261.

Long, R. R.: 1953, Some aspects of the flow of stratified fluids I. A theoretical investigation, *Tellus* **5**, 42–58.

Oliger, J. and Sundström, A.: 1978, Theoretical and practical aspects of some initial boundary value problems in fluid dynamics, *SIAM J. Appl. Math* **35**, 419–446.

Orlanski, I.: 1976, A simple boundary condition for unbounded hyperbolic flows, *J. Comp. Phys.* **21**, 251–269.

Pearson, R.: 1974, Consistent boundary conditions for numerical models of systems that admit dispersive waves, *J. Atmos. Sci.* **31**, 1481–1489.

Scinocca, J. F. and Peltier, W. R.: 1989, Pulsating downslope windstorms, *J. Atmos. Sci.* **46**, 2885–2914.

Sundström, A. and Elvius, T.: 1979, Computational problems related to limited area modelling, *Numerical Methods Used in Atmospheric Models*, Vol. II of *GARP Publication Series No. 17*, World Meteorological Organization, Geneva, pp. 379–416.

PARAMETERIZING EDDIES IN OCEAN CLIMATE MODELS

PETER R. GENT

National Center for Atmospheric Research,
Boulder, Colorado, U.S.A.

Abstract

Gent and McWilliams (1990) proposed a parameterization for the effects of mesoscale eddies on the large-scale flow for use in ocean climate models. It proposes that eddies advect tracers in addition to mixing them along isopycnal surfaces. This parameterization can be considered as a generalization of the residual-mean meridional circulation of Andrews and McIntyre (1976) to three dimensions. The resulting Eliassen-Palm fluxes have to be parameterized in order to determine the momentum equation used in non-eddy-resolving ocean climate models. The Antarctic Circumpolar Current region of the ocean is most like the atmosphere in that there are no continents to block the zonal jet. The nonacceleration theorem would suggest that the eddy-induced advection of tracers should oppose mean flow advection at the latitude of Drake Passage, and this is confirmed in global ocean model simulations.

1. Introduction

The ocean component of coupled climate models must be integrated for hundreds to thousands of years in both uncoupled and coupled modes. This has meant that these components have used non-eddy-resolving resolution, although some are now starting to enter the eddy-permitting regime, with a horizontal resolution of about 1°. However, in both these regimes it is very important to have a good parameterization of the effects of the unresolved mesoscale ocean eddies on the large-scale flow.

This problem will be related to the residual-mean meridional circulation of Andrews and McIntyre (1976), and to a generalization of Eliassen-Palm fluxes. These concepts were developed to analyse aspects of stratospheric flow, and are elegantly discussed by Andrews et al. (1987) in sections 3.5, 3.6 and elsewhere in that textbook. The material in this article comes mostly from the paper by Gent and McWilliams (1996), and the figures are taken from the paper by Danabasoglu and McWilliams (1995).

2. Generalized Eliassen-Palm Fluxes

The governing equations for incompressible, Boussinesq, hydrostatic, adiabatic flow in cartesian coordinates are

$$\frac{D}{Dt}\mathbf{u} + f\mathbf{k} \times \mathbf{u} + \nabla p = 0, \tag{1}$$

$$\frac{D}{Dt}\rho = 0, \tag{2}$$

$$p_z + g\rho/\rho_0 = 0, \tag{3}$$

$$\nabla \cdot \mathbf{u} + w_z = 0, \tag{4}$$

$$\frac{D}{Dt} = \frac{\partial}{\partial t} + \mathbf{u} \cdot \nabla + w\frac{\partial}{\partial z}. \tag{5}$$

$(\mathbf{u}, w)$ is the three-dimensional velocity, ∇ is the horizontal gradient operator, ρ is the density, and p is the pressure divided by a reference density ρ_0. The variables are now decomposed into low-pass components, denoted by an overbar, and eddy components by a low-pass filtering operator in time and space. This operator is difficult to define explicitly, but it cannot be either a time-average or a zonal-average. The reason is that non-eddy-resolving ocean models have both time and longitudinal variability in their solutions. Now define the filtered residual, denoted by a hat, and modified velocities by

$$\widehat{ab} = \overline{ab} - \overline{a}\overline{b}, \tag{6}$$

$$\mathbf{U} = \overline{\mathbf{u}} - (\widehat{\mathbf{u}\rho}/\overline{\rho}_z)_z, \tag{7}$$

$$W = \overline{w} + \nabla \cdot (\widehat{\mathbf{u}\rho}/\overline{\rho}_z). \tag{8}$$

The velocities in equations (7) and (8) generalize the usual definition of the zonally-averaged residual-mean circulation to three dimensions by including a zonal component. The modified velocity still obeys the usual continuity equation, and a modified substantial derivative can be defined that advects with $(\mathbf{U}, W)$. Thus,

$$\nabla \cdot \mathbf{U} + W_z = 0, \tag{9}$$

$$\frac{D^*}{Dt} = \frac{\partial}{\partial t} + \mathbf{U} \cdot \nabla + W \frac{\partial}{\partial z}. \tag{10}$$

The filtered equations from (1) - (3) can then be written exactly in the form

$$\frac{D^*}{Dt}\bar{u} - fV + \bar{p}_x = \nabla^{3D} \cdot \mathbf{E}, \tag{11}$$

$$\frac{D^*}{Dt}\bar{v} + fU + \bar{p}_y = \nabla^{3D} \cdot \mathbf{F}, \tag{12}$$

$$\frac{D^*}{Dt}\bar{\rho} = -G_z, \tag{13}$$

$$\bar{p}_z + g\bar{\rho}/\rho_0 = 0, \tag{14}$$

where ∇^{3D} is the three-dimensional gradient operator. Equations (11) - (14) extend the usual definition of the transformed Eulerian-mean equations, and $(\mathbf{E}, \mathbf{F}, G)$ are generalized Eliassen-Palm fluxes defined by

$$\begin{aligned}\mathbf{E} = \Big[& \bar{u}_z \widehat{u\rho}/\bar{\rho}_z - \widehat{uu}, \quad \bar{u}_z \widehat{v\rho}/\bar{\rho}_z - \widehat{vu}, \\ & -\bar{u}_x \widehat{u\rho}/\bar{\rho}_z + (f - \bar{u}_y)\widehat{v\rho}/\bar{\rho}_z - \widehat{wu} \Big],\end{aligned} \tag{15}$$

$$\begin{aligned}\mathbf{F} = \Big[& \bar{v}_z \widehat{u\rho}/\bar{\rho}_z - \widehat{uv}, \quad \bar{v}_z \widehat{v\rho}/\bar{\rho}_z - \widehat{vv}, \\ & -(f + \bar{v}_x)\widehat{u\rho}/\bar{\rho}_z - \bar{v}_y \widehat{v\rho}/\bar{\rho}_z - \widehat{wv} \Big],\end{aligned} \tag{16}$$

$$G = \bar{\rho}_x \widehat{u\rho}/\bar{\rho}_z + \bar{\rho}_y \widehat{v\rho}/\bar{\rho}_z + \widehat{w\rho}. \tag{17}$$

It is clear that if the filtered equations are to retain the adiabatic assumption of the original equation (2), then G_z in (13) must be zero. This is true if mesoscale eddy density fluxes are assumed to be aligned along mean isopycnals, which are surfaces of constant density. Making G equal zero could also be used as a definition of the low-pass filtering operator in time and space. For either reason, G is now set to zero, so that the filtered equations are also adiabatic.

3. Parameterizing the Eddy Density Fluxes

It is well known that baroclinic instability feeds off the potential energy of the mean flow and transfers energy to the mesoscale eddies. If an eddy-resolving ocean model is run adiabatically with only wind forcing, then there will be a domain-averaged transfer of mean to eddy potential energy by baroclinic instability. This will be balanced by a net conversion of mean kinetic to mean potential energy, even though the model is adiabatic. In a non-eddy-resolving model, this can be parameterized as follows. The equation for potential energy can be written in the form

$$\frac{D^*}{Dt}(g\overline{\rho}z) = g\overline{\rho w} + \nabla \cdot \left(g\overline{\rho}\,\widehat{\mathbf{u}\rho}/\overline{\rho}_z\right) - g\nabla\overline{\rho} \cdot \widehat{\mathbf{u}\rho}/\overline{\rho}_z. \tag{18}$$

The first term on the right-hand-side of equation (18) is the transfer term from mean kinetic energy. The second term integrates to zero in domain-average with suitable boundary conditions. The third term shows that a domain-averaged sink of potential energy can be assured if a simple downgradient form is assumed for the eddy density fluxes

$$\widehat{\mathbf{u}\rho} = -\kappa\nabla\overline{\rho}, \tag{19}$$

where κ is positive everywhere. Note that $\overline{\rho}_z$ is negative in statically stable flow. This choice was made in Gent and McWilliams (1990), but is justified and explained much more clearly in Gent et al. (1995).

4. Parameterizing the Eliassen-Palm Fluxes

One choice for the Eliassen-Palm fluxes was proposed in Gent and McWilliams (1996). The choice is

$$\mathbf{E} = \left[\nu_H(\overline{u}_x - \overline{v}_y),\ \nu_H(\overline{u}_y + \overline{v}_x),\ \nu_V\overline{u}_z + f\widehat{v\rho}/\overline{\rho}_z\right], \tag{20}$$

$$\mathbf{F} = \left[\nu_H(\overline{v}_x + \overline{u}_y),\ \nu_H(\overline{v}_y - \overline{u}_x),\ \nu_V\overline{v}_z - f\widehat{u\rho}/\overline{\rho}_z\right]. \tag{21}$$

The Eliassen-Palm flux divergences are then almost downgradient horizontal and vertical momentum diffusion, plus the appropriate Coriolis term. The resulting parameterized non-eddy-resolving momentum equation is

$$\frac{D^*}{Dt}\overline{\mathbf{u}} + f\mathbf{k}\times\overline{\mathbf{u}} + \nabla\overline{p} = \nabla\cdot(\nu_H\nabla\overline{\mathbf{u}}) + \mathrm{J}_{xy}(\nu_H,\mathbf{k}\times\overline{\mathbf{u}}) + (\nu_V\overline{\mathbf{u}}_z)_z, \tag{22}$$

where J is the two-dimensional Jacobian operator. Note that this Jacobian term is nonzero only when the horizontal viscosity has spatial dependence, and is necessary to ensure that no stress results from a uniform rotation.

The kinetic energy equation based on the momentum equation (22) is

$$\frac{D^*}{Dt}\left(\frac{\overline{\mathbf{u}} \cdot \overline{\mathbf{u}}}{2}\right) + \nabla^{3D} \cdot \mathbf{H} + g\overline{\rho w}/\rho_0 = \\ - \nu_H[(\overline{v}_x + \overline{u}_y)^2 + (\overline{u}_x - \overline{v}_y)^2] - \nu_V \overline{\mathbf{u}}_z \cdot \overline{\mathbf{u}}_z. \tag{23}$$

The normal component of the vector $\mathbf{H}$ is zero at the boundaries, so that this term integrates to zero in domain-average. The last term on the left-hand-side of (23) is the transfer to potential energy term, and the right-hand-side assures a domain-average sink of kinetic energy.

There is still much debate about, and no concensus on, the form of the momentum equation that should be used in non-eddy-resolving ocean models. Equation (22) has not been used in models so far; every ocean model uses the usual substantial derivative, defined by equation (5), and not the modified substantial derivative, defined by equation (10).

5. The Nonacceleration Theorem

The nonacceleration theorem was first described by Andrews and McIntyre (1976), although there is both earlier and later work, see section 3.6 of Andrews *et al.* (1987). The theorem was developed and applied mostly to flow in the stratosphere, which is often nearly adiabatic. First, the filtering must be zonal averaging which eliminates the pressure gradient term in equation (11). Andrews and McIntyre then proved that if the eddies are steady, linear waves then the divergences of the Eliassen-Palm fluxes in equations (11) and (13) are zero. Then there exists a steady state solution with $V = W = 0$. In this case, the eddy-induced velocity is equal and opposite to the mean velocity, and their advective effects on passive tracers, for example, cancel.

The only place in the world's ocean where the nonacceleration theorem is relevant is at the latitude of Drake Passage, where no continent blocks the zonal jet called the Antarctic Circumpolar Current (ACC). There are pressure differences across an enclosed ocean basin, so that the zonally-integrated pressure gradient term in (11) is nonzero. Recall that the filtering operator in this work is not zonal averaging, and that ocean eddies are not steady, linear waves. However, the nonacceleration theorem strongly suggests that the effects of the eddy-induced velocity should counteract those of the mean flow in the ACC. This is precisely what occurs in coarse resolution numerical models of the world ocean.

6. Global Ocean Model Results

These results come from a very early, coarse resolution version of the NCAR CSM Ocean Model (NCOM). CSM is the Climate System Model, and the ocean component was developed from the Modular Ocean Model (MOM) that was developed and maintained at the Geophysical Fluid Dynamics Laboratory. Details about this particular model setup can be found in Danabasoglu and McWilliams (1995). This version has a zonal resolution of 4°, a meridional resolution of 3°, and 20 levels in the vertical varying from 50m near the surface to 450m in the deep ocean. The prognostic model variables are horizontal velocity, potential temperature and salinity, with the density calculated from the equation of state for seawater. The eddy effects on potential temperature and salinity are parameterized as eddy-induced advection, as in equation (13), plus mixing terms oriented along and perpendicular to the mean isopycnals. Using the approximation of small isopycnal slopes gives the equation for potential temperature, T, in the form

$$\frac{D^*}{Dt}T = Q + R(\kappa_I, T) + (\kappa_V T_z)_z, \qquad (24)$$

where Q is the surface heat flux and R represents Laplacian mixing along isopycnals.

The results are from an equilibrium solution that has been integrated for several thousand years. Figure 1 shows the annual-mean, zonally integrated meridional overturning streamfunction in Sverdrups (Sv) from the mean velocity and from the eddy-induced velocity. The only region of the world ocean where the eddy-induced velocity transports a significant amount is in the region of the ACC between 45°S and 60°S where the transport is more than 25 Sv near the surface. Note also that, as anticipated from the nonacceleration theorem, the eddy-induced streamfunction is the same magnitude as that from the mean velocity and opposite in sign. This means that the meridional streamfunction from the residual-mean velocities (V, W) will be small in the ACC. This is confirmed in figure 2b which shows the annual-mean, zonally averaged meridional overturning streamfunction calculated from (V, W). Figure 2a shows the meridional overturning streamfunction from

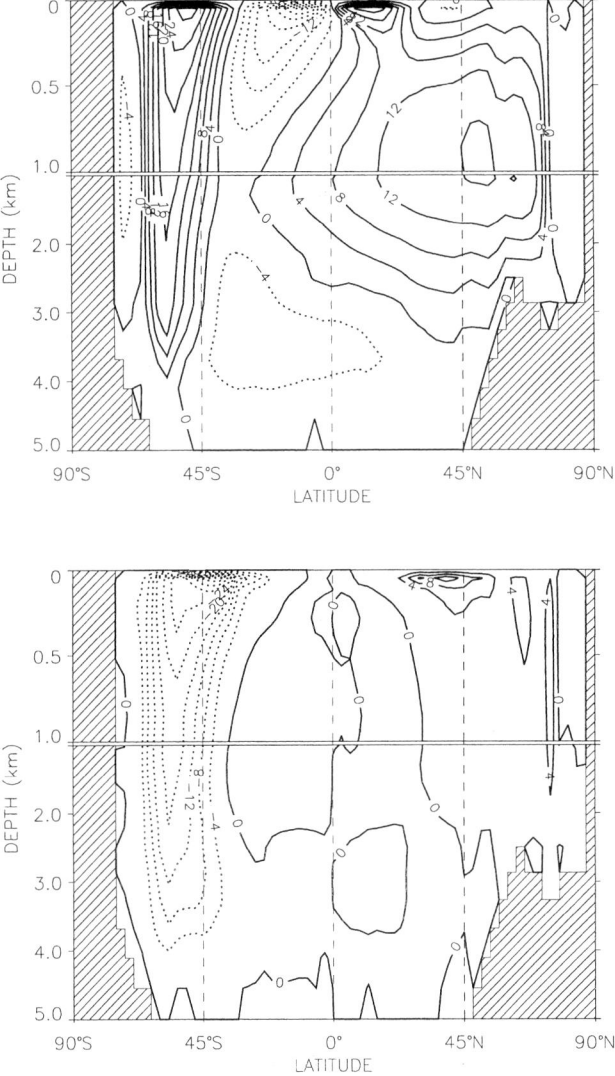

Figure 1. The annual-mean, zonally-averaged meridional overturning streamfunction in Sv from a) the mean velocity, and b) the eddy-induced velocity from an equilibrium solution using equation (24).

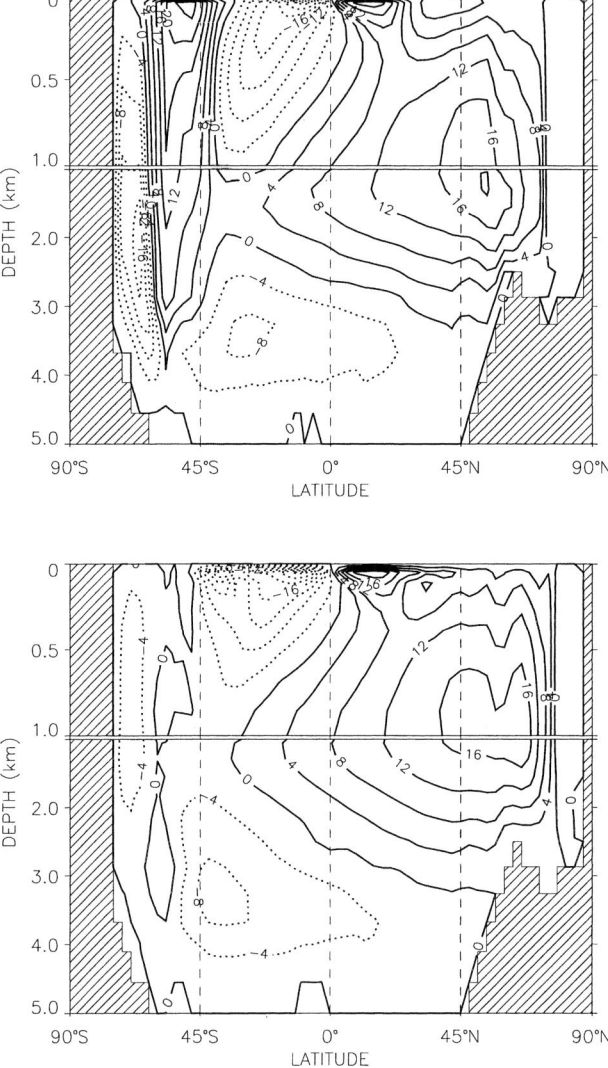

Figure 2. The annual-mean, zonally-averaged meridional overturning streamfunction in Sv from a) the mean velocity using horizontal tracer mixing, and b) the residual-mean velocity using equation (24).

a companion calculation using the earlier eddy parameterization that mixes T and S horizontally and has no additional advection. Figure 2a shows a strong cell in the ACC, similar to that in figure 1a.

The obvious conclusion from figure 2 is that in the ACC region, T and S will be advected very differently in the two companion calculations using horizontal tracer mixing and the tracer equation (24). One of the most important consequences of this different advection is the extent of convective adjustment in the ACC region in the two calculations. In a model using the hydrostatic approximation (14), vertical transfer of water is often accomplished by convective adjustment which is activated when the water column becomes statically unstable. In high latitudes, this can occur over most of the water column, and this is the model representation of deep water formation. Therefore, a measure of the location and frequency of deep water formation is easily diagnosed as when convective adjustment is triggered. This is shown in figure 3a,b from the calculations using horizontal tracer mixing and equation (24). In the first case, then is very extensive deep water formation over much of the ACC region, as well as in the North Atlantic and Arctic Oceans. In stark contrast, figure 3b shows deep water formation in rather few places, and is very much more realistic. Deep water formation is restricted to the Weddell and Ross Seas in the southern hemisphere and to the high North Atlantic and Arctic Oceans in the northern hemisphere. This represents a dramatic improvement in the ability of coarse resolution ocean models to simulate the global thermohaline circulation, which is very important for climate studies.

There are other improvements in results using equation (24) instead of horizontal tracer mixing. The deep water masses are much better represented because their density differences are not diffused away. One measure of this is shown in figure 4, which plots the globally averaged T and S from Levitus (1982) observations and from three equilibrium integrations of the global ocean model. The first two are those described above with different eddy parameterizations, which both use restoring boundary conditions at the surface on T and S. The last integration uses equation (24) and the bulk forcing described in Large et al. (1997). The globally averaged potential temperature in the horizontal tracer mixing integration is 5.7°C, which is much warmer than the Levitus value of 3.7°C. Using (24) improves this considerably to a value of 3.5°C. Note that the globally averaged

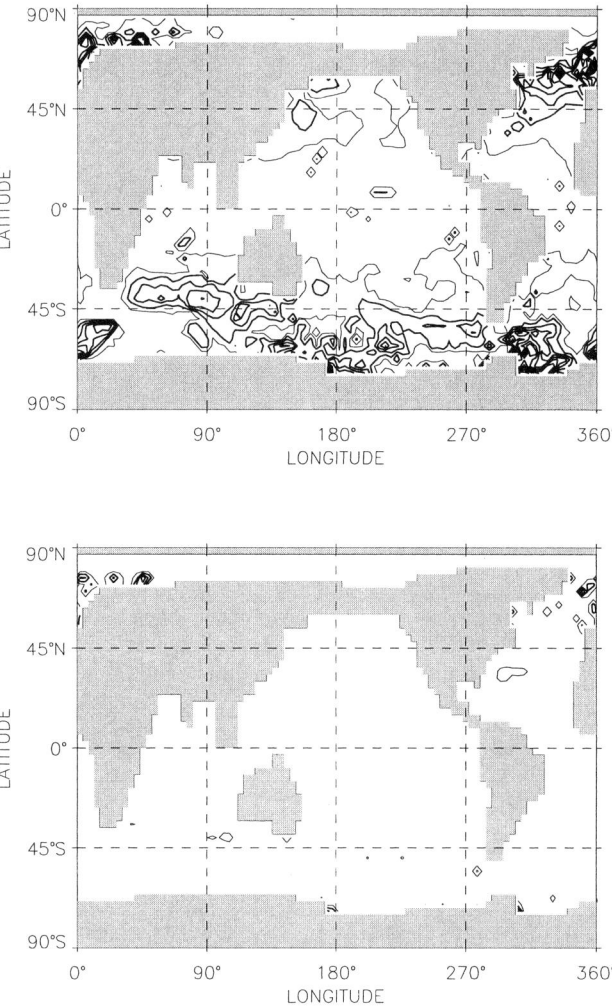

Figure 3. The areas of deep water formation from equilibrium solutions using a) horizontal tracer mixing, and b) using equation (24).

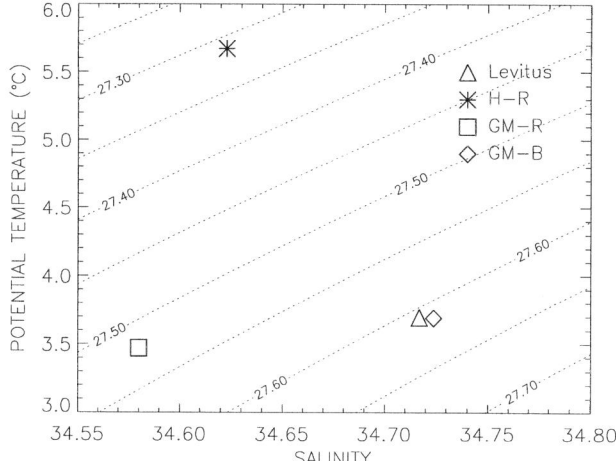

Figure 4. Values of globally-averaged potential temperature and salinity from observations and three equilibrium model solutions: H-R - horizontal tracer mixing and restoring boundary conditions; GM-R - equation (24) and restoring boundary conditions; GM-B - equation (24) and bulk forcing boundary conditions.

salinity changes much less, and remains too fresh compared to reality. This salinity bias is due to the restoring boundary condition on S, which gives a poor representation of the ocean surface fresh water flux. This bias is removed by going to the bulk forcing, which uses realistic precipitation and evaporation fields. The change to bulk forcing affects the globally-averaged potential temperature much less, but does improve it slightly. The values from the third experiment are very close to the Levitus values of 3.7°C and 34.72 ppt.

7. Conclusions

The parameterization of Gent and McWilliams (1990) is shown to be a generalization of the residual-mean circulation to three dimensions. Parameterizing the resulting Eliassen-Palm fluxes determines the momentum equation to use in non-eddy-resolving ocean models. The nonacceleration theorem of Andrews and McIntyre (1976) does not apply exactly to the ocean, but strongly suggests that the eddy-induced advection of tracers should oppose mean flow advection at the latitude of Drake Passage. Results from coarse resolution global ocean models confirm that considerable cancellation does occur at this latitude near the Antarctic Circumpolar Current when the Gent and McWilliams (1990) parameterization is used. This leads to a much more realistic pattern of deep water formation compared to results obtained using the earlier eddy parameterization of horizontal diffusion of tracers.

8. Limerick

> There once was an ocean model called MOM,
> That occasionally used to bomb,
> But eddy advection, and much less convection,
> Turned it into a stable NCOM.

References

Andrews, D. G. and M. E. McIntyre, 1976: Planetary waves in horizontal and vertical shear: The generalized Eliassen-Palm relation and the mean zonal acceleration. *J. Atmos. Sci.*, **33**, 2031–2048.

Andrews, D. G., J. R. Holton and C. B. Leovy, 1987: *Middle Atmosphere Dynamics.* Academic Press, 489pp.

Danabasoglu, G. and J. C. McWilliams, 1995: Sensitivity of the global ocean circulation to parameterizations of mesoscale tracer transports. *J. Climate*, **8**, 2967–2987.

Gent, P. R., and J. C. McWilliams, 1990: Isopycnal mixing in ocean circulation models. *J. Phys. Oceanogr.*, **20**, 150–155.

Gent, P. R., and J. C. McWilliams, 1996: Eliassen-Palm fluxes and the momentum equation in non-eddy-resolving ocean circulation models. *J. Phys. Oceanogr.*, **26**, 2539–2546.

Gent, P. R., J. Willebrand, T. J. McDougall, and J. C. McWilliams, 1995: Parameterizing eddy-induced tracer transports in ocean circulation models. *J. Phys. Oceanogr.*, **25**, 463–474.

Large, W. G., G. Danabasoglu, S. C. Doney and J. C. McWilliams, 1997: Sensitivity to surface forcing and boundary layer mixing in the NCAR CSM ocean model: annual-mean climatology. *J. Phys. Oceanogr.* **27**, 2418–2447.

Levitus, S., 1982: Climatological Atlas of the World Ocean, *NOAA Prof. Pap. 13*, U.S. Govt. Print. Office, 173pp.

BOUNDARY FILTER INITIALIZATION OF THE HIRLAM MODEL

PETER LYNCH and RAY MCGRATH
Peter.Lynch@met.ie, Ray.Mcgrath@met.ie
Met Éireann, Glasnevin Hill, Dublin 9, Ireland

1. Introduction

In this paper we establish the mathematical framework for the filtering problem on a bounded interval, and provide a motivation for the use of the 'half-sinc' function as a boundary filter. A boundary filter is one which yields output valid at the start or end of the interval on which the input is available. Such filters have obvious relevance to the problem of initialization for Numerical Weather Prediction. We also discuss some fundamental difficulties encountered when filtering time-limited functions.

Let us assume that the function $f(t)$ is specified on an interval $t_1 \leq t \leq t_2$. We seek a function $\bar{f}(t)$ which is as close as possible to the given function, subject to its being of low frequency. We shall show that, with an obvious definition of 'low frequency' the method results in an approximation of the function by itself. This paradoxical result arises from the difficulty of defining 'low frequency' on a bounded domain (Slepian, 1976). We circumvent the difficulty by modifying the approximation method, and obtain applicable results.

The first idea that might spring to mind is straightforward Fourier filtering: the function $f(t)$ is analysed into its spectral components, the high frequency components are removed and the remainder re-synthesized to produce a low frequency approximation to the function. Thus, if we express $f(t)$ as a Fourier series

$$f(t) = \sum_{n=-\infty}^{+\infty} \varphi_n \exp\left[2\pi i n \left(\frac{t-t_1}{t_2-t_1}\right)\right],$$

the high frequencies are easily removed by restricting the series to values of $|n|$ less than some value n_0 determined by the desired cutoff frequency. There are two difficulties with this technique. First, if the interval (t_1, t_2) on which $f(t)$ is given is shorter than the cutoff period, the longest-period non-constant component appearing in the Fourier analysis is above the cutoff, so that all that remains is the constant component, a poor approximation to a slowly varying function. Second, expansion into a Fourier series involves an assumption that the function $f(t)$ is periodic, with a period $\tau = t_2 - t_1$; in general, this is not so, although we are told nothing about $f(t)$ outside the range (t_1, t_2). Moreover, if $f(t_1) \neq f(t_2)$, the assumption of periodicity implies a discontinuity, with the consequence that Gibbs oscillations are produced in the Fourier representation of $f(t)$. Convergence of the Fourier series is nonuniform near points of discontinuity, and the accuracy of approximation is generally poor there.

The filtering method derived below may be simply described: the function $\bar{f}(t)$ is expressed as an integral over all sinusoidal components with frequencies less than the cutoff. The minimization of the distance between $\bar{f}(t)$ and $f(t)$ (defined precisely below) leads to an equation for the amplitude of the components: the amplitude function must satisfy a Fredholm integral equation of the first kind. This equation may be solved by standard numerical techniques but, because of the nature of the kernel, the resulting algebraic system is highly ill-conditioned. The equation can also be solved by use of the eigenfunctions of the corresponding homogeneous Fredholm equation of the second kind. It transpires that these eigenfunctions are complete in $L_2(-\frac{1}{2}T, +\frac{1}{2}T)$ so that the filtering procedure makes no change to the input. Considering the discrete case, we formulate an explicit expression for the filter matrix and devise a means of modifying it, using the special characteristics of the eigenvalues, so that it has the desired filtering effect.

2. Mathematical Framework

In this section we examine the mathematical framework for the filtering problem on a bounded interval, and demonstrate why the 'half-sinc' function is a natural choice for use as a boundary filter. Readers content to accept the appropriateness of using the half-sinc function may prefer to skip immediately to §3 (but they will miss the austere beauty of the Fredholm equation, Mercer's theorem, the Dirichlet kernel and other mathematical delicacies).

2.1. DERIVATION OF THE INTEGRAL EQUATION

Let a function $f(t)$ be given on a finite interval $t_1 \leq t \leq t_2$. The derivation is simplified if we assume, without loss of generality, that the interval is symmetric about the origin: $t_1 = -T/2$ and $t_2 = +T/2$, where $T = t_2 - t_1$ is the duration or *span*. The problem is to find a low frequency approximation to $f(t)$. We assume that a cutoff frequency $\omega_c = 2\pi\nu_c$ is specified, and require the approximation to vary more slowly than this.

Let us consider a function $\bar{f}(t)$ which is a super-position of low frequency components:

$$\bar{f}(t) = \frac{1}{2\pi} \int_{-\omega_c}^{+\omega_c} \varphi(\omega') \exp(i\omega' t) \, d\omega' . \tag{1}$$

The L_2-norm of $f(t)$ on the given interval is defined by

$$\|f\|^2 = \int_{-T/2}^{+T/2} |f(t)|^2 \, dt .$$

We may express the error E in the approximation of f by $\bar{f}$ as

$$E \equiv \|f - \bar{f}\|^2 = \int_{-T/2}^{+T/2} |f(t) - \bar{f}(t)|^2 \, dt \,.$$

The problem now is to choose $\varphi(\omega)$ in such a way as to minimize E. It is a problem in variational calculus. Let φ be replaced by $\varphi + \delta\varphi$; then $E \to E + \delta E$ where

$$\delta E = -2 \int_{-T/2}^{+T/2} \left(f(t) - \bar{f}(t)\right) \left(\frac{1}{2\pi} \int_{-\omega_c}^{+\omega_c} \delta\varphi(\omega') \exp(i\omega' t) \, d\omega'\right) dt \,.$$

Let us suppose that $\delta\varphi(\omega') = \delta(\omega + \omega')$, a delta-function centered at a fixed point $-\omega$ such that $\omega \in (-\omega_c, +\omega_c)$. The variation of E then reduces to

$$\delta E = -\frac{1}{\pi} \int_{-T/2}^{+T/2} \left(f(t) - \bar{f}(t)\right) \exp(-i\omega t) \, dt \,.$$

For an extremum of E the variation vanishes, $\delta E = 0$ so that, for $|\omega| \leq \omega_c$,

$$F(\omega) \equiv \int_{-T/2}^{+T/2} f(t) \exp(-i\omega t) \, dt = \int_{-T/2}^{+T/2} \bar{f}(t) \exp(-i\omega t) \, dt \,. \tag{2}$$

Using (1) to expand $\bar{f}$ yields

$$\begin{aligned}
F(\omega) &= \int_{-T/2}^{+T/2} \left(\frac{1}{2\pi} \int_{-\omega_c}^{+\omega_c} \varphi(\omega') \exp(i\omega' t) \, d\omega'\right) \exp(-i\omega t) \, dt \\
&= \int_{-\omega_c}^{+\omega_c} \left(\frac{1}{2\pi} \int_{-T/2}^{+T/2} \exp\bigl[i(\omega' - \omega)t\bigr] \, dt\right) \varphi(\omega') \, d\omega' \,.
\end{aligned}$$

If we define the kernel $K(\omega, \omega')$ by

$$K(\omega, \omega') \equiv \frac{1}{2\pi} \int_{-T/2}^{+T/2} \exp\bigl[i(\omega' - \omega)t\bigr] \, dt = \frac{\sin(\omega' - \omega)T/2}{\pi(\omega' - \omega)}, \tag{3}$$

the equation for $F(\omega)$ may be written as

$$\int_{-\omega_c}^{+\omega_c} K(\omega, \omega') \varphi(\omega') \, d\omega' = F(\omega) \,. \tag{4}$$

This is a Fredholm integral equation of the first kind (Tricomi, 1985). Given $f(t)$, the forcing term $F(\omega)$ may be evaluated and the equation solved for $\varphi(\omega)$.

The kernel (3) is symmetric and of convolution type:

$$K(\omega, \omega') = K(\omega', \omega), \qquad K(\omega, \omega') = k(\omega - \omega') \,. \tag{5}$$

The properties of the Fredholm equations having this kernel have been studied extensively; see, for example, Slepian and Pollak (1961). More general references on integral equations, in addition to Tricomi (1985), include Courant and Hilbert (1953) and Polyanin and Manzhirov (1998).

2.2. DIRECT NUMERICAL SOLUTION

The integral in (4) may be approximated by a finite sum using one of many methods, such as Gaussian quadrature or the trapezoidal rule. We assume the interval $(-\omega_c, +\omega_c)$ is divided into N sub-intervals, the n-th containing the point ω_n. Then a finite approximation to (4) is given by

$$\sum_{n=1}^{N} K(\omega_m, \omega_n) \varphi(\omega_n) w_n = F(\omega_m), \qquad (6)$$

where w_n is the weight for the n-th interval. If we define a matrix and two vectors by

$$\mathbf{K}_{mn} = K(\omega_m, \omega_n) w_n \qquad \mathbf{\Phi}_n = \varphi(\omega_n) \qquad \mathbf{F}_n = F(\omega_n),$$

equation (6) may be written as a standard system of algebraic equations

$$\mathbf{K}\mathbf{\Phi} = \mathbf{F} \qquad (7)$$

which can be solved for $\varphi(\omega_m)$ provided $\mathbf{K}$ is non-singular. We can use the chosen quadrature method for evaluating the forcing function $F(\omega)$ in (2), approximating the integral in (4) and evaluating the filtered result $\bar{f}(t)$ in (1).

The problem with this direct method is that the matrix $\mathbf{K}$ is ill-conditioned. We define the *Shannon Number*, Sh, as the duration-bandwidth product:

$$\text{Sh} = 2\nu_c T = \frac{\omega_c T}{\pi} = \frac{2T}{\tau_c}$$

For $n \ll \text{Sh}$, the eigenvalues of $\mathbf{K}$ are close to unity; for $n \gg \text{Sh}$, they decay rapidly with n (an example will be given below). Thus if, in the interests of accurate approximation, we choose n to be large, the condition number of $\mathbf{K}$ is large and the matrix is practically uninvertable.

2.3. EIGENFUNCTION ANALYSIS

Tricomi (*loc. cit.*, p143) writes: "Some mathematicians still have a kind of fear whenever they encounter a Fredholm integral equation of the *first* kind ... Today this fear is no longer justified, especially for the *symmetric case* $K(x,y) = K(y,x)$". So, we shall proceed fearlessly, but cautiously, to investigate (4). Caution is required because things are not quite as simple as they seem. For general $F(t)$, (4) may have no solution at all. If there is a solution, the fact that $F(t)$ can be expressed in the integral form of (4) guarantees, by the Hilbert-Schmidt theorem

(Tricomi, p110), that it can also be expanded in the eigenfunctions of the kernel. These are the solutions of the homogeneous equation of the *second* kind:

$$\int_{-\omega_c}^{+\omega_c} K(\omega, \omega')\varphi(\omega')\,d\omega' = \lambda\varphi(\omega). \tag{8}$$

Note that the historical convention, to write the equation $\lambda \int K(\omega, \omega')\varphi(\omega')\,d\omega' = \varphi(\omega)$, is used by many authors. But the convention here is more convenient as, in the discrete case λ corresponds to the matrix eigenvalues.

We shall give some properties of the solutions of (8). A more complete discussion can be found in Slepian and Pollak (1961). The eigenvalues λ_n form an infinite set

$$1 > \lambda_1 > \lambda_2 > \cdots > \lambda_k > \cdots > 0$$

whose only accumulation point is at zero. Because of the symmetry of the kernel, they are real. The eigenfunctions are doubly orthogonal. They may be assumed to be orthonormal on the interval $[-\omega_c, \omega_c]$:

$$\int_{-\omega_c}^{+\omega_c} \varphi_m(\omega)\varphi_n(\omega)\,d\omega = \delta_{mn}.$$

They are complete in the space of square-integrable functions on this interval. They are also orthogonal on the infinite interval:

$$\int_{-\infty}^{+\infty} \varphi_m(\omega)\varphi_n(\omega)\,d\omega = \frac{\delta_{mn}}{\lambda_n}.$$

The functions $\varphi_n(\omega)$ are bandlimited and span the space of bandlimited functions on the real line. The implication of these two orthogonality relationships is that, for small n ($n \ll$ Sh), $\varphi_n(\omega)$ is large in $[-\omega_c, \omega_c]$ and small outside, whereas for large n ($n \gg$ Sh) the opposite holds.

If the forcing function $F(\omega)$ in (4) is expanded in the eigenfunctions of (8) as

$$F(\omega) = \sum_{n=1}^{\infty} b_n \varphi_n(\omega)$$

and the solution is assumed to have a similar expansion with coefficients a_n it follows, *under certain convergence conditions*, that $a_n = b_n/\lambda_n$, so that

$$\varphi(\omega) = \sum_{n=1}^{\infty} (b_n/\lambda_n)\varphi_n(\omega). \tag{9}$$

Thus, if we can solve (8) for the eigenvalues and eigenfunctions, (9) gives a formal solution of our equation of the *first* kind, (4).

Now for the paradox: the eigenfunctions are complete in $L_2(-\frac{1}{2}T, +\frac{1}{2}T)$. Thus *any* L_2-function can be expanded in these eigenfunctions. But the eigenfunctions themselves are band-limited (Slepian and Pollak, 1961). Consequently, any L_2-function on a bounded interval can be approximated arbitrarily closely by a band-limited function. This is the surprise: our intuitive notion of a band-limited function on a time-limited or bounded domain is defective.

We are led to a strange conclusion: we have tried to filter $f(t)$, but it remains resolutely unchanged! How is this so? The problem is that the notion of bandlimitedness is unequivocally defined only for functions on an infinite domain. Our intuitive notion of bandlimited functions on a bounded interval is misleading: any L_2-function given on a finite interval can be approximated arbitrarily closely on that interval by a function which has a natural extension to the entire real line as a bandlimited function (Slepian and Pollak, 1961). For a general discussion of the paradox, see Slepian's (1976) paper *On Bandwidth*. It is possible to show explicitly that $\bar{f}(t) = f(t)$, but this is best done for the discrete case, which we shall examine after a brief digression.

2.4. THE BILINEAR SERIES

Before proceeding to the discrete case, we derive an expansion of the kernel in terms of its eigenfunctions, the so-called bilinear formula. Considering the kernel $K(\omega, \omega')$ for fixed ω as a function of ω', let us assume it can be expanded in the eigenfunctions:

$$K(\omega, \omega') = \sum_{m=1}^{\infty} c_m(\omega) \varphi_m(\omega'),$$

where $c_m(\omega)$ are to be found. Multiply by $\varphi_n(\omega')$ and integrate:

$$\int_{-\omega_c}^{+\omega_c} K(\omega, \omega') \varphi_n(\omega') \, d\omega' = \sum_{m=1}^{\infty} c_m(\omega) \int_{-\omega_c}^{+\omega_c} \varphi_m(\omega') \varphi_n(\omega') \, d\omega'.$$

But then, from the definition and orthonormality of the eigenfunctions, it follows that

$$c_n(\omega) = \lambda_n \varphi_n(\omega)$$

and this leads immediately to the bilinear series expansion:

$$K(\omega, \omega') = \sum_{n=1}^{\infty} \lambda_n \varphi_n(\omega) \varphi_n(\omega'). \tag{10}$$

This is Mercer's Theorem. For our kernel it can be shown that the series converges absolutely and uniformly. We shall see the convenience of abandoning the historical convention, and writing the eigenvalue on the right side of (8), when the discrete analogue of (10) appears below.

2.5. THE DISCRETE CASE

We now replace $f(t)$ by a function of a discrete variable, and assume that we have knowledge of it only on an index-limited set, $\{f_0, f_1, \ldots, f_{N-1}\}$. It will prove convenient to assume that N is an odd number and to re-index:

$$\{f_{-M}, f_{-M+1}, \ldots, f_{+M}\}$$

where $2M+1 = N$. We shall minimise the difference $E = \|f - \bar{f}\|^2 = \sum_{m=-M}^{M} |f_m - \bar{f}_m|^2$ where

$$\bar{f}_m = \frac{1}{2\pi} \int_{-\omega_c}^{\omega_c} \varphi(\omega) e^{im\omega} \, d\omega. \tag{11}$$

By the usual variational procedure, we find that E is a minimum if

$$\sum_{m=-M}^{M} \bar{f}_m e^{-im\omega} = \sum_{m=-M}^{M} f_m e^{-im\omega} \equiv F(\omega). \tag{12}$$

The astute reader will realize that $\bar{f}_m = f_m$ satisfies (12), but the question is: can $\bar{f}_m = f_m$ be expressed in the form (11)? This is not obvious. Substituting (11) in (12) and manipulating, we find that

$$\int_{-\omega_c}^{+\omega_c} K(\omega, \omega') \varphi(\omega') \, d\omega' = F(\omega), \tag{13}$$

where the kernel is now of the form

$$K(\omega, \omega') \equiv \frac{1}{2\pi} \sum_{m=-M}^{M} e^{-im\omega} e^{im\omega'} = \frac{N}{2\pi} \mathcal{D}_N(\omega - \omega') \tag{14}$$

where $\mathcal{D}_N$ is the Dirichlet kernel

$$\mathcal{D}_N(\omega) \equiv \frac{\sin N\omega/2}{N \sin \omega/2}.$$

Although (13) is formally similar to (4), there is a crucial difference: the kernel is now degenerate: it is a finite sum of products of the form $g(\omega)h(\omega')$. As a result, the integral equation of the second kind

$$\int_{-\omega_c}^{+\omega_c} K(\omega, \omega') \varphi(\omega') \, d\omega' = \lambda \varphi(\omega). \tag{15}$$

can be reduced to a classical algebraic eigenvalue problem. We define

$$\xi_m = \int_{-\omega_c}^{+\omega_c} e^{im\omega'} \varphi(\omega') \, d\omega', \quad m = -M, \ldots, +M$$

and substitute in (15), using (14) to obtain

$$\lambda \varphi(\omega) = \frac{1}{2\pi} \sum_{m=-M}^{M} e^{-im\omega} \xi_m . \qquad (16)$$

Now multiplying by $e^{in\omega}$ and integrating over ω, we arrive at

$$\lambda \xi_n = \sum_{m=-M}^{M} K_{nm} \xi_m \qquad (17)$$

where K_{nm} is an element of the symmetric matrix $\mathbf{K}$ defined by

$$[\mathbf{K}]_{mn} = K_{mn} = \frac{1}{2\pi} \int_{-\omega_c}^{+\omega_c} e^{-im\omega} e^{+in\omega} \, d\omega = \frac{\sin(m-n)\omega_c}{\pi(m-n)} . \qquad (18)$$

Thus, equation (15) is completely equivalent to the standard eigenvalue problem (17).

We denote the eigenvalue/eigenvector pairs of (17) by (λ_k, ξ^k). The eigenvector matrix $\mathbf{\Xi}$ is defined by $\Xi_{nk} = \xi^k_n$ and the eigenvalue matrix by $\mathbf{\Lambda} = \mathrm{diag}(\lambda_1, \ldots, \lambda_N)$. Then the eigenvector equation (17) may be written

$$\mathbf{K}\mathbf{\Xi} = \mathbf{\Xi}\mathbf{\Lambda} . \qquad (19)$$

The eigenvalues and eigenvectors corresponding to the Dirichlet kernel have been discussed in detail by Slepian (1978) and by Percival and Walden (1993). The eigenfunctions of (15) are called discrete prolate spheroidal wave functions, and the eigenvectors of (17) or (19) are known as discrete prolate spheroidal sequences (dpss). They have many interesting properties and a wide range of applications in digital signal processing. We shall assume a normalization different to that in the continuous case. For each pair (λ_k, ξ^k), there is a corresponding eigenfunction $\varphi^k(\omega)$ of (15) given by (16). We shall assume that the eigenvectors ξ^k are orthonormal; this means $\mathbf{\Xi}^T = \mathbf{\Xi}^{-1}$. It also constrains the norm of the eigenfunction $\varphi^k(\omega)$, as the relationship $\lambda_k \|\varphi^k\|^2 = \|\xi^k\|^2$ is easily shown to hold.

To illustrate the interesting properties of the eigenvalues, we set $N = 21$ and choose the parameters $T = 3\,\mathrm{hours}$ and $\tau_c = 0.6\,\mathrm{hours}$, so that the Shannon number Sh=10. We plot the eigenvalues in Fig. 1. For $k \ll \mathrm{Sh}$, the eigenvalues are close to unity, and for $k \gg \mathrm{Sh}$ they are very small. We may note that $\lambda_1 = 0.999999999999997$ and $\lambda_{21} = 2.88657986402541 \times 10^{-15}$, so the condition number of the matrix $\mathbf{K}$ is 3.5×10^{14}, a warning signal that caution is required in its numerical analysis.

We now use the eigenvectors to solve (13) and then use (11) to find $\bar{f}_m$. Let us expand $\varphi(\omega)$ and $F(\omega)$ as

$$\varphi(\omega) = \sum_{k=1}^{N} a_k \varphi^k(\omega), \qquad F(\omega) = \sum_{k=1}^{N} b_k \varphi^k(\omega) \qquad (20)$$

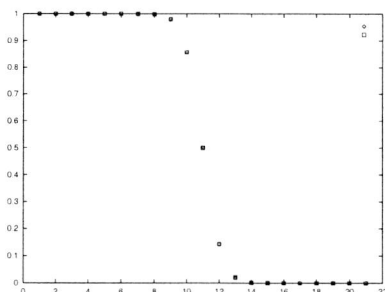 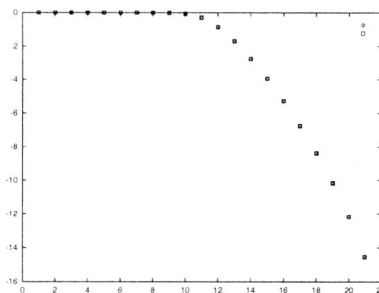

Fig. 1. Eigenvalues λ_k of the Matrix **K**. The parameter values are $N = 21$, $T = 3$ hours and $\tau_c = 0.6$ hours, so that the Shannon number Sh=10. Left: linear vertical scale. Right: logarithmic vertical scale.

It follows from (13) that $a_k = b_k/\lambda_k$. But now using (12) to express b_k in terms of f_n, and (16) to substitute ξ^k for $\varphi(\omega)$ we arrive, after some algebra, at the result

$$\bar{f}_m = \sum_{n=-M}^{M} \left\{ \sum_{k=1}^{N} \xi_m^k \xi_n^k \right\} f_n. \qquad (21)$$

This is a matrix product and the matrix in braces is just $\mathbf{M} = \Xi \Xi^T$. However, since the eigenvectors ξ^k are orthonormal we know that $\Xi^T \Xi = \mathbf{I}$. Since Ξ^T is a left inverse, it is also a right inverse of Ξ, and $\mathbf{M} = \mathbf{I}$. Thus, $\bar{f}_n = f_n$ and no filtering is achieved.

2.6. THE FLIGHT OF THE PHŒNIX

Our filtering strategy using (11) is unsuccessful. But all is not lost and we can still salvage something of value from the ashes. We know that the eigenfunctions $\varphi^k(\omega)$ oscillate more rapidly with increasing index: $\varphi_k(\omega)$ has k zeros in $(-\omega_c, \omega_c)$ and, similarly, the eigenvector ξ^k has k zeros. Since rapid oscillations run counter to our intuitive notion of low frequency, it makes sense to eliminate them. We can do this by truncating the expansion (21) in ξ^k. A convenient way to do this is to exploit the properties of the eigenvalues by using them as weighting coefficients. The natural truncation point is given by the Shannon number. We know that λ_k is near unity for $k < $ Sh and small for $k > $ Sh. So the sum in (21) is replaced by

$$\sum_{k=1}^{[\text{Sh}]} \xi_m^k \xi_n^k \approx \sum_{k=1}^{N} \lambda_k \xi_m^k \xi_n^k.$$

But this is just the (m,n)-th element of the matrix $\Xi\Lambda\Xi^T$ which, by (19), is nothing other than the original matrix:

$$\mathbf{K} = \Xi\Lambda\Xi^T \qquad (22)$$

This expansion is just a discrete analogue of the bilinear series (10). We reach the delightful conclusion that $\mathbf{K}$ can be used directly to act on the input signal f_n to produce a filtered output $\bar{f}_n$. Despite our long mathematical peregrinations in the frequency domain and our sojourn amongst the integral equations, we arrive back in the time domain and find that all that is needed to filter the signal is the matrix $\mathbf{K}$ defined by (18). We do not need to solve any integral equations or calculate any eigenvalues or eigenfunctions explicitly.

3. Properties of the Half-sinc Function

An ideal low-pass filter with cut-off θ_c has frequency response

$$H(\theta) = \begin{cases} 1 & \text{for } |\theta| < \theta_c \\ 0 & \text{for } |\theta| > \theta_c \end{cases}$$

The corresponding impulse response is that of a non-causal FIR (Finite Impulse Response filter).

$$h_n = \frac{\sin n\theta_c}{n\pi} = \left(\frac{\theta_c}{\pi}\right)\operatorname{sinc}\left(\frac{n\theta_c}{\pi}\right)$$

where $\operatorname{sinc}\alpha = \sin(\pi\alpha)/\pi\alpha$ and $n \in \mathbb{Z}$. In practice, we must restrict n to some finite range. Moreover, for a causal filter we require $n \geq 0$. Then the coefficients are

$$h_n = \frac{\sin n\theta_c}{n\pi}, \qquad n = 0, 1, \ldots, N-1. \qquad (23)$$

We refer to this sequence as a *half-sinc* sequence. We note that the elements of the matrix $\mathbf{K}$ defined by (18) above are $\mathbf{K}_{mn} = h_{m-n}$. Boundary filters, that is, those which yield output valid at the extremity of the time-range, must be either causal or anti-causal. They are represented by the first and last rows of $\mathbf{K}$. We examine these cases below.

The frequency response may be written

$$\sum_{n=0}^{N-1} h_n e^{in\theta} = H(\theta) = M(\theta)e^{i\varphi(\theta)} . \qquad (24)$$

The response to a signal of vanishingly small frequency (the DC component) is $H(0)$. We always normalize so that $H(0) = 1$. Thus, the sum of the filter weights h_n is unity. The *group delay* is defined as $\delta = -d\varphi/d\theta$. Taking the derivative of (24) and recalling that h_n are real, we easily see that $\delta_0 = \delta(0) = \sum n h_n$. For the output of a boundary filter to apply at the start of the time interval it must be

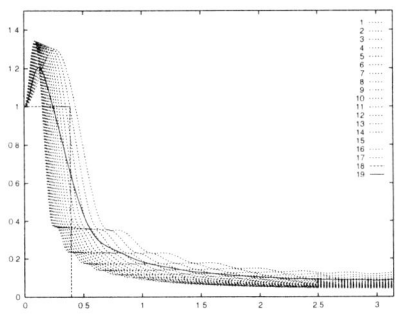 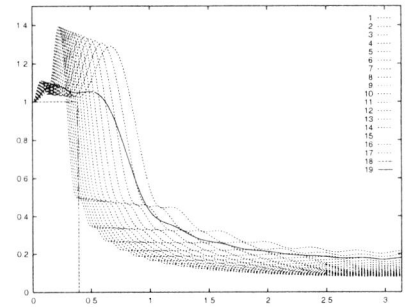

Fig. 2. Dashed curves: Frequency responses $H(\theta)$ for seventeen half-sincs with varying spans. Parameters are $\Delta t = 225\,\text{s}$, $N \in \{17, 19, \ldots, 49\}$, $T_S = (N-1)\Delta t \in \{1\text{h}, 1\tfrac{1}{8}\text{h}, \ldots, 3\text{h}\}$ and $\tau_c = T_S$. Solid curve: weighted sum of seventeen half-sincs, to reduce intermediate frequency boost. Left panel: $K = 1$. Right panel: $K = 2$.

'zero-delay'. That is, we require $\delta_0 = 0$. For the half-sinc sequence, this can be satisfied if we truncate after an exact number of wavelengths:

$$\sum_{n=0}^{N-1} n h_n = \frac{1}{\pi} \sum_{n=0}^{N-1} \sin n\theta_c = 0 \tag{25}$$

provided $(N-1)\theta_c = 2\pi K$ for some integer K.

We now examine the frequency response of some boundary filters based on zero-delay half-sinc sequences. In Fig. 2, the responses $H(\theta)$ are shown for a selection of (seventeen) half-sincs with $K = 1$ and varying spans. The parameters are $\Delta t = 225\,\text{s} = \frac{1}{16}$ hours, filter order $N \in \{17, 19, \ldots, 49\}$, span $T_S = (N-1)\Delta t \in \{1\text{h}, 1\tfrac{1}{8}\text{h}, \ldots, 3\text{h}\}$ and cut-off $\tau_c = T_S$. All are normalized so that $H(0) = 1$. The square wave shows an ideal response with cut-off period of one hour. All responses are low-pass, as high frequencies are strongly attenuated. However, all half-sincs have an over-shoot or *boost* for some intermediate frequencies. To reduce this boost, we construct a weighted combination, the response of which is shown by the solid curve. The corresponding response curves for $K = 2$ are shown in the right-hand panel of Fig. 2. The boost is further reduced in this case, but at the expense of widening the pass-band.

In Fig. 3, the responses are plotted with horizontal axis $\mu = \log_{10}(\tau/\tau_c)$ where $\tau_c = 3\,\text{h}$ and $\tau = 2\pi\Delta t/\theta$ is the period. Again, the square response is for an ideal one-hour cut-off filter. The left panel is for $K = 1$ and the right one for $K = 2$; the choice of K represents a compromise between reducing boost and optimising high frequency cut-off. So far, we have not succeeded in obtaining a boundary filter which is free from some boost at intermediate frequencies. A number of possibilities to achieve this are being explored.

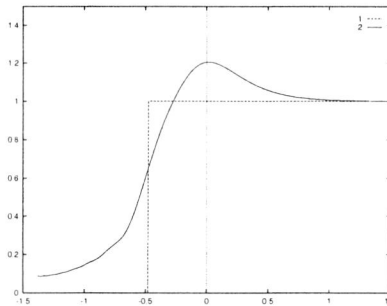 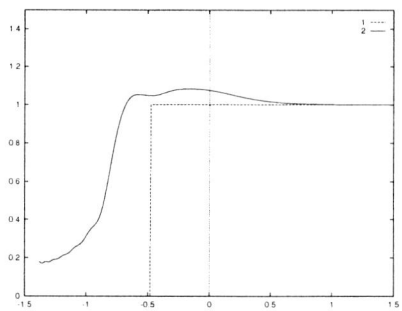

Fig. 3. Frequency response of weighted sums, as in Fig. 2, plotted against a horizontal axis $\mu = \log_{10}(\tau/\tau_c)$ where $\tau_c = 3\,\mathrm{h}$ and $\tau = 2\pi\Delta t/\theta$ is the period. The square response is for an ideal one-hour cut-off filter. Left panel: $K = 1$. Right panel: $K = 2$.

4. Implementation in HIRLAM Model

The forecast model used to test the efficiency of the filter is a development of the international HIRLAM (HIgh Resolution Limited Area Model) Project (see http://www.knmi.nl/hirlam for further details and a list of meteorological institutes involved in the project). The model was configured to run on an area covering approximately 45°–60°N and 20°W–0°E, with a horizontal resolution of approximately 0.15° (98×97 points in the E–W and N–S directions on an Arakawa C-grid; a rotated coordinate system is used with the the South Pole at 38°S, 9°E) with 24 levels in the vertical. A two time-level, three dimensional semi-Lagrangian semi-implicit scheme with a time step of 225 seconds was used in the integration. The STRACO condensation scheme was used in the diabatic stage. Global analysis fields from the European Centre for Medium-range Weather Forecasts (ECMWF) were used as boundary fields.

4.1. PRACTICAL PROCEDURE

The boundary filter initialization is performed by applying the filter to time series of model variables. The coefficients of the half-sinc filter, $\{h_n : 0 \leq n \leq N\}$, are chosen such that the output is valid at the initial point of the span, $t = 0$, so the filter produces output valid at that time. The filtered values are calculated by performing a model integration from $t = 0$ to $t = T_S$ and accumulating sums of the form

$$\bar{x} = \sum_{n=0}^{n=N} h_n x_n,$$

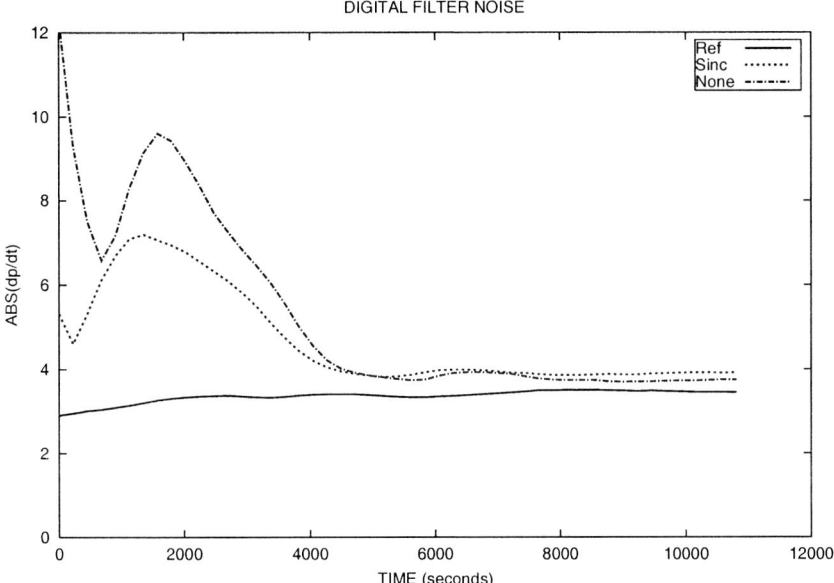

Fig. 4. Time evolution, during a 3-hour forecast, of the area-averaged absolute value of the surface pressure tendency (units: hPa per 3 hours) for three forecasts. Dot-dashed line ('None') is for the uninitialized forecast. Dotted line ('Sinc') is for the BFI scheme. Solid line ('Ref') is for the reference DFI scheme.

where x is a particular prognostic variable at a particular grid point and level (the same sum is accumulated for all prognostic variables). The output $\bar{x}$ is valid at time $t = 0$ and comprises the initialized data. The values of the prognostic variables at the lateral boundaries are left unchanged during the digital filtering process.

4.2. CASE STUDY

The HIRLAM analysis system (based on an Optimum Interpolation scheme) was used to generate weather analyses for the forecast model using observational data from ECMWF. A 1-hour assimilation cycle was used, the forecast model producing 'first-guess' fields for the analysis in the following cycle. The test data covered the period 23–24 December, 1997 during which a particularly severe storm moved across Ireland from the South-west. The assimilation cycle was initiated from 12 UTC, 23 December, using the reference digital filtering for initialization during the cycle. The analysis for 12 UTC, 24 December was then used for testing the boundary filter.

The boundary filter initialization was compared to the reference digital filter initialization (DFI) and to forecasts with no initialization. Fig. 4 shows the time evolution, during a 3-hour forecast, of the area-averaged absolute value of the surface pressure tendency for three forecasts. The dot-dashed line (marked 'None') is for the uninitialized forecast. The mean tendency starts at a spuriously high level of 12 hPa/3h. the BFI scheme (dotted line, marked 'Sinc') reduces the noise drastically, to about half the value of the uninitialized run. However, the reference scheme (solid line, marked 'Ref') yields a forecast with essentially no high-frequency noise: the mean surface pressure tendency starts at a low level (about 3 hPa/3h) and remains small throughout the run. We conclude that the BFI scheme is partially effective in removing gravity-wave noise but, as currently formulated and implemented, is clearly not as effective as the reference DFI scheme. Further modifications to improve the efficacy of the BFI scheme are under investigation.

References

Courant, R and D Hilbert, 1953: *Methods of Mathematical Physics, Vol I*. Interscience Publishers, New York, 560pp.

Farhang-Boroujeny, B, 1998: *Adaptive Filters: Theory and Applications*. John Wiley & Sons, 529pp.

Percival, D B and A T Walden, 1993: *Spectral Analysis for Physical Applications*. Cambridge Univ. Press, 583pp.

Polyanin, A D and A V Manzhirov, 1998: *Handbook of Integral Equations*. CRC Press, 787pp.

Slepian, D, 1976: On Bandwidth. *Proc. IEEE*, **64**, 292–300.

Slepian, D, 1978: Prolate spheroidal wave functions, Fourier analysis and uncertainty — V: the discrete case. *Bell Sys. Tech. J.*, **57**, 1371–1430.

Slepian, D and H O Pollak, 1961: Prolate spheroidal wave functions, Fourier analysis and uncertainty — I. *Bell Sys. Tech. J.*, **40**, 43–63.

Tricomi, F G, 1985: *Integral Equations*. (First published by Interscience, 1957). Dover Publications Inc., 238pp.

BALANCE, POTENTIAL-VORTICITY INVERSION, LIGHTHILL RADIATION, AND THE SLOW QUASIMANIFOLD

M. E. McINTYRE
Centre for Atmospheric Science at the
Department of Applied Mathematics and Theoretical Physics,[†]
Silver Street, Cambridge CB3 9EW, England.
http://www.atm.damtp.cam.ac.uk/people/mem/

Abstract

Practically our entire understanding of large-scale atmosphere–ocean dynamics depends on the notions of balance and potential-vorticity inversion. These are essential, for instance, for a clear understanding of the basic Rossby-wave propagation mechanism, or quasi-elasticity, that underlies almost every large-scale fluid-dynamical phenomenon of meteorological and oceanographical interest, from the global-scale transport of terrestrial greenhouse gases (and similar problems in the solar interior) to Rossby-wave-mediated global teleconnection, baroclinic and barotropic shear instability, vortex coherence, and vortex-core isolation. The ideas involved in understanding balance and inversion continue to hold special fascination because of their central importance both for theory and for applications, such as data assimilation, and the fact that complete mathematical understanding is still elusive. The importance for applications was adumbrated by Richardson in his pioneering study of numerical weather prediction. The importance for theory — and the exquisite subtlety involved — was adumbrated by Poincaré in his discovery of the homoclinic tangle, and by Lighthill in his discovery of the quadrupole nature of acoustic radiation by unsteady vortical motion.

1. Introduction

I'll begin by quoting Lewis Fry Richardson's classic vignette — which *isn't* a limerick — following Jonathan Swift and Augustus De Morgan and summarizing the nature of three-dimensional turbulence:

[†]The Centre for Atmospheric Science is a joint initiative of the Department of Chemistry and the Department of Applied Mathematics and Theoretical Physics (http://www.atm.damtp.cam.ac.uk/).

> Big whirls have little whirls
> That feed on their velocity;
> Little whirls have lesser whirls,
> And so on to viscosity.

This is what buffets us, and brings us Atlantic clear air, on a walk in the beautiful Irish hills; and Richardson was well aware of what can be learnt by casual observation of one's surroundings together with acute thinking. I think many of us who find fluid dynamics fascinating share a delight in such learning by observation, of anything from the power of a great storm seen from space or felt on the ground, to the progression of waves on an otherwise calm pond to the wavelike and vortical phenomena hints of which we see in stirred tea or coffee. Something that thrills me personally is that one can learn about fluid phenomena on a grand scale, up to global scales on our home planet and even vaster scales in the Sun's interior, from theory together with simple kitchen-sink experiments that even a theoretician like me can easily perform. It's all about what J. E. Littlewood called the "impudence" of creative thinking, or Erasmus Darwin the "damn fool experiment", which often fails but sometimes comes off brilliantly.

Jonathan Swift added, to the original well-known lines that inspired De Morgan and Richardson (about fleas, as you'll recall) the aphorism "Thus every poet, in his kind, is bit by him that comes behind." One might say that the same thing happens to scientists, even though for the sake of the scientific ethic we try to be more polite about it. I don't think Richardson's vignette of turbulence would be disputed as far as it goes, despite all the sophistication of fractal modelling and data analysis that refines the picture today. But the nature of large-scale atmosphere–ocean turbulence is in some ways at an opposite extreme:

> Big whirls meet bigger whirls,
> And so it tends to go on:
> By merging they grow bigger yet,
> And bigger yet, and so on.

And that's not all because, co-existing and competing with the tendencies toward vortex merging and upscale energy cascade, there are yet more, very different processes. One of these, as Rhines (1975) may have been the first to point out in a turbulence-related context, is Rossby-wave propagation. The eddying flows conspicuous in today's satellite images of the atmosphere and ocean do indeed, very often, look turbulent in a perfectly real sense. The flow is manifestly chaotic, and it stirs and mixes tracers, properties often quite reasonably taken to be characteristic of turbulence. But as soon as you have an important wave propagation mechanism — and there are several such mechanisms in stratified rotating fluid systems — you have a very profound difference by comparison with any classical turbulence picture, even if lots of tracer mixing is going on at the same time and in the same locations.

That difference is central to understanding global-scale atmospheric circulations, and hence to understanding, for instance, the whole business of chlorofluorocarbon (CFC) lifetimes and other problems concerning the ozone layer, the distribution of greenhouse gases, and their role in the entire Earth System.

Why should the presence of wave propagation mechanisms make such a "profound difference", and what's wrong with classical pictures of turbulence? The answer is fundamental: wave propagation changes everything qualitatively. (This is changing our ideas about the Sun's differential rotation as well; see Gough and McIntyre (1997).) As soon as you have a wave propagation mechanism you are liable to have, by that very fact, systematic correlations between fluctuating fields, of a kind that are completely neglected in classical turbulence theories, and in most modern turbulence theories as well. Such correlations profoundly alter the nature of the fluctuation-induced momentum transport, and with it the global circulation dynamics. Instead of momentum transport over the relatively short ranges characterizing eddy-induced material displacements, of the order of classical "mixing lengths", you suddenly have a long-range momentum-transport mechanism, limited only by the distances over which waves can propagate.

You can then have phenomena like anti-frictional behaviour. This means fluctuations that on average drive the system not toward, but away from, solid rotation: upgradient momentum transport, if you will. The most conspicuous illustrations are the quasi-biennial oscillation (QBO) of the zonal winds in the equatorial stratosphere, and the celebrated laboratory experiment of Plumb and McEwan (1978), which produces a similar phenomenon in a very simple way.

The Plumb–McEwan experiment illustrates anti-frictional behaviour in the most clearcut and unequivocal way imaginable: stratified flow in an annulus with symmetry axis vertical is driven away from solid — in this case zero — rotation, entirely by fluctuations of relatively high frequency, though still low enough to feel the internal gravity wave propagation mechanism. The fluctuations are excited simply by standing oscillations of the top or bottom boundary. The imposed conditions are mirror-symmetric (to reflection in a vertical mirror), but that symmetry is spontaneously broken and a strongly sheared differential rotation arises, first one way and then the other, reversing again and again on a much longer timescale and displaying a characteristic spacetime signature in which shear zones move systematically upward or downward, toward the fluctuating boundary that is the sole means of driving the system.[1]

Such anti-frictional behaviour is impossible in a classical turbulence scenario. But it is perfectly easy to understand in terms of the long-range momentum-transport — reaching far beyond mixing-length range — associated with the generation of waves in one place and their dissipation in another. Further discussion would be too much of a digression here; let me just say that the tendency toward anti-frictional behaviour, and the relevance to it of wave breaking, wave dissipation, critical layers and so on, have been discussed systematically in a recent review

---

[1] The experiment has recently been repeated at Kyoto University, with results available in the form of a movie at http://www.gfd-dennou.org/library/gfd_exp/ where, in addition, "technical tips" are given on how to run the experiment successfully.

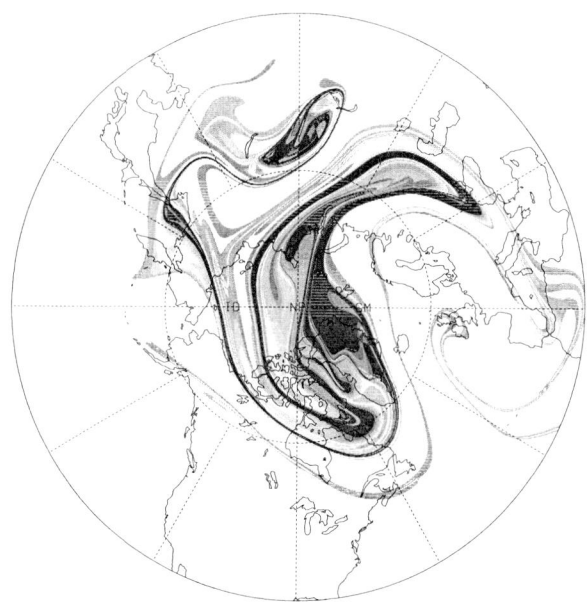

Figure 1. Contour-advective plot of polar-vortex air in the real lower stratosphere on 28 January 1992, from the work of Plumb *et al.* (1994) and D. W. Waugh (personal communication). Similar plots emphasizing midlatitude instead of polar air show strong mixing along the isentropic (stratification) surfaces, very like that seen in Figure 2.

(McIntyre 2000) in which special care is taken to distinguish, among other things, critical-layer myths from critical-layer realities.

And, as already hinted, there is more to the whole story than just the QBO and similar phenomena. Wave-induced momentum transport is now recognized as central to the fluid dynamics of greenhouse-gas-transporting atmospheric circulations, a crucial part of the Earth system, or climate system if you will. On a rapidly rotating planet such as the Earth you can, and do, have an almost ubiquitous, global-scale "gyroscopic pumping" remotely driven by wave-induced momentum transport. This is a persistent mechanical pumping action that drives, for instance, the stratospheric Brewer–Dobson circulation. Wave-induced mean forces push stratospheric air persistently westward, and Coriolis effects persistently turn the air poleward. Air is pulled up in the tropics and pushed back down in higher latitudes, against thermal radiative relaxation.

The Brewer–Dobson circulation is basic to understanding the behaviour of the ozone layer, and its part in the Earth system and in shaping the detectable and attributable patterns of climate change. The strength and persistence of the circulation accounts for many aspects of greenhouse-gas behaviour, for instance explaining why CFC lifetimes are of the order of 10^2 years. CFCs, which would otherwise take millennia or more to be removed from the atmosphere, are transported by the Brewer–Dobson circulation to altitudes above about 25 km, where they are destroyed relatively rapidly by hard solar ultraviolet.

And in the case of the stratospheric Brewer–Dobson circulation, where our understanding and its observational underpinning are now very secure, it is clear that the wave propagation mechanism chiefly responsible is the Rossby-wave mechanism.

2. The Rossby-wave jigsaw

In view of the above, I was astonished to hear someone say earlier this week, at this Symposium, that some proposal or paper had been rejected on the grounds that "Rossby waves do not exist in the real atmosphere or ocean". We actually know quite a lot about the real Rossby waves that propagate or diffract up from the real troposphere into the real stratosphere, where they dissipate through breaking and infrared radiative damping and drive the real Brewer–Dobson circulation — whose reality is itself consistent with many lines of chemical tracer evidence, all the way from Brewer's historic water-vapour measurements of 1949 (Norton et al. 2000) to today's beautiful visualizations of the tropical upwelling branch through the "tape recorder effect", in which the annual water-vapour cycle is imprinted on the air as it rises at about $0.2\,\mathrm{mm\,s}^{-1}$ (Mote et al. 1996, 1998). The Rossby mechanism is also central to our understanding of such phenomena as vortex coherence, barotropic and baroclinic shear instabilities and their nonlinear saturation, dynamical teleconnections, and indeed nearly all large-scale fluid motions of oceanographic and meteorological interest (e.g. Hoskins et al. 1985).

Much of today's knowledge of real, finite-amplitude Rossby waves began with the theoretical work of Rossby (1936), Ertel (1942), Charney (1948) and Kleinschmidt (1950–1) on the concept of potential vorticity (PV) and its inversion, together with observational studies of isentropic distributions of PV in the real stratosphere, beginning with a "damn fool experiment", in Erasmus Darwin's sense, that was done in the early 1980s at the UK Meteorological Office.

Having been working on some relevant theory at the time (McIntyre 1982, Killworth and McIntyre 1985), I became closely involved in interpreting that experiment together with one of the Met Office's young luminaries, Dr Tim Palmer (McIntyre and Palmer 1983–5). The experiment was foolish in Erasmus' sense because it attempted to use satellite data to visualize isentropic distributions of Rossby–Ertel PV in the middle stratosphere. In those days atmospheric researchers were already well aware of the problems with satellite data retrieval, and there was one version of the folklore saying that anyone with the temerity — the impudence — to try to compute, from satellite data, such a highly differentiated quantity as the Rossby–Ertel PV was, not to put too fine a point on it, a fool.

Nevertheless, the experiment defied the folklore and worked. It gave us our first glimpses of the PV distributions at altitudes around 30 km, in the midwinter extratropical stratosphere. We likened what we saw to "a blurred view of reality seen through... knobbly glass." Idealized theoretical models, especially the Stewartson–Warn–Warn (SWW) model of a Rossby-wave critical layer, helped us to make sense of that blurred view; and we were able to do enough cross-checks, for instance from Lagrangian trajectory computations — I remember the labour of doing some of

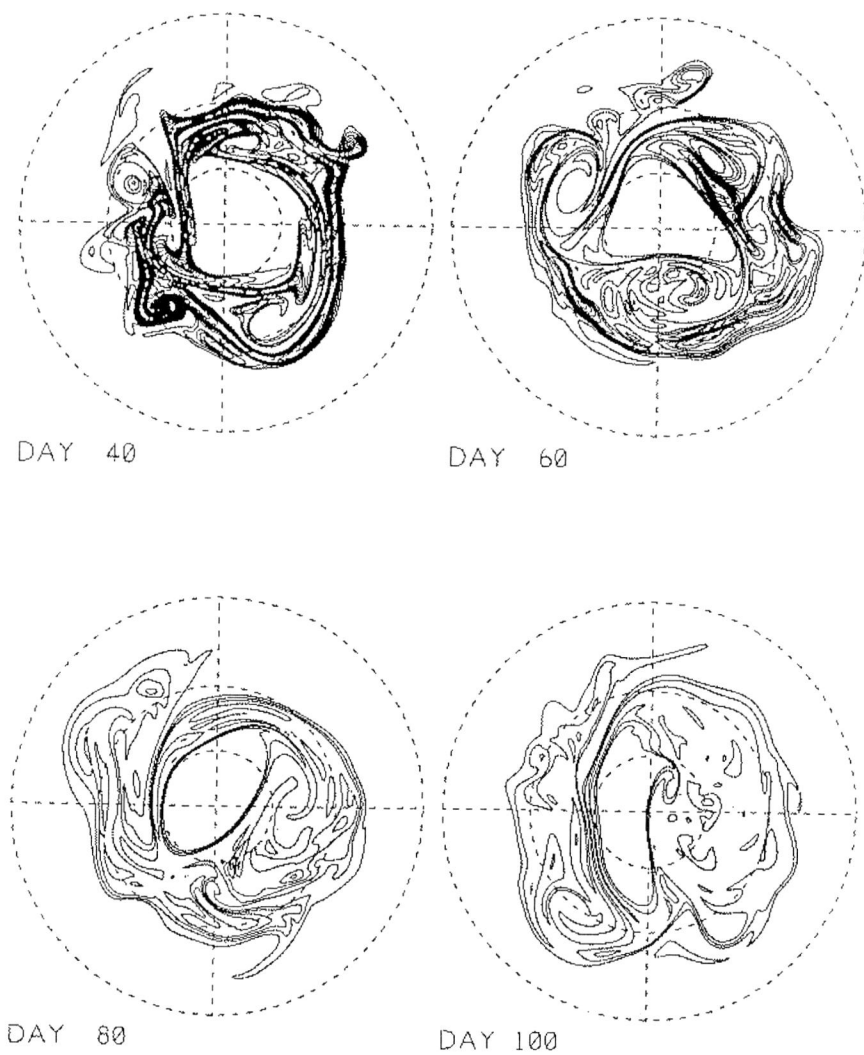

Figure 2. Shallow-water model flow on the sphere, an animation of which was shown at the Symposium, closely resembling flow in the real wintertime stratosphere at altitudes around 25 or 30 km. The map projection is conformal (polar stereographic), with the equator and the 30°N and 60°N latitude circles shown dashed. The flow is visualized by passive tracer released as a compact blob into the midlatitude stratospheric surf zone, clearly showing the fast two-dimensional turbulent mixing in that region, despite which the stratospheric polar vortex remains almost completely isolated from its surroundings, and likewise, to a lesser extent, the tropics (cf. Mote *et al.* 1998). The isolation of the (core of the) polar vortex recalls classic smoke rings and is of great importance to stratospheric polar chemistry, including the Antarctic ozone hole and its (so far less severe) Arctic counterpart. The isolation is due to the combined effects of the Rossby-wave restoring mechanism and the strong shear just outside the edge. Courtesy of Dr W. A. Norton, from whom an animated video of the model run is available (Dept of Atmospheric, Oceanic, and Planetary Physics, Clarendon Laboratory, Parks Road, Oxford, OX1 3PU, UK). Details of the model and the model run are given in Norton (1994). The mean depth is $h_0 = 4$ km, giving behaviour qualitatively like that implied by Equations (20) and (25) below, with the Rossby length $L_0 \sim 2000$ km in middle latitudes, and quantitatively very close to that implied by Equations (10)–(19) below.

them by hand — to add up to a convincing picture. The rest is history. The original experiment was enough to to make conspicuous the highly inhomogeneous "wave–turbulence jigsaw puzzle" that is typical of real, finite-amplitude Rossby-wave fields. The midlatitude stratosphere at altitudes around 30 km revealed itself as a gigantic "surf zone" driven by the "world's largest breaking waves." The surf zone corresponds to the theoretician's critical layer, whose supposed narrowness was thus revealed as one example of a critical-layer myth.

The same gigantic surf zone and spatial inhomogeneity is now seen routinely, and much more clearly, through clever combinations of data analysis and adaptive Lagrangian computational techniques such as "contour advection" (Norton 1994, Waugh and Plumb 1994), or "domain-filling trajectories" (Manney et al. 1994, Lahoz et al. 1996), applied to state-of-the-art meteorological wind analyses, themselves the result of operational data assimilation into large numerical models. In addition, the picture has been confirmed again and again by global-scale images of chemical tracer distributions remotely sensed from space, an outstanding recent example being some remarkably detailed CFC-11, methane, and nitrous oxide images from the helium-cooled CRISTA spectrometer flown on the Space Shuttle (Riese et al. 2001).

Figure 1 shows a contour-advection example from the work of Plumb et al. (1994), for the lower stratosphere at around 18 km altitude. This is a passive-tracer picture emphasizing polar-vortex air, constructed by contour advection over 4 days (see figure caption) on the 450 K isentropic surface. Some of the fine detail in pictures like these has been directly verified by *in situ* airborne chemical measurements (Waugh et al. 1994). Figure 2 shows another example, this time from a shallow-water model (Norton 1994), a full animation of which was shown as a video at the Symposium. This model behaves in a manner astonishingly like the real wintertime stratosphere at altitudes anywhere between about 20 and 40 km, as revealed for instance by the CRISTA images, which cover a similar range and an example of which is shown in Figure 3; note the different map projection emphasizing polar regions.

In the model example of Figure 2, whose bottom-right panel may be qualitatively compared to Figure 3, a small blob of passive tracer released in middle latitudes quickly fills the well developed surf zone, a layerwise-two-dimensional turbulent region sandwiched between relatively isolated polar-vortex and tropical regions. The video animation of Figure 2 makes especially vivid the wavelike, quasi-elastic behaviour of the polar-vortex edge under the Rossby-wave mechanism.[2] It is a clear example of what is now called an "eddy transport barrier", almost completely preventing surf-zone air from penetrating into the polar vortex region, a matter of some chemical importance.

From a dynamical perspective the wavelike and turbulent regions — the quasi-elastic vortex edge and the adjacent surf zone — are closely interdependent, as the term "jigsaw" is meant to suggest. That interdependence, a manifestation of nonlinear dynamics, is illustrated most plainly and explicitly by the SWW critical-

---

[2] Copies of the video are obtainable from Dr W. A. Norton at the Department of Atmospheric Physics, Clarendon Laboratory, Parks Road, Oxford, OX1 3PU, UK; see also Norton (1994).

layer model, of which our understanding is comprehensive (Killworth and McIntyre 1985, Haynes 1989). The model applies in a different parameter regime — in which the surf zone or critical layer is, in fact, narrow — but has the virtue that one can precisely quantify the dynamical interdependence of the surf zone and its more wavelike surroundings through matched asymptotic expansions.

The upshot of all these observational and theoretical studies, then, is that we may think of the main mass of polar-vortex air as being held together quasi-elastically by the Rossby mechanism,[3] with help from differential advection by the shear just outside — a good example of what is also called "vortex coherence" — even though some of the polar airmass is being eroded and mixed into the surrounding surf zone. The erosion and mixing involve the irreversible deformation of "otherwise-wavy" material contours, *i.e.* those contours, also PV contours in this case, that would undulate reversibly under the conditions implicitly assumed by linearized wave theory.

Irreversible deformation of such otherwise-wavy contours violates the nonacceleration theorem of wave–mean interaction theory, a consequence of Kelvin's circulation theorem applied to such contours. This makes the contour-deformation process fundamental to understanding the circumstances under which irreversible wave-induced momentum transport occurs. When the contour deformation is irreversible the process may therefore, very reasonably in this context, be designated as "Rossby-wave breaking", and a comparison drawn with the more familiar breaking of ocean-beach waves, which leads to the convergence of wave-induced momentum transport and its well-known consequence near ocean beaches, the generation of longshore currents. The big long tongue of polar air in Figure 1, curving away from the main polar airmass toward Mediterranean Europe, is quite like what Palmer and I originally saw "through knobbly glass" and is one of the typical large-scale Rossby wave-breaking patterns seen in practice. By contrast, the quasi-elastic region near the edge of the vortex core — showing up especially clearly in the case of Figure 2 — is marked by a set of material and PV contours that to good approximation undulate reversibly, as Norton (1994) showed in detail through high precision contour-advection calculations. Such reversible undulations are in stark contrast with the irreversible behaviour of the contours just outside.[4]

It will have been noticed that the extreme spatial inhomogeneity that shows up so vividly in all these cases, including the SWW model, represents another big departure from classical turbulence-theoretic scenarios. Those scenarios assume statistical homogeneity or near-homogeneity. The real spatial inhomogeneity is a very robust feature, showing up not only again and again in observations such

---

[3] See Figure 4. I am aware that "Rossby waves" might more aptly be called "Kelvin waves", especially in the context of vortex coherence. Though no historian of science, I have the impression that if you see someone's name on something there is a more than even chance that someone else thought of it first. There are always good reasons. "Kelvin waves" are usually understood to refer to Coriolis-trapped gravity waves, a different animal entirely. So in speaking of "Rossby waves" I am only using what has become the standard terminology, tending to displace the more logical alternatives "vorticity waves" and "potential-vorticity waves".

[4] McIntyre and Palmer (1984) discuss the time-reversal "paradox" involved here in the use of the term "irreversible".

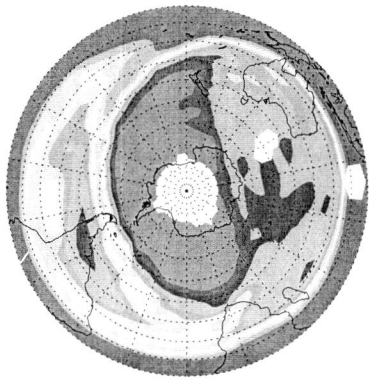 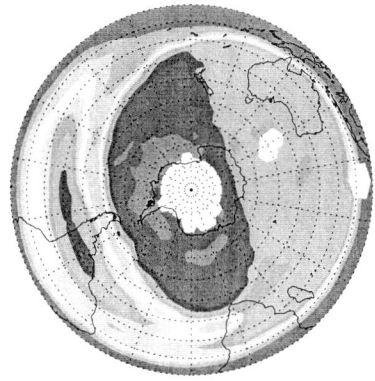

Figure 3. N_2O (nitrous oxide) mixing ratios observed at two stratospheric altitudes on 11 August 1997 by the CRISTA instrument, from Riese *et al.* (2001), for the southern hemisphere but mirror-flipped, in this display, for qualitative comparison to Figure 2 bottom right, which has positive Coriolis parameter $f > 0$ as appropriate to the northern hemisphere. Notice the different map projections, with polar regions emphasized here but tropical regions in Figure 2. Irregular white areas are data gaps. On Rossby-wave timescales of days and weeks N_2O is an accurate passive tracer, though destroyed photochemically on Brewer–Dobson timescales of years. In the right half of each picture the N_2O mixing ratios increase nearly monotonically or stepwise monotonically outward (being nearly constant over large areas in the surf zone). They increase from polar-vortex values close to zero to large tropical values imported from the troposphere by the Brewer–Dobson upwelling. **At left and right respectively,** pressure-altitudes are 4.64 hPa and 10 hPa, roughly 37 km and 31 km; ranges of mixing ratios in parts per billion by volume are 0–90+ and 0–150+ with contour intervals 10 and 16.67, where "+" signifies that values may slightly overshoot the plotted range. The lightest band at the subtropical edge of the surf zone highlights the ranges 60–70 and 100–116.67 ppbv. CRISTA (Cryogenic Infrared Spectrometers and Telescopes for the Atmosphere) detects a number of chemical species through their infrared spectral signatures and is a large (1350 kg) helium-cooled instrument flown from the Space Shuttle.

as Figure 3 but also in model simulations, all the way from idealized shallow-water models to the big three-dimensional models now widely used in studies of stratospheric chemistry and ozone depletion.

3. PV inversion

V. I. Arnol'd once stated that "Hamiltonian mechanics cannot be understood without differential forms" (Arnol'd 1978). In the same spirit I would say that Rossby waves cannot be understood without PV inversion. An understanding worth having will include, of course, an understanding of the nonlinear effects, such as the "jigsaw dynamics", the dynamical interdependence of vortex-edge undulation and surf-zone turbulence.

Rossby waves are often discussed without mentioning the idea of PV inversion at all — and if you are content with an idealized background state and infinitesimal wave amplitude then it's easy enough to calculate wavelike solutions and say that's all there is to it — but the idea of PV inversion is always there implicitly, right from the moment one tries to go further and understand even the linearized wave dynamics intuitively. I think it clarifies one's thinking to make the inversion

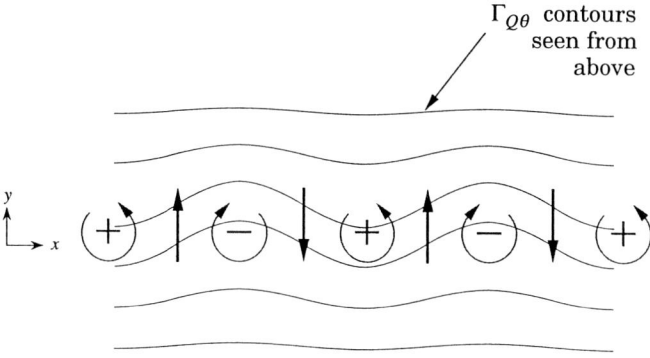

Figure 4. Visualization of the Rossby-wave propagation mechanism, redrawn from Hoskins *et al.* (1985) and McIntyre (2000).

idea explicit. Moreover, the resulting insight does help to bridge the gulf between idealized models and complex, nonlinear reality. I feel it's part of one's job as a theoretician to try to discern what is robust about an idealized model, and hence what aspects are likely to carry over to the real world and what aspects are not.

Thus when, for instance, the edge of the polar vortex is distorted, as illustrated in the video corresponding to Figure 2, this means that the PV, which has steep gradients across the vortex edge and is approximately a materially conserved, or advected, quantity, develops a pattern of anomalies relative to the undistorted circular shape of the edge. This pattern of PV anomalies is qualitatively as shown by the curved arrows and ± signs in Figure 4, if we read the coordinate x as azimuthal distance around the edge, periodically repeating as appropriate. The figure assumes that the background PV gradient is positive in the positive y-direction. The sideways displacement of material and PV contours on an isentropic surface — the edge distortion itself — is like the displacement of the wavy contours marked $\Gamma_{Q\theta}$, in the central part of Figure 4.

PV invertibility says that PV inversion is possible. This means that, under reasonable assumptions about boundary conditions, you can *deduce* the velocity field from the PV field. Equations (1)–(2) below provide the simplest illustration of this. In particular, any x-periodic pattern of PV anomalies like that in Figure 4 *implies* an x-periodic pattern of velocity anomalies a quarter wavelength out of phase with the displacement pattern, as suggested by the straight arrows. But if you have velocities a quarter wavelength out of phase with displacements, then you can directly see, by making a mental movie of the way in which the $\Gamma_{Q\theta}$ contours are advected, that the velocity field *makes* the undulations propagate, in this case in the $-x$ direction, and in all cases to the left of the background isentropic gradient of PV.

In summary, displacements produce PV anomalies, which (via PV inversion) imply velocities a quarter wavelength out of phase with displacements, which produce propagation in whichever direction has high PV values on the right. This is the basic Rossby-wave mechanism. It is basic to any stratified fluid motion that has gradients of PV on stratification surfaces (or buoyancy-acceleration gradients acting as concentrated PV gradients at upper or lower boundaries), and for which invertibility holds (Hoskins *et al.* 1985, & refs.). It is therefore basic to most large-scale fluid motions of oceanographic and meteorological interest; for instance, if you put opposite-signed PV gradients next to each other you will often get barotropic or baroclinic shear instabilities, in which counterpropagating Rossby waves phase-lock and make each other grow.

I think everyone is familar with the world's simplest example of the Rossby-wave mechanism, in the context of two-dimensional, nondivergent barotropic vortex dynamics in the original Rossby beta-plane or "approximately flat Earth" model. In this model, invertibility reduces simply to solvability of a Poisson equation $\nabla^2 \psi = (Q - f)$, under boundary conditions such as evanescence at infinity. Symbolically,

$$\psi = \nabla^{-2}(Q - f) . \qquad (1)$$

Here f is the Coriolis parameter, and $Q(x, y, t)$ is the PV, which in this example is the same as the two-dimensional absolute vorticity. The stream function $\psi(x, y, t)$ describes a strictly nondivergent velocity field $\mathbf{u}(x, y, t)$,

$$\mathbf{u} = (u, v) , \qquad u = -\partial \psi / \partial y , \qquad v = \partial \psi / \partial x , \qquad (2)$$

where (x, y) are eastward and northward Cartesian coordinates and (u, v) the corresponding components of the velocity vector $\mathbf{u}(x, y, t)$. The inversion is a diagnostic process; no time derivatives appear in (1)–(2), and so one can talk about inversion at a single instant t, independently of neighbouring t values.

In this dynamical system there is just one evolution equation,

$$DQ/Dt = 0 , \qquad (3)$$

where D/Dt is the two-dimensional material derivative,

$$D/Dt := \partial / \partial t + \mathbf{u} \cdot \nabla = \partial / \partial t + u \partial / \partial x + v \partial / \partial y . \qquad (4)$$

We can think of the dynamical system as a limiting case of real stratified flow in which friction goes to zero and the stratification becomes infinitely strong, constraining the motion to be exactly horizontal. This picks out the Coriolis parameter f as precisely the vertical component of twice the Earth's angular velocity, implying that f increases monotonically with northward distance y like the sine of the latitude. In taking such a limit, one must assume that vertical scales of motion stay finite. Then the dependence on the vertical coordinate z becomes a mere parametric dependence, with no vertical derivatives $\partial/\partial z$ appearing in the problem. The equations at each altitude z then reduce to Equations (1)–(4) above.

The standard Rossby-wave theory for this system — linearize equation (3) about rest, set $\beta = df/dy =$ constant, look for solutions $\psi = \Re \hat{\psi} \exp(ikx + ily - i\omega t)$ ($\hat{\psi}$ and ω constant, k and l real, constant) so that $\nabla^{-2} = -(k^2 + l^2)^{-1}$ and

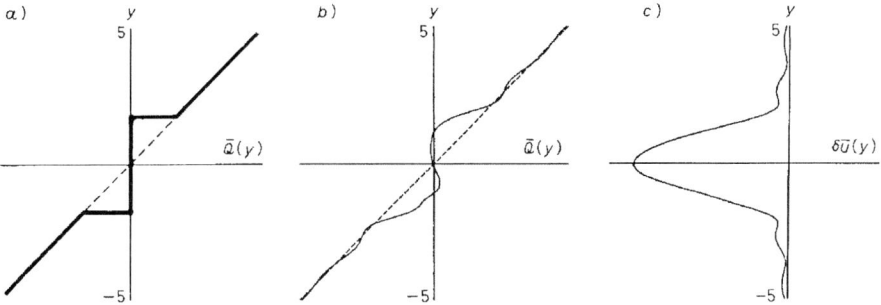

Figure 5. The relation between mean wave-induced force and Q-rearrangement by a breaking Rossby wave, in the simplest relevant model system, the dynamical system (1)–(4). Courtesy P. H. Haynes; for mathematical details see Killworth and McIntyre (1985) and Haynes (1989). Plot (a) shows idealized Q distributions before and after mixing Q in some y-interval or latitude band; (b) shows the x-averaged Q distribution in an actual model simulation using equations (1)–(4); (c) shows the resulting mean momentum deficit, equation (5), whose profile would take on a simple parabolic shape in the idealized case corresponding to (a).

$\omega = -\beta k/(k^2 + l^2)$ (real) — confirms the qualitative picture suggested by Figure 4. The y-velocity $\partial \psi/\partial x$, with complex amplitude $ik\hat{\psi}$, is a quarter wavelength out of phase with the displacement, with complex amplitude $(-i\omega)^{-1}ik\hat{\psi}$. The chirality — the one-way character of the propagation, tied to the sense of the Earth's rotation — is expressed by the single power of ω in the dispersion relation $\omega = -\beta k/(k^2 + l^2)$.

The association between Rossby-wave breaking and irreversible wave-induced angular momentum transport, and the dynamical interdependence of vortex-edge undulation and surf-zone turbulence in general, are likewise well illustrated by the dynamical system (1)–(4). The PV inversion operator (1) is non-local, so it's obvious from the outset that there has to be some such dynamical interdependence. And for angular momentum transport we may consider, for instance, an idealized thought experiment in which Rossby waves arrive and form a surf zone by perfectly mixing the PV in some band of y-values, as shown on the left of Figure 5. The mean change $\delta \bar{Q}(y)$ in $\bar{Q}$ is a linear function within that band, integrating to zero across the band. Inversion according to (1) with reasonable boundary conditions gives an x-averaged mean flow change

$$\delta \bar{u}(y) = \int_y^\infty \delta \bar{Q}(\tilde{y}) \, d\tilde{y} \leqslant 0 \, . \tag{5}$$

This is a parabolic profile qualitatively like the $\delta \bar{u}(y)$ profile shown on the right of Figure 5, which is from a less idealized, fully detailed chaotic critical-layer calculation due to Haynes (1989), with the $\delta \bar{Q}(y)$ profile shown in the middle. In both cases there is a robustly negative net momentum change. This is a phenomenon in which wavelike and turbulent motion are both inextricably involved.[5]

It is also an example of anti-frictional behaviour, because $\delta \bar{u}(y)$ has to be added to a background flow $\bar{u}(y) \propto y$ with positive shear $\partial \bar{u}(y)/\partial y > 0$ (not shown in the figure), and the wave source is located at positive y, outside the domain of the figure. Thus the momentum has been transported against its own mean gradient.

The point of all this is that more realistic models have just the same generic structure, and qualitative properties, provided only that we have PV invertibility. In particular, if we go back to realistic finite stratification, then the dynamical system still has the single evolution equation (3) with its single time derivative, expressing chirality, together with an inversion operator that behaves qualitatively like that in (1). It implies the same qualitative description of Rossby-wave propagation as before, including the one-way character associated with the single time derivative and the single power of ω in the dispersion relation. It implies the same qualitative association between surf-zone formation and irreversible wave-induced momentum transport, robustly one-signed. This has been verified explicitly by the work of Robinson (1988), on stratified flow in spherical geometry.

PV invertibility is helpful, moreover, in coping with the central difficulty in fluid dynamics. This is the advective nonlinearity, as expressed by the $\mathbf{u} \cdot \nabla$ terms in the Euler equations. If the main effects of $\mathbf{u} \cdot \nabla$ are captured by the D/Dt in a single evolution equation of the simple form (3), then we have the simplest conceivable way of visualizing and understanding the effects of the advective nonlinearity. It is in all these ways that the qualitative insight from the world's simplest model does, indeed, carry over to more sophisticated models of reality.

The discussion up to this point is one way of showing why the ideas of PV inversion and PV invertibility are important, and therefore, by implication, why the ideas of balance and "slow manifold" are important. As is well known, the same ideas are crucial to an understanding of how to initialize weather-prediction models (*e.g.* Lynch and McGrath, this Proceedings), a matter whose importance was first revealed by Richardson's pioneering attempt to integrate the equations numerically.

4. Accurate balance and PV inversion

As is well known, PV invertibility, and weather-prediction initialization, both depend on imposing some balance condition. Imposing such a condition amounts to artificially constraining the dynamical system to some prescribed "slow manifold" within its full phase space. In the system (1)–(4), the balance condition is simply the incompressibility condition (2). It implies an absence of sound waves. In more

---

[5] Results like (5) have sometimes been thought to show that the PV mixing illustrated at left and middle of Figure 5 is impossible (*e.g.* Stewart and Thomson 1977) since, the argument goes, momentum conservation would be violated. This underlines the danger of neglecting the wavelike aspects of the jigsaw puzzle, *i.e.* of thinking solely in terms of classical turbulence paradigms and hence missing the possibility of momentum being imported from elswhere via wave-induced momentum transport. It is just this possibility, or rather actuality, that is so cogently illustrated by the SWW and related critical-layer solutions such as those of Haynes (1989).

realistic systems, from shallow-water to fully-stratified, condition (2) is replaced by a functional relation of the form

$$\mathbf{u}(\mathbf{x}) = \mathbf{u}^B(\mathbf{x}; h(\cdot)) \ , \tag{6}$$

where explicit reference to time t is suppressed for the moment, and where $h(\cdot)$ symbolically represents the mass configuration of the fluid system at a given instant t. In a shallow-water system, for instance, $h(\cdot)$ is shorthand for the height $h(\mathbf{x}) = h(x, y)$ of the free surface above some horizontal reference level. The balance condition (6) says that the velocity field is completely determined, at each instant t, by a knowledge of the mass configuration — the spatial distribution of fluid mass throughout the physical domain at that instant. It says that the velocity field no longer represents independent degrees of freedom in phase space. The condition (6) confines the system to a submanifold within phase space. A simple albeit crude illustration is provided by the familiar geostrophic relation, defined for the shallow-water case by

$$\mathbf{u}(\mathbf{x}) = \mathbf{u}^B(\mathbf{x}; h(\cdot)) = \mathbf{u}^G(\mathbf{x}; h(\cdot)) := \frac{g}{f}\left(-\frac{\partial h(\mathbf{x})}{\partial y}, \frac{\partial h(\mathbf{x})}{\partial x}\right) , \tag{7}$$

where g is the gravitational acceleration. The associated PV inversion operator is defined by (7) together with a definition of the PV and suitable boundary conditions, such as evanescence at infinity. Once again, no time derivatives appear: we have a diagnostic process, defined at each instant t. For the definition of PV we may use for instance the shallow-water PV discovered by Rossby (1936), who showed it to be an exact material invariant for frictionless flow (*op. cit.*, Eq. (75)). In the simplest case of a flat bottom identified with the horizontal reference level, the Q of §3 is then replaced by the exact Q defined by Rossby's formula[6]

$$Q = \frac{1}{h}\left(f + \frac{\partial v}{\partial x} - \frac{\partial u}{\partial y}\right) = \frac{1}{h}(f + \zeta) , \quad \text{say.} \tag{8}$$

The resulting boundary-value problem has variable coefficients but is elliptic and robustly invertible, like (1) above, at least when Froude and Rossby numbers are sufficiently small:

$$F = \sup(|\mathbf{u}|/c) \ll 1 \ ; \qquad R = \sup(|\zeta|/f) \ll 1 \ . \tag{9}$$

Here $c = \sqrt{(gh)}$, the shallow-water gravity-wave speed. Replacing Q by a linearized counterpart based on writing $h' = h - h_0$ and neglecting squares $(h'/h_0)^2$ and products $(h'/h_0)(\zeta/f)$, with h_0 constant, gives the standard "quasi-geostrophic" version of the theory in which the PV inversion problem has constant coefficients.

Now balance conditions far more accurate than (7) are known. For instance we may replace (7) by the following set of diagnostic equations, to be solved (at

---

[6]Rossby's formula is still valid for a sloping bottom, of course, provided that h is temporarily redefined as layer depth instead of surface elevation, in which case (7) is modified appropriately.

each instant t) for $\mathbf{u}^{\mathrm{B}}$ when a suitable $h'(\mathbf{x})$ field is given. It is convenient to introduce the Helmholtz decomposition of $\mathbf{u}^{\mathrm{B}}$ into its solenoidal and irrotational contributions, under suitable boundary conditions,

$$\mathbf{u}^{\mathrm{B}} = \mathrm{curl}^{-1}\zeta^{\mathrm{B}} + \mathrm{div}^{-1}\delta^{\mathrm{B}} \qquad (10)$$

where ζ^{B} and δ^{B} are the corresponding vorticity and divergence defined respectively as $\partial v^{\mathrm{B}}/\partial x - \partial u^{\mathrm{B}}/\partial y$ and $\partial u^{\mathrm{B}}/\partial x + \partial v^{\mathrm{B}}/\partial y$. In the remaining equations, $c_0 := \sqrt{(gh_0)}$; the quantities with subscripts 1 and 2 are auxiliary fields to be explained shortly:

$$f\zeta^{\mathrm{B}} = g\nabla^2 h' + \nabla \cdot (\mathbf{u}^{\mathrm{B}} \cdot \nabla \mathbf{u}^{\mathrm{B}}) + \delta_1^{\mathrm{B}}, \qquad (11)$$

$$\delta^{\mathrm{B}} = \mathcal{L}^{-1}\nabla \cdot \left\{f\zeta^{\mathrm{B}}\mathbf{u}^{\mathrm{B}} + \mathbf{u}_1^{\mathrm{B}} \cdot \nabla \mathbf{u}^{\mathrm{B}} + \mathbf{u}^{\mathrm{B}} \cdot \nabla \mathbf{u}_1^{\mathrm{B}} - g\nabla^2(h'\mathbf{u}^{\mathrm{B}})\right\}, \qquad (12)$$

$$\delta_1^{\mathrm{B}} = \mathcal{L}^{-1}\nabla \cdot \left\{f\zeta_1^{\mathrm{B}}\mathbf{u}^{\mathrm{B}} + f\zeta^{\mathrm{B}}\mathbf{u}_1^{\mathrm{B}} + \mathbf{u}_2^{\mathrm{B}} \cdot \nabla \mathbf{u}^{\mathrm{B}} + 2\mathbf{u}_1^{\mathrm{B}} \cdot \nabla \mathbf{u}_1^{\mathrm{B}}\right.$$
$$\left. + \mathbf{u}^{\mathrm{B}} \cdot \nabla \mathbf{u}_2^{\mathrm{B}} - g\nabla^2(h_1^{\mathrm{B}}\mathbf{u}^{\mathrm{B}} + h'\mathbf{u}_1^{\mathrm{B}})\right\}, \qquad (13)$$

$$\mathbf{u}_1^{\mathrm{B}} = \mathrm{curl}^{-1}\zeta_1^{\mathrm{B}} + \mathrm{div}^{-1}\delta_1^{\mathrm{B}}, \qquad (14)$$

$$\mathbf{u}_2^{\mathrm{B}} = \mathrm{curl}^{-1}\zeta_2^{\mathrm{B}}, \qquad (15)$$

$$\zeta_1^{\mathrm{B}} + f\delta^{\mathrm{B}} = -\nabla \cdot (\mathbf{u}^{\mathrm{B}}\zeta^{\mathrm{B}}), \qquad (16)$$

$$\zeta_2^{\mathrm{B}} + f\delta_1^{\mathrm{B}} = -\nabla \cdot (\mathbf{u}_1^{\mathrm{B}}\zeta^{\mathrm{B}} + \mathbf{u}^{\mathrm{B}}\zeta_1^{\mathrm{B}}), \qquad (17)$$

$$h_1^{\mathrm{B}} + h_0\delta^{\mathrm{B}} = -\nabla \cdot (\mathbf{u}^{\mathrm{B}}h'). \qquad (18)$$

Here

$$\mathcal{L} := c_0^2\left(\nabla^2 - L_0^{-2}\right), \qquad (19)$$

a modified Helmholtz operator depending on the natural length scale

$$L_0 := c_0/f, \qquad (20)$$

the Rossby length or "radius" based on c_0. No time derivatives appear anywhere in (10)–(18). When $h'(\mathbf{x})$ is given, these are nine equations (six scalar and three vector) to determine, diagnostically, the nine unknown functions $\zeta_2^{\mathrm{B}}(\mathbf{x})$, $\zeta_1^{\mathrm{B}}(\mathbf{x})$, $\zeta^{\mathrm{B}}(\mathbf{x})$, $\delta_1^{\mathrm{B}}(\mathbf{x})$, $\delta^{\mathrm{B}}(\mathbf{x})$, $h_1^{\mathrm{B}}(\mathbf{x})$, $\mathbf{u}_2^{\mathrm{B}}(\mathbf{x})$, $\mathbf{u}_1^{\mathrm{B}}(\mathbf{x})$, and $\mathbf{u}^{\mathrm{B}}(\mathbf{x})$. The quantities with subscripts 1 and 2 are diagnostic estimates of first and second partial time derivatives. Being diagnostic quantities, they must be sharply distinguished from actual rates of change in a model integration. The point is disussed more fully in my paper with Norton (2000, hereafter MN00), along with related issues concerning local mass conservation and Galilean invariance. If $\delta_1^{\mathrm{B}}(\mathbf{x})$ were to be replaced by $\partial\delta/\partial t$ and the superscripts B deleted from the other variables, then (11) would become the divergence equation of the exact shallow-water equations; (16) similarly corresponds to the exact vorticity equation and (18) to the exact mass-conservation equation. Note also that if every term on the right of (11) were to be deleted except the first, then (11) would reduce to the curl of the geostrophic relation (7). Equations (12) and (13) are derived from the first and second time derivatives of the divergence equation with the leading time derivatives deleted. Standard scale-analytic considerations would argue that these deleted terms are relatively small if F and R are small.

When (8) is appended to (10)–(19), one obtains the PV inversion operator that MN00 called a "third order direct" inversion operator. It is exquisitely accurate, over an astonishingly wide range of values of F and R, as MN00 demonstrate for complicated shallow-water vortical flows on a hemisphere. (See Appendix A of MN00 for notes on the numerical procedures and for the counterpart of (10)–(19) in spherical coordinates, taking account of variable f.) Not only the vorticity but also the divergence field are reconstructed in considerable detail from a knowledge of the PV alone. Almost incredibly, this accuracy is obtained despite R being infinite at the equator, and F reaching values in excess of 0.7 in some cases. One can hardly say that ∞ and 0.7 are small. Trying to carry out a PV inversion for such parameter values certainly counts as another "damn fool experiment" — and most of the credit for it is due to Dr Warwick Norton, who was my research student at the time, and who showed marvellous intellectual courage as well as considerable computational ingenuity. The exquisite accuracy comes at a price, of course: it depends on subtle, weakly nonlinear corrections that vitiate the superposition principle and demand elaborate iterative numerical methods.

Even more accurate inversion operators can be defined, based on normal mode expansions. For further discussion see MN00 and Mohebalhojeh and Dritschel (2001). Figure 6 shows an example taken from MN00 — which still astonishes me even though it was first obtained a decade ago — again in a hemispherical domain and with F again exceeding 0.7. This is a cumulative accuracy test, using the PV-conserving balanced model defined by a normal-mode-based PV inversion operator together with the single prognostic equation $DQ/Dt = 0$, for shallow-water flow on the hemisphere, slightly modified with a ∇^6 hyperdiffusion to control numerical noise at discretization scales. The top half of Figure 6 shows two PV fields from a 10-day run of the balanced model, and the bottom half the corresponding fields from a carefully initialized run of the exact shallow-water equations, serving as the benchmark of accuracy. This is a complicated, highly unsteady vortical flow exhibiting hyperbolicity or phase-space sensitivity. See MN00 for evidence of that sensitivity and for the precise specification of the inversion operator. Even after 10 days or several eddy times, and some nontrivial vortex interactions including merging, the two PV distributions track each other almost perfectly.

Now there is indeed something truly mysterious about such accuracy. Standard order-of-magnitude arguments say that we have no right to expect the concepts of balance and inversion, and the resulting balanced models, to be accurate unless F and R are small. Accurate balance and inversion involve nonlocal functional relations, through operators such as curl^{-1}, div^{-1}, and $\mathcal{L}^{-1}$, as the notation $\mathbf{u}^B(\mathbf{x}; h(\cdot))$ in (6) was meant to suggest. Just as with (1), such relations imply, so to speak, action-at-a-distance. A change in the Q value here influences the velocity over there; and it does so instantaneously. In the simple system (1)–(4) this is reasonable, because the balance condition (2) makes the speed of sound infinite. But $F > 0.7$ means that even the fastest inertia–gravity waves are barely faster, in terms of group velocity, than relative fluid velocities. PV inversion is indeed

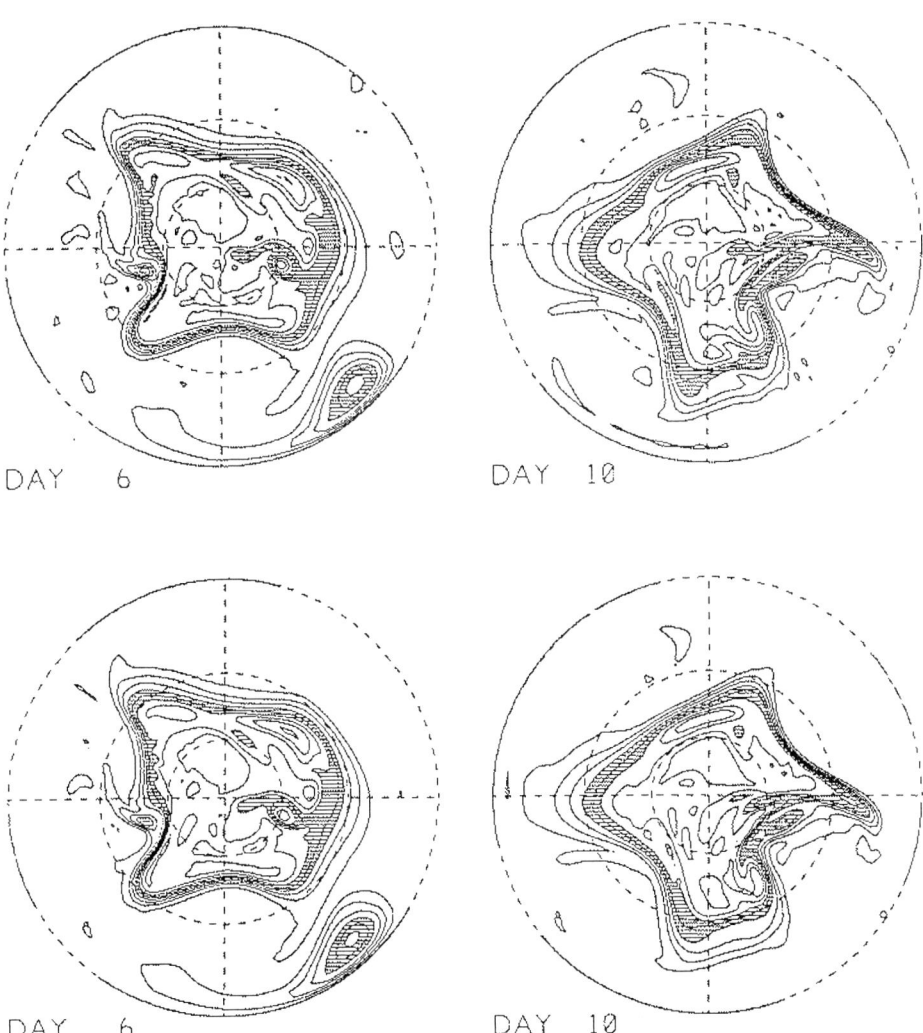

Figure 6. PV fields showing cumulative accuracy over 10 days (several eddy times) of a sophisticated "nonlinear normal mode" PV inverter for shallow-water flow on a hemisphere. The top pair are from a 10-day run of a PV-conserving balanced model based on this PV inverter; the bottom pair are from the corresponding run of the full shallow-water equations. Parameter regime is realistic for upper-tropospheric flow, with velocities in the tens of m s^{-1} and gravity-wave speed $c_0 = \sqrt{(gh_0)} = 100 \text{ m s}^{-1}$, with h_0 defined as area-averaged layer depth. The PV is defined as in (8). The contour interval is $3 \times 10^{-8} \text{ m}^{-1}\text{s}^{-1}$; the shading marks values lying between 4 and 6 of these units. The two sets of PV fields are astonishingly close to each other, almost down to the discretization scales that feel the model ∇^6 hyperdiffusion, despite the infinite Rossby number R at the equator and the value, just over 0.7, reached by the Froude number F as defined by (9).

5. Poincaré's homoclinic tangle and the slow quasimanifold

So let us draw a deep breath and stand back a moment. As has often been pointed out, balanced motion has its counterparts in simpler dynamical systems. One of these is the "springy pendulum" consisting of a mass suspended from a pivot by a stiff elastic spring. Such a pendulum has slow, rotating or swinging, modes of motion in which the relatively fast, compressional, modes of the mass and spring are hardly excited. A first approximation in describing such slow motions is simply to set the length of the spring to be constant. This might be compared to the incompressibility condition (2). More accurate approximations would allow the spring to change its length in a quasi-static way. The fast modes are then, so to speak, "slaved" to the slow modes. In the fluid system this is another way of viewing the effect of a balance condition $\mathbf{u}(\mathbf{x}) = \mathbf{u}^B(\mathbf{x}; h(\cdot))$, be it the geostrophic relation (7) or any of its more accurate counterparts such as (10)–(19). Such approximations and their ultimate limitations can be studied mathematically via techniques ranging all the way from formal two-timing (multiple scales) to bounded-derivative theory and, in the pendulum case at least, KAM (Kolmogorov–Arnol'd–Moser) theory and other dynamical-systems techniques for finite phase spaces. There is an enormous literature (*e.g.* Bokhove and Shepherd 1996 & refs.).

Basic to understanding what is involved is Henri Poincaré's picture of the "homoclinic tangle". This is a dynamical-systems classic that is now textbook material; it applies unequivocally, for instance, to the pendulum started slowly from its upside-down equilibrium position. In the phase space of a simple, idealized *rigid* pendulum, there is a homoclinic orbit representing one complete circuit of the pendulum, taking a logarithmically infinite time. Poincaré's homoclinic tangle (between perturbed stable and unstable manifolds, whose formidable fine-grain complexity Poincaré himself did not even attempt to draw, and which there is no room to discuss adequately here) tells us, in effect, that if we nonlinearly couple the simple rigid pendulum to practically any other oscillator — such as that associated with the springiness — or even just persistently jiggle the suspension point in a prescribed, deterministic way, then the single homoclinic orbit will break up into a "stochastic layer" or "chaotic layer" having finite though possibly small thickness. In KAM language, this homoclinic orbit is a torus that always breaks up under perturbation. This is intuitively reasonable. A logarithmically infinite time is available, which is plenty of time for practically any slight disturbance to nudge the pendulum into neighbouring parts of phase space, and enormously large numbers of ways for it to do so — depending on initial conditions on the fast modes, such as the states of the spring, or on the imposed jiggling. So if one replaces the exact springy-pendulum motion by a balanced model of it (in which information about the fast initial conditions or the jiggling is lost) then one is replacing the stochastic layer — which we might call a *slow quasimanifold* to emphasize that it is *not* a single invariant manifold of the system — by an artificially imposed, infinitesimally thin slow manifold. In effect, this artifice suppresses all the uncertainty due to the lost information.

By analogy, we expect that the accuracy seen in Figure 6 cannot mean that the exact shallow-water equations naturally have a true "slow manifold", a precisely invariant manifold within the full phase space whose existence implies the possibility of exact "superbalance", exact PV inversion, and therefore (incredibly) exact action-at-a-distance despite finite inertia–gravity wave speeds. Rather, even though strict mathematical proof in this infinite-dimensional problem seems well beyond anyone's capability, the pendulum analogy leads us to expect that exact shallow-water motion close to balance — such as the motion corresponding to the bottom half of Figure 6 — must have a trajectory within a thin, though not infinitesimally thin, slow quasimanifold or infinite-dimensional stochastic layer within the infinite-dimensional phase space of the exact shallow-water equations.

6. Lighthill radiation — geostrophic it's not!

It is here that James Lighthill's work enters the picture. Lighthill's pioneering thinking about aerodynamic sound generation, the emission of sound waves by, in his case, three-dimensional vortical motion such as turbulent jets, can be adapted to our problem even though there are some nontrivial technicalities (Ford *et al.* 2000). Lighthill's thinking beautifully complements Poincaré's homoclinic-tangle argument. It gives us a profound physical insight (a) showing, quite independently of Poincaré's argument, why one expects a slow quasimanifold rather than a true slow manifold (in the infinite-dimensional phase space of the exact fluid equations), and (b) some idea of why the slow quasimanifold should be so remarkably thin in the fluid case, as Figure 6 emphasizes, certainly thinner than any simple order-of-magnitude consideration could possibly predict.

Lighthill's essential insight is contained in the phrases "quadrupole radiation" and "destructive interference". It tells us (a) that unsteady vortical motions, like those illustrated in Figures 1–3 and 6 of this paper, will, in continually adjusting toward balance, emit inertia–gravity waves spontaneously but (b) that such "spontaneous-adjustment emission" is weak because of destructive interference. This begins to explain why PV inversion can work better than it has a right to, and why any approach via simple order-of-magnitude analysis, or any other approach neglecting the full subtlety of the problem, will tend to overestimate the strength of spontaneous-adjustment emission.

The adjective *spontaneous*, incidentally, is essential for clarity here. Spontaneous-adjustment emission is not to be confused, as sometimes occurs in the literature, with inertia–gravity wave emission due to what is today called "Rossby adjustment". Rossby or initial-condition adjustment is the process famously encountered by Richardson, in his first attempt at numerical weather prediction. It is simply and straightforwardly the emission of inertia–gravity waves due to imbalanced initial conditions, artificially imposed. There is nothing subtle here. If you kick the system, you will excite inertia–gravity waves. I am also avoiding the term "geostrophic adjustment" because — when the adjustment is spontaneous — it is more likely, if anything, to be adjustment away from geostrophic balance (7), $\mathbf{u} = \mathbf{u}^{\mathrm{G}}(\mathbf{x}; h(\cdot))$, as adjustment toward it. It is likely to be toward a more accurate balance $\mathbf{u} = \mathbf{u}^B(\mathbf{x}; h(\cdot))$. Part of the trouble may be that the term "geostrophic" is often used not in its precise technical sense, $\mathbf{u} = \mathbf{u}^G(\mathbf{x}; h(\cdot))$, but rather as a word to mean anything vaguely to do with the general concept of balance. The next thing that happens, of course, is that the word gets used in both senses at once!

For our rotating systems, the appropriate generalization of Lighthill's original argument is as follows. Again we use the shallow-water equations with constant f. The momentum and mass-conservation equations are taken in flux form:

$$\frac{\partial}{\partial t}(hu_i) + \frac{\partial}{\partial x_j}(hu_i u_j) - \varepsilon_{ij} f h u_j + \frac{g}{2}\frac{\partial}{\partial x_i}(h^2) = 0, \qquad (21)$$

$$\frac{\partial h}{\partial t} + \frac{\partial}{\partial x_i}(hu_i) = 0, \qquad (22)$$

where u_i is the ith Cartesian component of $\mathbf{u}$ ($i = 1, 2$), and ε_{ij} is the two-dimensional alternating tensor, defined by $\varepsilon_{12} = -\varepsilon_{21} = 1$ and $\varepsilon_{11} = \varepsilon_{22} = 0$. A little manipulation (Ford et al. 2000) produces

$$\left(\mathcal{L} - \frac{\partial^2}{\partial t^2}\right)\frac{\partial h}{\partial t} = -\frac{\partial^2}{\partial x_i \partial x_j} T_{ij}, \qquad (23)$$

where $\mathcal{L}$ is the modified Helmholtz operator (19) and where

$$T_{ij} = \frac{\partial}{\partial t}(hu_i u_j) + \frac{f}{2}(\varepsilon_{ik} h u_j u_k + \varepsilon_{jk} h u_i u_k) + \frac{g}{2}\frac{\partial}{\partial t}(h'^2)\delta_{ij}. \qquad (24)$$

Equation (23) has the form of the linear inertia–gravity wave operator $\mathcal{L} - \partial^2/\partial t^2$ acting on $\partial h/\partial t$, on the left-hand side, and an *apparent wave source* consisting of the nonlinear terms on the right-hand side.

The essential point noted in Lighthill (1952) is that, in the case $F \ll 1$, the right-hand side is known to good approximation from the vortical flow alone, and can therefore be regarded as a *given* source of inertia–gravity waves.

The only essential assumption is that corrections $O(F)$ or weaker cannot change the qualitative character of the right-hand side. It is crucial, however — and this was Lighthill's most important point — that any approximate representation of the vortical flow be first substituted into T_{ij} before the differentiations $\partial^2/\partial x_i \partial x_j$ are carried out. This is because the weakness of the radiation depends on the cancellation, or destructive interference, already mentioned. It is the second-derivative form of the right-hand side of (23), rather than the precise form of T_{ij} itself, that is crucial to the cancellation and corresponds to the celebrated "quadrupole radiation". It is the weakness of quadrupole radiation, in other words, that begins to account for the possible accuracy of PV inversion. It can be added that the introduction of Coriolis effects should weaken the radiation still further, because of the cutoff at the inertia frequency $\omega = f$. Further discussion may be found in Ford et al. (2000) and in Saujani and Shepherd (2001).

Of course Lighthill's argument leaves unanswered the question "How does the wave emission remain weak even when the Froude number F is not small?" It is not obvious from Lighthill's argument why such a thing should come about. Here there is still considerable mystery. Part of the answer seems to be that the PV inversion operator takes on more and more of a *short-range character* as the Froude number F increases. Quasi-geostrophic theory, though quantitatively inaccurate, is enough to give a qualitative feel for this point. Within that theory — based, as mentioned earlier, not only on approximating $\mathbf{u}^B(\mathbf{x}; h(\cdot))$ by (7) but also on neglecting squares $(h'/h_0)^2$ and products $(h'/h_0)(\zeta/f)$ in Q, replacing it by a linearized counterpart Q_{qg} say — it is well-known that the PV inversion operator becomes essentially $\mathcal{L}^{-1}$, where $\mathcal{L}$ is again the modified Helmholtz operator defined in (19), relevant to cases in which the Coriolis parameter f is exactly or approximately constant. The resulting PV inversion sets the divergence to zero and can be written explicitly, in an unbounded xy-domain, in terms of the stream function

$$\psi = -\frac{1}{2\pi} \iint K_0\Big(\frac{|\mathbf{x}-\mathbf{x}'|}{L_0}\Big) Q_{qg}(\mathbf{x}')\, d^2\mathbf{x}' \; ; \qquad (25)$$

$K_0(\cdot)$ is the modified Bessel function with exponential decay at large argument. This represents a short-range interaction, because of the exponential tail of the Bessel function, and is such that the range decreases in proportion to L_0 and in inverse proportion to F, for given typical velocities $|\mathbf{u}|$, as F increases.

So even though the inertia–gravity waves cannot travel infinitely fast in order to mediate the action-at-a-distance implied by balance and PV inversion, the influence has a shorter and shorter distance to travel, as F increases and L_0 decreases.

Another consequence of this shortening interaction distance is that the vortex dynamics becomes distinctly more sluggish or slow moving. Thus the spontaneous-adjustment emission, which depends on unsteadiness of the vortex dynamics, is in this problem weaker, in most circumstances, than Lighthill's scaling laws would suggest. This must be another part of the explanation of the near-miracle of Figure 6, in which it should be noted that the Rossby length $L_0 \sim 1000\,\mathrm{km}$ in middle latitudes. For further insight the reader is referred to a paper by Ford (1994), presenting some very careful numerical experiments on Lighthill radiation.

7. Concluding remarks

It has sometimes been argued that spontaneous-adjustment emission is not properly described as "Lighthill emission" or "Lighthill radiation" because, for instance, a domain such as that of Figure 6 isn't really infinite, as in Lighthill's original problem or its rotating counterpart studied by Ford (1994) and Ford *et al.* — or, again, because in a rotating system the emission is a lot weaker than Lighthill's original (non-rotating) scaling laws would say. I would argue that that misses Lighthill's main point. It comes down to saying that in *any* situation where the spontaneous-adjustment emission of inertia–gravity waves is in fact weak, for any reason, one can regard the right-hand side of (23) as known to leading order and thus make a conceptual separation between vortical motion and wave emission. This point seems to me to be quite independent of whether the domain is bounded or unbounded or whether the Coriolis parameter f is zero or nonzero. It is important because of the help it gives in beginning to understand the ubiquity yet weakness of spontaneous-adjustment emission, hence the physical cause of slow-quasimanifold fuzziness and the remarkable thinness of the slow quasimanifold in at least some cases of interest such as that of Figure 6.

I want to end by mentioning briefly some recent developments in the Hamiltonian theory of balanced motion that I have been involved in. The unapproximated shallow water system, and analogous stratified, rotating fluid systems, are Hamiltonian in the classical sense. Ways of making this explicit mathematically are now very well known. In three pioneering papers, Salmon (1983, 1985, 1988) took the first steps toward developing systematic procedures to construct balanced models from their exact "parents" that inherit the parent Hamiltonian structure, including all the associated conservation relations such as that for PV. Recently, Ian Roulstone of the U.K. Meteorological Office and I have succeeded in simplifying and clarifying these procedures in such a way as to make plain a number of generic or universal properties of such Hamiltonian balanced models (McIntyre and Roulstone 1996, 2001). One of these is a property we call "velocity splitting", which can be viewed as an immediate consequence of imposing the balance condition $\mathbf{u} = \mathbf{u}^B(\mathbf{x}; h(\cdot))$.

We find it mnemonically useful to say that the imposition of the balance condition — for any choice of the functional $\mathbf{u}^B(\mathbf{x}; h(\cdot))$, accurate or inaccurate — splits the parent velocity field into two distinct velocity fields, whose difference provides a natural intrinsic measure of the inaccuracy of the model. Because of Lighthill radiation, this inaccuracy can never be zero, no matter how delicately one tries to refine the balance condition, save in a tiny set of exceptional special cases where the vortex dynamics is steady and the Lighthill radiation vanishes.

Recently, Mohebalhojeh and I have shown that velocity splitting is not peculiar to Hamiltonian balanced models (paper in preparation). It is a generic property of all accurate *non-Hamiltonian* balanced models as well, with just one class of exceptions. The most important member of that class — an exception that has long diverted attention from what is generic — is the balanced model called the Bolin–Charney "balance equations" in the form described by Gent and McWilliams (1984) and Whitaker (1993). Our result is essentially that all PV-conserving balanced models significantly more accurate than the Bolin–Charney model, including for instance the balanced model defined by (10)–(19) above, must suffer velocity splitting *as a direct price for their accuracy.*

And now I am out of time and out of space! How can I capture something of all this in a limerick, as seems to be *de rigueur* at this Symposium? Well, here goes:

> On balance, consider the angle
> Of the pendulum's upside-down dangle:
>   Geostrophic it's not,
>   But I don't care a lot,
> 'Cause it's all in a bit of a tangle.

Acknowledgements: Special thanks are due to my colleagues Rupert Ford, Peter Haynes, Björn Haßler, Ali Mohebalhojeh, Philip Mote, Warwick Norton, Dirk Offermann, Alan Plumb, Martin Riese, Ian Roulstone, Jürgen Theiss, Darryn Waugh, and Christof Wunderer, all of whom have contributed in important ways to the most recent developments described or mentioned in this all-too-brief survey, which barely does justice to recent advances. Darryn Waugh kindly allowed me to reproduce Figure 1 from his ozone-layer research, and Martin Riese Figure 3. This work received generous support from the Natural Environment Research Council through the UK Universities' Global Atmospheric Modelling Programme, from the Engineering and Physical Sciences Research Council through the award of a Senior Research Fellowship, and from the Isaac Newton Institute's Programme in the Mathematics of Atmosphere and Ocean Dynamics. Finally, I want to thank Frank Hodnett and his Organizing Committee for inviting me here and for an unusual, delightful, and interesting Symposium.

References

Arnol'd, V. I., 1978: *Mathematical Methods of Classical Mechanics*, New York: Springer-Verlag, 462 pp.
Bokhove, O., and Shepherd, T. G., 1996: On Hamiltonian balanced dynamics and the slowest invariant manifold. *J. Atmos. Sci.*, **53**, 276–297.
Charney, J. G., 1948: On the scale of atmospheric motions. *Geofysiske Publ.*, **17(2)**, 3–17.
Ertel, H., 1942: Ein Neuer hydrodynamischer Wirbelsatz. *Met. Z.*, **59**, 271–281.
Ford, R., 1994: Gravity wave radiation from vortex trains in rotating shallow water, *J. Fluid Mech.*, **281**, 81–118.
Ford, R., *et al.*, 2000: Balance and the slow quasimanifold: some explicit results, *J. Atmos. Sci.*, **57**, 1236–1254.
Gent, P. R., McWilliams, J. C., 1984: Balanced models in isentropic coordinates and the shallow water equations. *Tellus*, **36A**, 166–171.
Gough, D. O., McIntyre, M. E., 1998: Inevitability of a magnetic field in the Sun's radiative interior. *Nature*, **394**, 755–757.
Haynes, P. H., 1989: The effect of barotropic instability on the nonlinear evolution of a Rossby-wave critical layer, *J. Fluid Mech.*, **207**, 231–266.
Hoskins, B. J., *et al.*, 1985: On the use and significance of isentropic potential-vorticity maps, *Q. J. Roy. Meteorol. Soc.*, **111**, 877–946; *Corrigendum*, **113**, 402–404.
Killworth, P. D., McIntyre, M. E., 1985: Do Rossby-wave critical layers absorb, reflect or over-reflect? *J. Fluid Mech.*, **161**, 449–492.
Kleinschmidt, E., 1950–1: Über Aufbau und Entstehung von Zyklonen (1–3 Teil). *Met. Runds.*, **3**, 1–6; **3**, 54–61; **4**, 89–96.
Lahoz, W. A., *et al.*, 1996: Vortex dynamics and the evolution of water vapour in the stratosphere of the southern hemisphere. *Q. J. Roy. Meteorol. Soc.*, **122**, 423–450.
Lighthill, M. J., 1952: On sound generated aerodynamically, *Proc. Roy. Soc. Lond.*, **A 211**, 564–587.
Manney, G. L., *et al.*, 1994: Simulations of the February 1979 stratospheric sudden warming: model comparisons and three-dimensional evolution. *Mon. Wea. Rev.*, **122**, 1115–1140.
McIntyre, M. E., 1982: How well do we understand the dynamics of stratospheric warmings? *J. Meteorol. Soc. Japan*, **60**, 37–65 [Special middle-atmosphere issue, ed. K. Ninomiya. *NB:* "latitude" should be "altitude" on page 39a line 2.]
McIntyre, M. E., 2000: On global-scale atmospheric circulations. In: *Perspectives in Fluid Dynamics: A Collective Introduction to Current Research*, ed. G. K. Batchelor, H. K. Moffatt, M. G. Worster, 557–624. Cambridge, University Press, 631 pp. *Corrigenda:* (1) The symbols × for 3-dimensional vector products were all misprinted as ∧; (2) the evaporation feedback at the tropical sea surface (p. 621) may be strongly compensated radiatively; see Held and Soden 2000, Ann. Rev. Energy Environment, **25**, 441–475.
McIntyre, M. E., Norton, W. A., 2000: Potential-vorticity inversion on a hemisphere. *J. Atmos.*

Sci., **57**, 1214–1235. *Corrigendum* **58**, 949 and on www.atm.damtp.cam.ac.uk/people/mem/

McIntyre, M. E., Palmer, T. N., 1983: Breaking planetary waves in the stratosphere. *Nature*, **305**, 593–600.

McIntyre, M. E., Palmer, T. N., 1984: The "surf zone" in the stratosphere. *J. Atm. Terr. Phys.*, **46**, 825–849.

McIntyre, M. E., Palmer, T. N., 1985: A note on the general concept of wave breaking for Rossby and gravity waves. *Pure Appl. Geophys.*, **123**, 964–975.

McIntyre, M. E., Roulstone, I., 1996: Hamiltonian balanced models: constraints, slow manifolds, and velocity splitting. *Forecasting Research Scientific Paper* **41**, *Meteorological Office, U.K.*. Shortened and corrected version in revision for *J. Fluid Mech.*; full text and corrections available at http://www.atm.damtp.cam.ac.uk/people/mem/

McIntyre, M. E., Roulstone, I., 2001: Are there higher-accuracy analogues of semigeostrophic theory? In: *Large-scale Atmosphere–Ocean Dynamics: II: Geometric Methods and Models* (Proc. Newton Inst. Programme on Mathematics of Atmosphere and Ocean Dynamics), ed. I. Roulstone and J. Norbury. Cambridge, University Press, in press.

Mohebalhojeh, A. R., Dritschel, D. G., 2001: Hierarchies of balance conditions for the f-plane shallow water equations. **J. Atmos. Sci.**, in press.

Mote, P. W., *et al.*, 1996: An atmospheric tape recorder: the imprint of tropical tropopause temperatures on stratospheric water vapor. *J. Geophys. Res.*, **101**, 3989–4006.

Mote, P. W., *et al.*, 1998: Vertical velocity, vertical diffusion, and dilution by midlatitude air in the tropical lower stratosphere. *J. Geophys. Res.*, **103**, 8651–8666.

Norton, W. A., 1994: Breaking Rossby waves in a model stratosphere diagnosed by a vortex-following coordinate system and a technique for advecting material contours. J. Atmos. Sci., 51, 654–673.

Norton, W. A., *et al.*, 2000: Brewer–Dobson Workshop (report plus transcripts of invited lectures by J. R. Holton and A. W. Brewer). *SPARC Newsletter* **15**, 25–32. Available from sparc.office@aerov.jussieu.fr, also http://www.aero.jussieu.fr/~sparc/ [sic].

Plumb, R. A., McEwan, A. D., 1978: The instability of a forced standing wave in a viscous stratified fluid: a laboratory analogue of the quasi-biennial oscillation. *J. Atmos. Sci.*, **35**, 1827–1839.

Plumb, R. A., *et al.*, 1994: Intrusions into the lower stratospheric Arctic vortex during the winter of 1991–92. *J. Geophys. Res.*, **99**, 1089–1105.

Rhines, P. B., 1975: Waves and turbulence on a beta-plane. *J. Fluid Mech.*, **69**, 417–443.

Riese, M., *et al.*, 2001: Stratospheric transport by planetary wave mixing as observed during CRISTA-2. *J. Geophys. Res.*, submitted. See also Riese, M., 2000: *Remote Sensing and Modeling of the Earth's Middle Atmosphere: Results of the CRISTA Experiment*, Habilitation thesis, Bergische Universität, Fachbereich 8, 42097 Wuppertal 1, Germany.

Robinson, W. A., 1988: Analysis of LIMS data by potential vorticity inversion. *J. Atmos. Sci.*, **45**, 2319–2342.

Rossby, C. G., 1936: Dynamics of steady ocean currents in the light of experimental fluid mechanics. Mass. Inst. of Technology and Woods Hole Oc. Instn. *Papers in Physical Oceanography and Meteorology*, **5(1)**, 1–43. [Material PV conservation: Equation (75).]

Salmon, R., 1983: Practical use of Hamilton's principle. *J. Fluid Mech.*, **132**, 431–444.

Salmon, R., 1985: New equations for nearly geostrophic flow. *J. Fluid Mech.*, **153**, 461–477.

Salmon, R., 1988: Semigeostrophic theory as a Dirac-bracket projection. *J. Fluid Mech.*, **196**, 345–358.

Saujani, S., and Shepherd, T.G., 2001: Comments on "Balance and the Slow Quasimanifold: Some Explicit Results". *J. Atmos. Sci.* **58**, submitted; FORD, R., *et al.*, 2001: Reply, *J. Atmos. Sci.* **58**, submitted.

Stewart, R. W., Thomson, R. E., 1977: Re-examination of vorticity transfer theory. *Proc. Roy. Soc. Lond.*, **A354**, 1–8.

Waugh, D. W., Plumb, R. A., 1994: Contour advection with surgery: a technique for investigating finescale structure in tracer transport. *J. Atmos. Sci.*, **51**, 530–540.

Waugh, D. W., *et al.*, 1994: Transport of material out of the stratospheric Arctic vortex by Rossby wave breaking. *J. Geophys. Res.*, **99**, 1071–1088.

Whitaker, J. S., 1993: A comparison of primitive and balance equation simulations of baroclinic waves. *J. Atmos. Sci.*, **50**, 1519–1530.

KELVIN'S THEOREM AND THE OCEANIC CIRCULATION IN THE PRESENCE OF ISLANDS AND BROKEN RIDGES

JOSEPH PEDLOSKY
WOODS HOLE OCEANOGRAPHIC INSTITUTION
CLARK LABORATORY
WOODS HOLE, MA 02543
e-mail: jpedlosky@whoi.edu

ABSTRACT

The fluid mechanics of the large-scale ocean circulation when planetary scale islands and barriers are present introduces new features to the problem of the ocean circulation. Islands introduce non simply-connected domains for the ocean circulation while barriers, representing either large island arcs or mid-ocean ridges would seem to impose strong limits on the communication between oceanic sub-basins. It is shown that the application of Kelvin's circulation theorem is an illuminating instrument in understanding both the steady and unsteady flows in the presence of such geometry. The constraint imposed by a proper application of Kelvin's theorem leads to non intuitive predictions for the circulation around such large islands and the production of strong zonal jets in the ocean circulation. A similar application of Kelvin's theorem to the problem of the propagation of large-scale Rossby waves yields, for example, the surprising result that barriers with just two small gaps become essentially transparent to the passage of large-scale wave energy.

The presence of planetary scale islands (such as Australia) renders the ocean basins non simply connected and introduces new features to the problem of the ocean circulation. For the abyssal flow of the ocean, the presence of steep topography in the form of the mid-ocean ridge system with its long linear features broken by transform faults, also introduces island-like elements to the circulation of the deep ocean. The interesting physical features associated with the presence of these "islands" is illuminated particularly well by an application of Kelvin's theorem. In this short review I will briefly discuss a number of phenomena connected to the fundamental constraint of the circulation theorem. Most of the work discussed has either been previously published [see references] or is the work of my colleagues at Woods Hole and their contributions will be cited in the appropriate location.

For the purpose of conceptual simplicity the geometry will be ruthlessly idealized and will generally consist of a square basin containing an island of regular shape, usually rectangular and often extremely thin in one of its two lateral dimensions. These are not essential to the physics.

Again, for simplicity we usually consider the motion to be gentle enough so that quasi geostrophic theory applies. To begin with we consider a barotropic model in which the flow is contained in a single layer for which the horizontal motion is described by a stream function $\psi(x,y,t)$. On the outer boundary of the basin the stream function must be constant and without loss of generality this constant can be taken to be zero although for a baroclinic fluid this is no longer the case and the function can then be a non-trivial function of z and t. However, that is a story for another day. For the barotropic case, the constant can be set to zero. On the island the stream function must also be a constant but now that constant is, in general, not zero but some value Ψ_I whose value determines the total flux of fluid from around the island. If the island a thin barrier with two gaps separating it from the basin boundary the island constant yields the flux through the gaps of the flow entering one sub-basin from the adjacent basin.

The determination of Ψ_I is a key goal of the theory and the extra scalar boundary condition to determine it is provided by Kelvin's theorem.

Consider a fluid, which satisfies:

$$\frac{\partial \vec{u}}{\partial t} + (f+\zeta)k \times \vec{u} = -\nabla(p/\rho_o + \vec{u}^2/2) + Diss(\vec{u}) + \vec{T} \qquad (1)$$

where the symbols have their usual meaning and where $Diss(u)$ represents the parameterization of frictional dissipation in the model and T is the applied external force. A line integral of (1) around the perimeter of the island of the component of (1) tangent to the island, along with the condition that the flow normal to the island vanish yields the fundamental constraint of the circulation theorem:

$$\frac{\partial}{\partial t}\oint_{C_I} \vec{u} \cdot d\vec{s} = \oint_{C_I} Diss(\vec{u}) \cdot d\vec{s} + \oint_{C_I} \vec{T} \cdot d\vec{s} \qquad (2)$$

where C_I is the contours girdling the island. We will consider dissipation of the form

$$Diss(\vec{u}) = -r\zeta + A\nabla^2 \zeta \qquad (3)$$

where ζ is the relative vorticity.

STEADY FLOWS:

The work discussed here is described in more detail in the paper by Pedlosky, Pratt, Spall, and Helfrich (JMR 1997).

Consider a wind driven barotropic flow in a basin, which contains a meridional barrier, which nearly splits the basin in two. What is the flow from one sub basin to the next and how effective is the barrier in isolating one barrier from the adjacent barrier? Suppose the flow, which would occur in the absence of the barrier, is gyral so that fluid would endlessly recirculate from the western to eastern sides of the total basin. What is the effect of emplacing the barrier island?

We know that for the pertinent β-plane dynamics that the strongest currents for steady flows will occur on the western sides of oceanic basins (or sub-basins) hence, the strongest currents on a long meridional island will occur on the *eastern* side of the island, compared with which, the flow on the western side of the island will be negligible. Thus to satisfy the circulation theorem:

$$\oint_{C_I} Diss(\vec{u}) \cdot d\vec{s} \Rightarrow \int_{y_s}^{y_n} Diss(v_{east}) dy = 0 \qquad (4)$$

where y_s and y_n are the southern and northern extremities of the meridional barrier and v_{east} is the meridional velocity on the eastern side of the island. Generally speaking this means that the flow on the eastern side of the island must be of two signs so that the integral in (4) will self cancel.

Figure 1a shows the calculated streamline pattern for a flow driven by a wind stress curl of the form $\sin(\Box y)$. The barrier is indicated by the crosses. The dissipation is of the Stommel type, i.e., $A=0$ in (4) and the Stommel boundary layer thickness is $\delta_s = r/(\beta)$. In the figure this is chosen to be 0.2% of the basin width.

The dashed contour in the figure outlines a region of *recirculation*, that is, a zone in which the fluid is isolated from the general, large-scale gyre flow in which the fluid flows from one sub-basin to the next. Indeed, within the recirculation region the flow along the boundary is in a direction opposite to that of the flow outside the zone. The extent of the zone is determined by the streamline of the Sverdrup flow, which joins the two stagnation points on the boundary indicated in the figure by the two circles on the island. The amount of fluid endlessly recirculating in the zone to the east of the barrier is a measure of the fluid impeded by the barrier from flowing from one sub-basin to the adjacent basin. In the absence of the barrier the maximum amount of fluid flowing across the line where the barrier is placed, x_t, is equal to Ψ_{max} which is the maximum value of the Sverdrup stream function on evaluated at the longitude of the barrier while the amount of fluid in the recirculation region is-($\Psi_I - \Psi_{max}$) the difference being Ψ_I the amount of fluid flowing through the gaps. The reduction of the flow is therefore exactly equal to the transport recirculating in the zone east of the ridge. In the example of Figure 1a, in spite of the narrowness of the gaps almost 50% of the fluid is able to flow from one sub-basin to the other. This is in part due to the narrowing of the flow at the gaps into two jets clearly seen in the figure. The flow in these jets forms an internal boundary layer whose structure is similar to the Stommel boundary layer when placed along potential vorticity contours. Figure 1b shows the profile of the zonal velocity in the jets at a position west of the barrier. Note how the concentrated flow is centered at

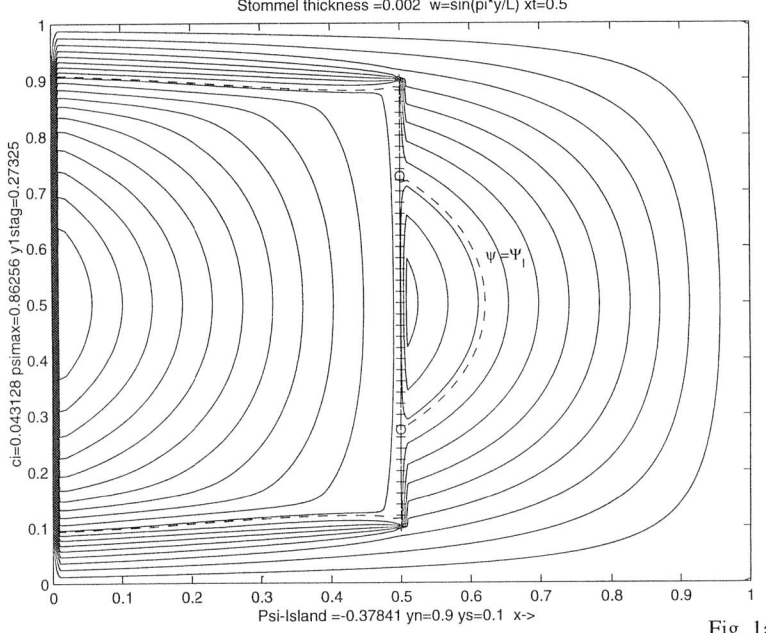

Fig. 1a

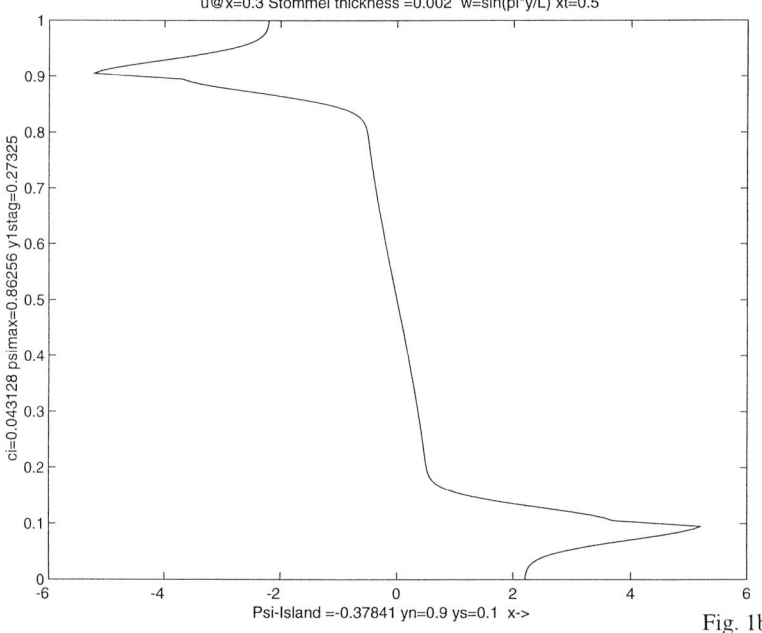

Fig. 1b

the latitudes of the gaps. Note also that for this problem the boundary layers containing these jets thicken westward whether the flow in the jets is westward or eastward.

We have found it immensely profitable to examine these ideas in the laboratory. With the help of Karl Helfrich an apparatus of the same type described by Pedlosky and Greenspan (1967) has been set up, the so-called sliced cylinder experiment. The reader is referred to the Pedlosky, Pratt, Spall, and Helfrich paper for details. Suffice it to say here that the cylindrical apparatus, driven by differential rotation of the upper surface is able to mimic the dynamics of the beta plane by introducing a slight slope to the lower boundary of the rotating cylinder. The island is formed by a plexiglas slab oriented, at first, in the north south direction. A detailed analysis of the flow expected in the linear limit in analogy to the case shown in Figure 1 predicts a slender but conspicuous recirculation region to the east of the barrier.

Figure 2 shows the result of the experiment.

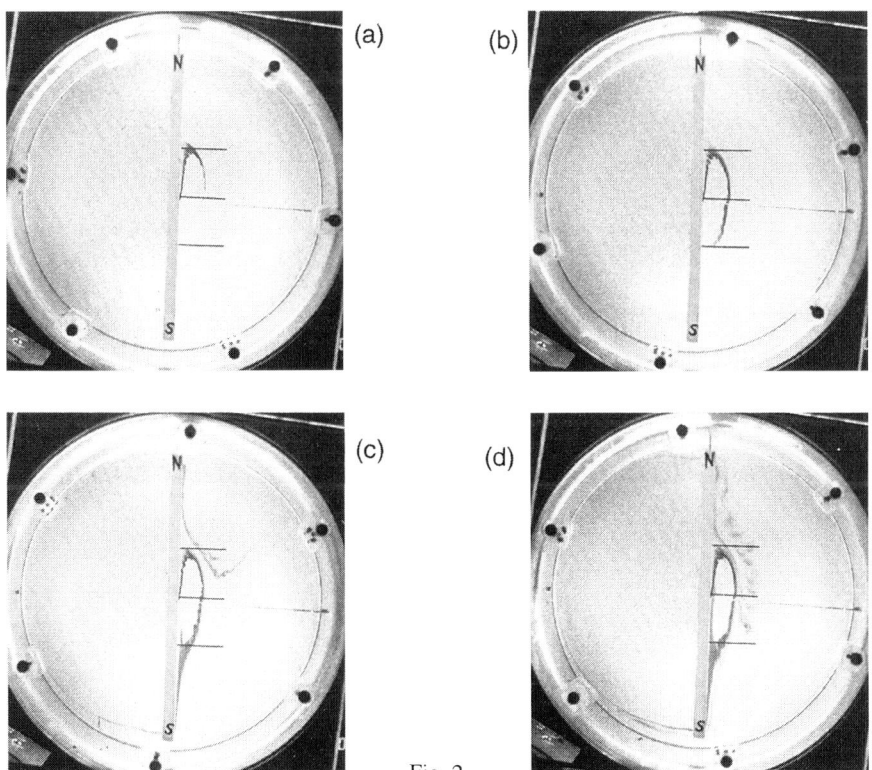

Fig. 2

The four panels show the path of dye inserted along the eastern boundary of the barrier and the sequence (a) to (d) shows the path of the dye at four later times in order of increasing time. It is evident that the dye traces out the boundary of the recirculation zone and detailed comparison of the zone with that predicted by the theory is excellent even though the experiment is performed in a parameter range where the boundary layer on the island is substantially non-linear. It is also obvious from the figure, especially in Figures, 2(c) and 2(d) that there is a small leak in the recirculation zone. This is as it must be since Ekman pumping from the differentially rotating upper lid, which drives the flow, will pump fluid downward into the recirculation zone. Mass conservation requires that flow, which is small, compared to the horizontal transport to leak out the southern end of the zone. The absence of the leak in quasi-geostrophic theory occurs because qg theory sees the Ekman pumping as a pure vorticity source and neglects the mass source, which is an order $(\beta L/f)$ smaller than the effect of the vorticity source. Recourse to the *planetary* geostrophic equations is sufficient to explain and predict the leak, which is consistent with the observations. Again, the details are in the *Journal of Marine Research* paper quoted above.

When additional barriers, in the form of peninsulas or other islands are place to the *east* of the island so far considered, in such a way as to cast a shadow on the gaps through which the fluid must circulate, the theory predicts that the flow through the gaps will be reduced. Since the strength of the recirculation is itself a measure of the impedance we can expect such shadowing barriers to produced enhanced regions of recirculation. Such phenomena were studied by Helfrich, et al. (1999) and Figure 3 shows the recirculation zone for the sliced cylinder apparatus when a peninsula attached to the boundary shadows the southern gap. Note the greatly expanded size of the recirculation.

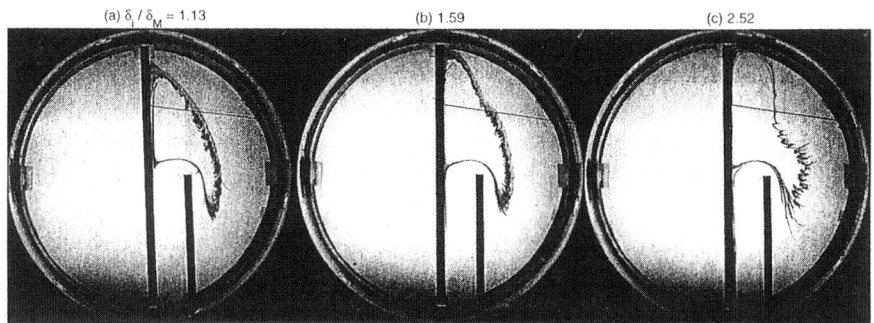

Fig. 3

The three figure panels show the recirculation for increasing strength of the non-linearity. Although there are detailed alterations in the flow the basic pattern is robust. The parameter, which measures the non-linearity here, is the ratio of the inertial western boundary layer thickness, based on the Rossby number of the flow to the Munk boundary layer thickness. That ratio, related to the Reynolds number, increases from about unity in the panel on the left to 2.5 for the panel on the right.

When the island is oriented along latitude circle features of special interest occur. First of all, the determination of the island constant really must be carried out by a direct application of Kelvin's theorem to the flow around the island. The ingenious alternative scheme devised by Godfrey (1989) is no longer valid. Again the details of the required calculation are discussed in Pedlosky et al (1997) but a key point of interest is the fact that, even though for a thin island the north-south extent of the island is very small, the flow still recognizes the primacy of the eastern side of the island as the position for the major "western" boundary current and the larger part of the southward flowing fluid impinging on the barrier flows around the eastern end of the island. Again, experiments have been invaluable in examining such non intuitive predictions of simple theory.

Figure 4 shows the result of an experiment by Wells and Helfrich (2000) in which the island barrier is placed in the center of the "sliced cylinder" along a "latitude circle" and the oncoming meridional flow {driven by uniform Ekman pumping from the upper boundary) impinges on the barrier. The upper panel of the figure shows the streamlines of the flow around the island. Note the strong "western" boundary layer flow on the eastern side of the island and in agreement with the theory, most of the flow is diverted to the east of the island. Panels b and c show the flow around the western and eastern edges of the island in more detail and illustrate the intense boundary current on the eastern side very clearly.

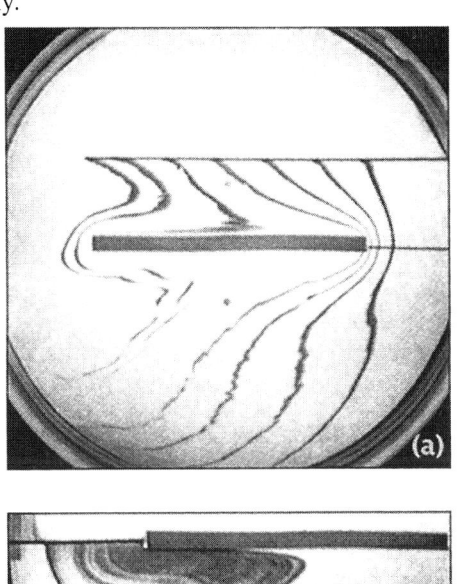

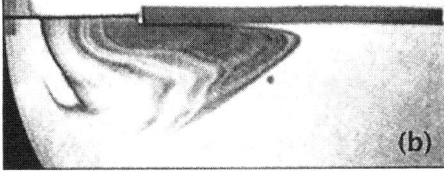

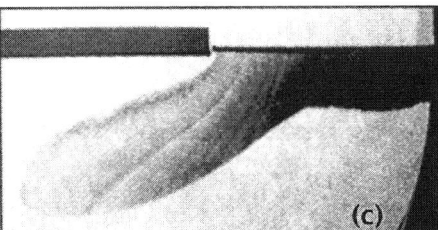

Fig. 4

UNSTEADY FLOWS

Consider the alternative limit of the circulation theorem (2) when forcing and dissipation provide time scales that are long compared to some characteristic interaction time for some time dependent motion. For example, suppose we consider the interaction of a Rossby wave of frequency ω with the island. The form of Kelvin's theorem in this case is the classic one, e.g., that circulation is conserved,

$$\frac{\partial}{\partial t} \oint_{C_I} \vec{u} \cdot d\vec{s} = 0 \qquad (5)$$

Note that the circuit C_p, once placed *exactly* on the boundary of the island, must remain on the island since there is no velocity perpendicular to the island and hence the contour. Thus (5) always applies to the tangential velocity at the island.

Suppose that the island consists of a long meridional barrier, which nearly divides the basin into two sub-basins except for the presence of two small gaps of width d at the locations y_s and y_n where the former is the southern gap and the latter the northern one (see Figure 5).

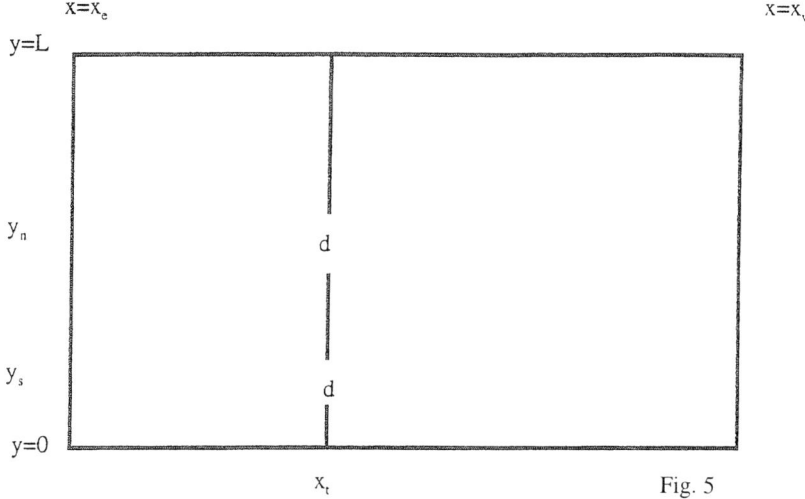

Fig. 5

For flows either starting from rest or for flow oscillating periodically the integral of (6) with time implies that the circulation itself must be zero.

If the island is thin enough the integral in (6) reduces to the requirement that the *average* meridional velocity on each side of the meridional barrier, at $x=x_t$, be equal. Thus if a Rossby wave impinging on the barrier from the east has a meridional scale that is large enough so that at each time t

$$\int_{y_s}^{y_n} v(x_t, y) dy \neq 0 \qquad (6)$$

then Kelvin's theorem would be violated. This would occur, for example, if a large vortical motion that would arise naturally in a Rossby normal mode of the lowest meridional modal form impinged on the boundary and left the eastern basin largely unperturbed. The impossibility of doing that (i.e.,) limiting the motion to one side of the barrier because of Kelvin's theorem means that there must be an (oscillating) meridional motion on the *western* side of the boundary. In quasi-geostrophic theory this happens immediately and this turns the entire western side of the island barrier into an antenna radiating Rossby wave energy to the western sub-basin. Thus, the theoretical prediction is that the barrier with two small gaps allows large meridional scale Rossby wave energy to penetrate through the barrier and produce an O(1) transmission of energy. The barrier with two gaps becomes transparent.

The issue has been studied in detail, first by Pedlosky and Spall (1999) where the Rossby normal modes of a basin with a dividing barrier pierced by two small gaps were studied in detail both analytically and numerically and the reader is referred to that paper for details.

Figure 6, below, shows an example of such wave transmission. The figure's four panels shows the response of a barotropic fluid in a basin to periodic forcing in the eastern basin at a frequency close to a Rossby normal mode of the full basin and at four different phases of the oscillation. The forcing itself is not evident since the form of the response is dominated by the form of the natural mode of oscillation. Of interest is the clear ability of the large vortical motion to squeeze through the two small gaps, extrude itself into the western basin and continue on its Rossby journey westward apparently unimpeded by the barrier.

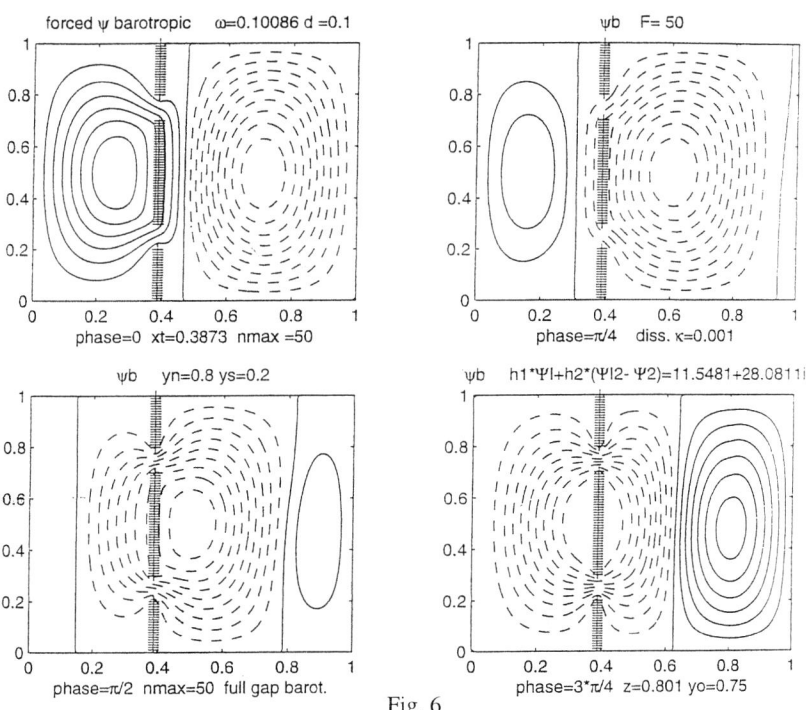

Fig. 6

The transparency of such barriers to Rossby wave transmission is not limited to normal modes or resonant excitation. In the references (Pedlosky 2000 a,b,c) it is shown that non-resonant motions, wave packets and baroclinic Rossby waves all share the transparency. What is important is the structure of the impinging motion at the barrier. In particular, if the integral of the velocity of the impinging wave along the boundary has a non-zero average Kelvin's theorem *demands* the appearance of a compensating oscillatory motion on the opposite side of the barrier turning the island from a barrier into an antenna. As Pedlosky and Spall (1999) show, if the impinging motion is such that the integrated tangential velocity is zero along the barrier between the two gaps, then no transmission takes place. It is clearly the satisfaction of Kelvin's theorem (or lack of it) that determines the degree of transmission.

If there is a single gap in the barrier so that Kelvin's theorem is not applicable it is not difficult to show that the amount of transmitted energy is small in the limit of small gap width as would be expected from the standard scattering and transmission wave problem (McKee, 1972). Similarly, as Pedlosky and Spall have shown the effect is not limited to the case of meridional barriers. Barriers, which are zonally oriented, also share the same properties although the details of the transmission change.

Finally, the reader is reminded that Kelvin's theorem is not limited to linear theory although the examples described here are largely linear. Again, a laboratory experiment is instructive.

Figure 7, below, shows a snapshot sequence of the interaction of an isolated eddy approaching and interacting with a circular island on the beta plane.

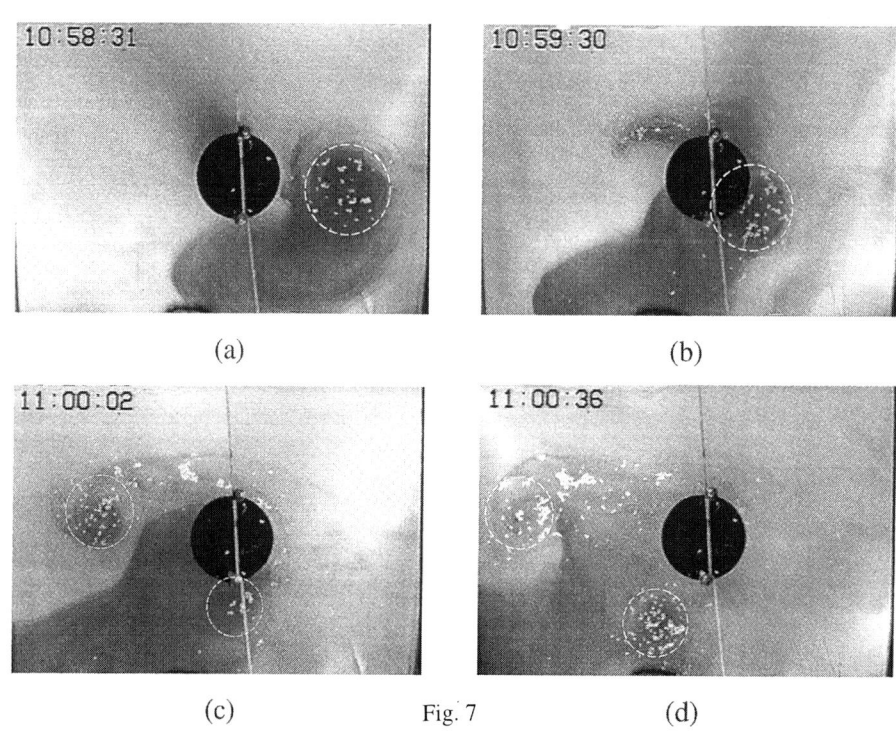

Fig. 7

These photos were generously made available to me by Claudia Cenedese of the Physical Oceanography department at Woods Hole who carried out the experiment. The self-propelled eddy is seen in the upper left-hand panel approaching the island. The cloud of fluid disturbed by the eddy is shown as an inky shaded region. Most of the initial disturbance appears to develop around the south-east sector of the island but in the second panel on the upper right it is clear that fluid shoots around the northern edge of the island. That fluid is marked by surface reflecting flecks of tracer and as time goes by it is the northern rim current, which produces a strong down stream, eddy which matches a similar eddy to the south. This splitting of the original eddy by the island due to the two currents, which develop on either side of the island, can be understood as a response to the requirement that Kelvin's theorem be satisfied. Before the eddy impinges on the island the circulation around the island is zero. To preserve the circulation at zero compensating currents on both sides of the island are necessary leading to the eddy splitting and the production of the downstream pair.

SUMMARY AND CONCLUSIONS

The application of Kelvin's theorem to the oceanographically pertinent problem of flow around islands and island-like barriers has illustrated the powerful consequences attendant on the application of Kelvin's circulation theorem. The theorem easily explains the appearance of recirculation regions in the steady flow problem and the transparency of nearly compete barriers to Rossby wave transmission. Both of these surprising results are (*a posteriori*) easily understood consequences of Kelvin's theorem.

The utility of laboratory experiments to a problem of general circulation scale is also very clearly demonstrated. The ability to flexibly examine a wide parameter range in the laboratory, far more easily than either analytical or numerical theory allows, is a fundamental tool of great power, too long overlooked in oceanography.

It will be of interest to extend the studies described above to more fundamentally nonlinear regimes, including stratification and more realistic topographic and geographic geometry. Such problems are currently under study.

ACKNOWLEDGMENT

It is a pleasure to acknowledge my colleagues Karl Helfrich and Claudia Cenedese who have generously provided images from their superb experiments for this paper. This research was supported in part by a grant (OCE 99–01654) from the US National Science Foundation.

REFERENCES:

Godfrey, J.S. 1989. A Sverdrup model of the depth-integrated flow from the world ocean allowing for island circulations. *Geophysical Astrophysical Fluid Dynamics, 45*, 89–112.

Helfrich, K.R., J. Pedlosky and E. Carter. 1999. The shadowed island. *Journal of Physical Oceanography, 29*, 2559–2577.

McKee, W.D. 1972 Scattering of Rossby waves by partial barriers. *Geophysical Fluid Dynamics*, 4, 83–89.

Pedlosky, J. 2000a The transmission of Rossby Waves through basin barriers. *Journal of Physical Oceanography,* 30, 495–511

Pedlosky, J. 2000b The transparency of Ocean Barriers to Rossby waves: The Rossby slit problem. *Journal of Physical Oceanography,* (to appear)

Pedlosky, J. 2000c. The transmission and transformation of baroclinic Rossby waves by topography. *Journal of Physical Oceanography,* (to appear)

Pedlosky, J and H. P. Greenspan, 1967 . A simple laboratory model for the oceanic circulation. *Journal of Fluid Mechanics*, 27, 291–304.

Pedlosky, J. and M. Spall. 1999. Rossby normal modes in basins with barriers. *Journal of Physical Oceanography,* 29, 2332–2348.

Pedlosky, J., L.J. Pratt, M.A. Spall, and K.R. Helfrich, 1997 Circulation around islands and ridges. *Journal of Marine Research,* 55, 1199–1251.

Wells, J.R. and K. R. Helfrich 2000, Circulation around a thin zonal island. *Journal of Physical Oceanography*, (submitted).

GLOBAL AND REGIONAL ATMOSPHERIC MODELING USING SPECTRAL ELEMENTS

F. BAER
Department of Meteorology
University of Maryland
College Park, MD 20742 USA
JOE TRIBBIA
National Center for Atmospheric Research
MARK TAYLOR
Los Alamos National Laboratory

1. Introduction

To meet some of the outstanding issues associated with climate modeling such as exploiting new methodology for computation, utilizing the latest computing hardware, increasing the speed of computations and incorporating various space scales into one comprehensive model, we have developed a new global climate model entitled a Spectral Element Atmospheric Model, (SEAM), which can perform regional integrations concurrently with the global scale in a very straightforward manner. The model offers a number of distinct advantages over other global/regional models and is highly flexible. It utilizes the geometric properties of finite element methods, it allows for very convenient local mesh refinement and regional detail, it is ideally suited to parallel processing by minimizing communication amongst the processors, it is very efficient computationally, and last but not least, it has no pole problems. A number of tests with this model are described which will demonstrate the efficacy of the model's advantageous features.

2. Characteristics of the Model

The method for generating this model is now well documented and details on it can be found in Taylor, et al. (1997). Using sigma coordinates in the vertical, we tile all selected spherical surfaces in the atmosphere with an arbitrary number and size of rectangular elements. These elements are generated by first inscribing a polyhedron with rectangular faces inside the sphere and mapping the surface of the polyhedron to the surface of the sphere with a gnomic projection. Each face is then arbitrarily subdivided as desired and independent of uniformity, yielding a set of elements that cover the surface. The elements can be made as uniform as desired similar to most global climate models (GCMs) in use today, (see Figure 1) or can be distributed with high resolution in selected regions (Figure 3 shows such a selection). Each element is subdivided into a two-dimensional grid array and using finite element methodology (Cullen, 1979), a set of basis functions is selected. The dependent variables appropriate to the model are then expanded in these basis functions with time dependent coefficients. Note that the basis functions vanish everywhere but at the point at which they are applied. Global test functions are then selected and they are here chosen identical to the basis functions. The model equations in the basis function format are then multiplied by a test function and integrated over the spherical surface; this is done at each point over the entire global domain. Careful consideration is given to the boundaries where the elements meet and continuity of functions between elements is preserved. The resulting equations for the time dependent basis function coefficients are unique and define the tendencies of the dependent variables at each point within an element and for all elements. The resulting equations are advanced in time by using Adams-Bashforth 3rd order time stepping and 4th order Runge-Kutta as a start. The vertical representation follows the NCAR/CCM3 (Kiehl, et al., 1998) using sigma coordinates and finite difference discretization.

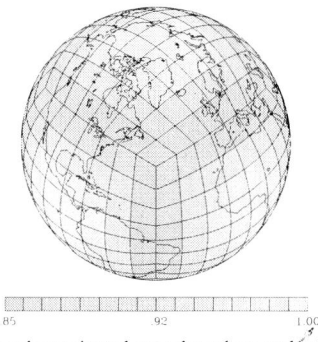

Figure 1. The cube projected onto the sphere and subdivided.

The basis functions we selected are the Legendre cardinal functions (an example of an eighth order function is seen on Figure 2) and we use the same functions in each of the two dimensions over the grid. Moreover, these functions lend themselves to Gauss-Lobotto quadrature provided that the grid in each element is selected to conform to this quadrature (these points are also depicted on Figure 2).

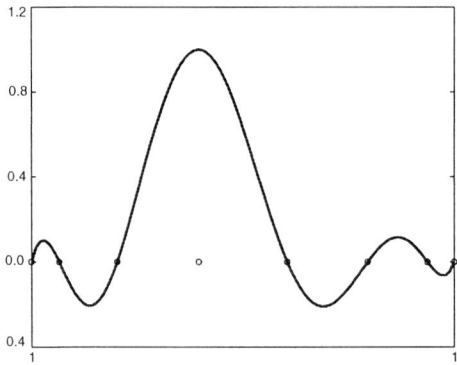

Figure 2. A Legendre cardinal function and the Gauss Lobotto quadrature points on an element.

Thus the integral equations defined above are reduced to a set of summations over the quadrature points and the simple Legendre spectral transform method may be employed. These choices result in an extremely simple finite element method with a diagonal mass matrix. It is our observation that the spectral element method appears to be a most efficient and natural way to achieve a high order finite-element discretization and in particular, to simplify regional modeling.

3. Experiments with the Model

We have tested the model (SEAM) under various representations and have found that in all circumstances it performs well, superior or equivalent to competitive models. It has been successful when applied to a three-dimensional dynamical core, to be discussed below, and we are currently introducing the dynamical core version into the NCAR/CCM3, which contains a full physics package. This form of the model should give insight into the ability of the model to predict a more realistic climate.

3.1. THE SHALLOW WATER MODEL

It is now commonplace to run new models on a standard test suite of initial conditions using the shallow water equations (Williamson, et al., 1992). All the recommended cases were tested with SEAM and our results were compared with those from the NCAR CCM3, the CSU TWIG model (Heikes and Randall, 1995) and the Arakawa-Lamb finite-difference model. In all cases, the SEAM was able to give results, which were at least comparable to the experiments discussed by Williamson, et al. (loc. cit.). Case 7, which describes an initial asymmetric polar vortex, is one of the more interesting of the cases and we present our results with it; they may be seen on Figure 3.

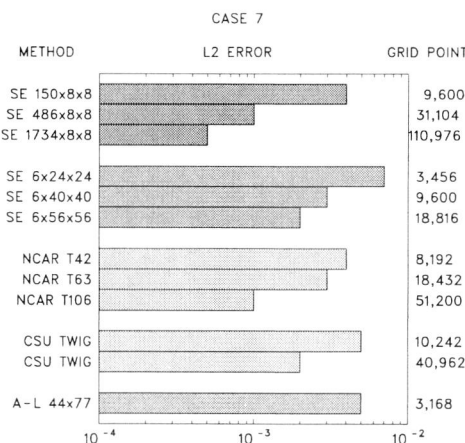

Figure 3. Errors generated by running the shallow water equations for various truncations and computational methods using the initial conditions for case 7 (Williamson et al., loc. cit.). Shown are the SEAM (SE), the NCAR spectral model, an icosahedral grid model (TWIG), and a finite-difference grid point model (A-L). Numbers of grid points involved for each experiment are listed on the right ordinate axis.

It is evident that SEAM (SE on the figure) has errors that are smaller than those from most of the other models tested, which include the NCAR spectral model, a twisted icosahedral grid model (TWIG) and a finite difference model (A-L), and is noted most clearly when comparing the different models for the same the number of grid points. In particular, SEAM is most successful when using an 8x8 grid in the elements. Additional experiments confirmed the efficiency of this gridding, and we have therefore used the 8x8 grid for subsequent experiments.

We introduced local topography into the SWE SEAM for an idealized mountain, and ran experiments with a more highly resolved grid both globally and regionally (mesh refined). Figure 4 shows the difference between local and global grid refinement. For comparable prediction results in the vicinity of the topography, the global reduction required 6936 total elements whereas the local refined version needed but a total of 310 elements. This gives a clue as to the virtues of the SEAM in calculating regional events embedded in a global domain and doing so with great computational efficiency.

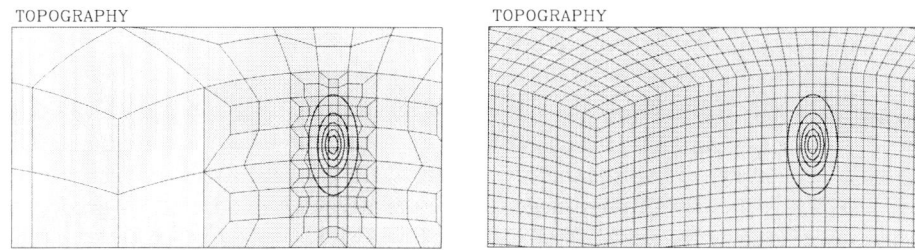

Figure 4. Mesh refinement around a mountain done locally (left) and globally (right).

3.2. THE THREE DIMENSIONAL MODEL

To test the three-dimensional dynamical core of SEAM with simple Newtonian physics as suggested by Held and Suarez (HS, 1994), we used 384 elements distributed uniformly over the globe and with an 8x8grid in each. This is roughly comparable to a T85 truncation in a spectral model. As with HS, we used 20 equally spaced sigma levels. Comparison of our model output (SEAM) with that given by HS for both a spectral model (T63) and a finite-difference grid point model (144 points on a latitude circle) show that all three models produce essentially the same results. It was not possible to make a direct comparison of the models' computational efficiency because no such data was available from HS, but we were able to demonstrate from this experiment that SEAM (in three dimensions) runs very efficiently on a MPP and generates the same number of mflops per processor independently of the number of processors used.

3.3. A POLAR VORTEX

To assess the impact of model forcing greater than simple Newtonian physics used in HS, we ran a primitive equation version of SEAM to determine the structure of Rossby wave breaking in the polar wintertime stratosphere (see Polvani and Saravanan, 1999). The model was dry, forced at the lower boundary to simulate an upward propagating

Rossby wave and with various resolutions, ranging from 50-200 vertical levels and horizontal scales of T85 and T181, the latter representing a grid length of about 70 km. After 25 days of integration, the model was able to depict in great detail the polar vortices that developed. As the resolution was increased the sharpening of the vortices became more evident. The high resolution encountered by this experiment and its successful prediction demonstrates the robustness of the method.

3.4. A BLOCKING EXPERIMENT

Using the three-dimensional dynamical core version of the SEAM with HS conditions at T85 resolution with uniformly distributed elements each having an 8x8 grid, and 10 sigma levels, we carried out ten year integrations with and without a surface topography field of T42 resolution to search for persistent ridges. The ridges found showed a frequency distribution in time similar to those produced by more realistic models (those with more complex physics), although SEAM produced fewer events. Given the model's simplicity, this result was very encouraging.

3.5. JETS ON JUPITER

Using the SWE version of the SEAM, we carried out integrations of decaying turbulence on planet Jupiter with very high resolution. Classical Jupiter dimensions were used (g = 23 ms^{-2}, radius = 7 x 10^4 km, Jupiter day = 9 Earth hours) and an equivalent depth of 20000m was selected, with a very weak dissipation. Four different truncations were used ranging from T170 to a maximum of T1033. This latter truncation represents 60000 8x8 elements or roughly 3000 equatorial latitude circle points. The model ran very efficiently on a Cray T3E with 128 processors, and at that resolution produced multiple jets from pole to pole as anticipated by Rhines. The intensity of the jet streaks as well as the developed vortices may be seen on Figure 5, which shows the potential vorticity after 276 Jovian days for T533 truncation.

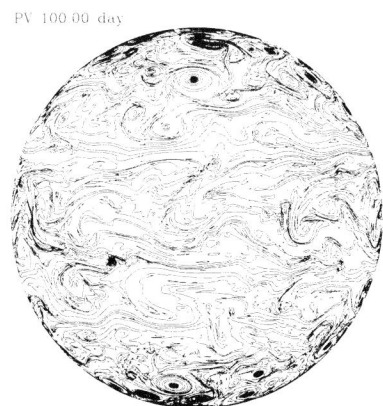

Figure 5. Potential vorticity on Jupiter for T533 truncation.

The number of jet streaks and the intensity of the vortices are greater still for T1033 truncation. It is evident from this experiment that the SEAM is capable of running efficiently at exceptionally high resolution on a MPP. Indeed there are few if any reports of other global models capable of running at the high resolution reported herein.

4. Conclusions

The experiments described in this report indicate that SEAM has the potential to be exceptionally useful for fast high-resolution climate studies and in particular for imbedded regional calculations both for weather prediction and for climate. The shallow water tests highlight the efficiency of local mesh refinement, the potential accuracy of the method and its ability to take advantage of MPP technology to speed up calculations. The Jupiter study highlights the model's ability to accommodate exceptionally small-scale events over the entire scale range. The studies using the three-dimensional dynamical core indicate that there is no difficulty in extending the method to higher dimensionality. Given these positive results, we are currently introducing state-of-the-art physics by using the NCAR/CCM3 with its complex physics package as a base and implanting the dynamical core version of the SEAM for dynamics. The model is so designed that no special programming is required to accomplish this procedure, although some efficiency may be lost. It should be emphasized that from an efficiency point of view, the method is almost ideal. If a column of elements is stored in a processor of a MPP, only the immediate neighbors are needed to complete a calculation over the grid, thus reducing the search for data in other processors of the computer to a minimum. This makes optimum use of the MPP architecture and minimizes the cost of the computation cycle. It is for this reason that the method is competitive with models using semi-Lagrange and semi-implicit integration methods. Although these methods have not yet been incorporated in a SEAM model, we plan to do so in the near future.

Acknowledgements. The research reported herein was supported by grant #DEFG0298ER62612 from DOE/CCPP to the University of Maryland. Both NERSC/DOE and NCAR/CGD and SCD provided computing resources.

References

Cullen, M., J., P., 1979: The finite element method, *Numerical methods used in atmospheric models* **II**. World Meteorological Organization, Geneva, 302-337.
Heikes, R. P., and Randall, D. A., 1995: Numerical integration of the shallow-water equations on a twisted icosahedral grid. Part I: Basic design and results of tests, *Mon. Wea. Rev.*, **123**, 1862-1887.
Held, I., and M. Suarez, 1994: A proposal for the intercomparison of the dynamical cores of atmospheric general circulation models. *Bull. Amer. Meteor. Soc.*, **75**, 1825-1830.
Kiehl, J. T., J. J. Hack, G. B. Bonan, B. A. Boville, D. L. Williamson, and P. J. Rasch, 1998: The National Center for Atmospheric Research Community Climate Model: CCM3. *J. Climate*, **11**, 1131-1149.
Polvani, L. M., and R. Saravanan, 1999: The three-dimensional structure of breaking Rossby waves in the polar wintertime stratosphere. *J. Atmos. Sci.*, submitted.
Taylor, M., J. J. Tribbia and M. Iskandarani, 1997: The spectral element method for the shallow water equations on the sphere. *J. Comput. Phys.* **130**, 92-108.
Williamson, D., et al., 1992: A standard test set for numerical approximations to the shallow water equations in spherical geometry. *J. Comput. Phys.*, **102**, 211-224.

PIECEWISE-CONSTANT VORTICES IN A TWO-LAYER SHALLOW-WATER FLOW

JEAN-MICHEL BAEY (1) AND XAVIER J. CARTON (2,3)
(1) ATLANTIDE, Brest, France
(2) Centre Militaire d'Océanographie, EPSHOM, Brest, France
(3) Laboratoire de Physique des Océans, IFREMER, Brest, France

1. Introduction

Vortices are a main feature of oceanic circulation, and thus have often been studied, mostly with a quasi-geostrophic model: stationary solutions such as tripoles have been found (Carton and Legras, 1994; Corréard and Carton, 1999). Recently, vortex stability has been investigated in a shallow-water framework (Dewar and Killworth, 1995; Carton and Baey, 2000; Stegner and Dritschel, 2000). These studies showed that the potential vorticity profile of the vortex, as well as its size (Burger number, see Benilov et al., 1998) and its intensity (Rossby number) are crucial for its stability. But they defined the vortex via its velocity profile, thus changing potential vorticity with stratification. In the present study, the potential vorticity profile of the vortex is given independently of the other parameters and is reduced to a very simple form: piecewise-constant.

After presenting the model equations and initial conditions (section 2), we study their non-linear evolutions numerically (section 3). Finally, the stability of tripoles is presented (section 4).

2. The mathematical model and the mean flow

Our model is described by the two-layer shallow-water equations which are written in polar coordinates

$$\partial_t u_j + u_j \partial_r u_j + \frac{v_j}{r} \partial_\theta u_j - \frac{v_j^2}{r} - f v_j = \frac{-1}{\rho_0} \partial_r p_j \qquad (1)$$

$$\partial_t v_j + u_j \partial_r v_j + \frac{v_j}{r}\partial_\theta v_j + \frac{u_j v_j}{r} + fu_j = \frac{-1}{\rho_0 r}\partial_\theta p_j \qquad (2)$$

$$\partial_t h_j + u_j \partial_r h_j + \frac{v_j}{r}\partial_\theta h_j + \frac{h_j}{r}[\partial_r(ru_j) + \partial_\theta v_j] = 0 \qquad (3)$$

where u_j, v_j are the radial, azimuthal velocities, p_j the pressure, h_j the local thickness of layer $j = 1, 2$ (upper and lower layers) and $f = f_0 + \beta y$ is the Coriolis parameter; beta-effect is introduced only at the end of this study. Here the flow is bounded above and below by rigid, flat surfaces so that $h_j = H_j + (-1)^j \eta$, where H_j is the thickness of the layer at rest, and η is the interface deviation. The hydrostatic balance is $p_2 = p_1 + g'\eta$, where $g' = g\delta\rho/\rho_0$ ($\delta\rho$ is the density jump between layers and ρ_0 is an average density).

These equations lead to the conservation of potential vorticity

$$[\partial_t + u_j \partial_r + \frac{v_j}{r}\partial_\theta]Q_j = 0, \quad Q_j = \frac{\zeta_j + f}{h_j}, \quad \zeta_j = \frac{1}{r}[\partial_r(rv_j) - \partial_\theta u_j] \qquad (4)$$

and the potential vorticity anomaly is

$$q_j = Q_j - \frac{f}{H_j}$$

The mean flow is a circular vortex defined by a piecewise-constant potential vorticity anomaly (see Figure 1): α is the baroclinicity of the core vortex (q_{lower}/q_{upper}) and the vortex is isolated (zero volume integral of potential vorticity anomaly). This mean flow is in hydrostatic and cyclogeostrophic balances and the basic velocities and thicknesses satisfy

$$\frac{d\bar{h}_2}{dr} = \frac{1}{g'}(\frac{\bar{v}_2^2}{r} - \frac{\bar{v}_1^2}{r} + f.\bar{v}_2 - f.\bar{v}_1) \qquad (5)$$

This problem is solved by a Newton-Raphson method.

This problem is rendered dimensionless with length scale R ($= R1$) and horizontal velocity scale U ($= q1.R.H$) so that the Rossby and Burger numbers are $Ro = U/fR, Bu = g'H/f^2R^2$ with $H = H_1 + H_2$. Thus we have $q1 = 1, R1 = 1$, and we set $R2 = 1.6, R3 = 2.6$ for application to oceanic eddies in the Bay of Biscay (swoddies). The layer thicknesses are equal. Note that thin vortices are usually more stable (Benilov, 2000).

3. Non-linear evolutions

The non-linear shallow-water equations are solved with a pseudo-spectral code on a biperiodic square grid ($L = 2\pi$) with 128^2 nodes. The time-step is

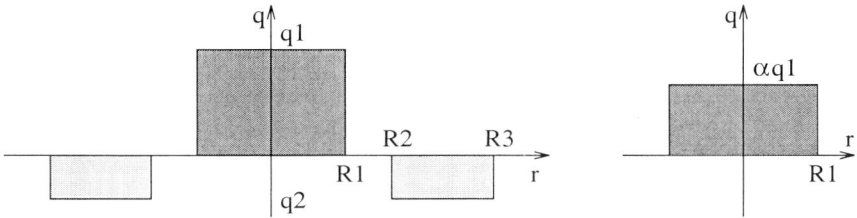

Figure 1. Radial profiles of potential vorticity anomaly for the mean isolated circular vortices; left-hand (right-hand) panel shows the upper (lower) layer.

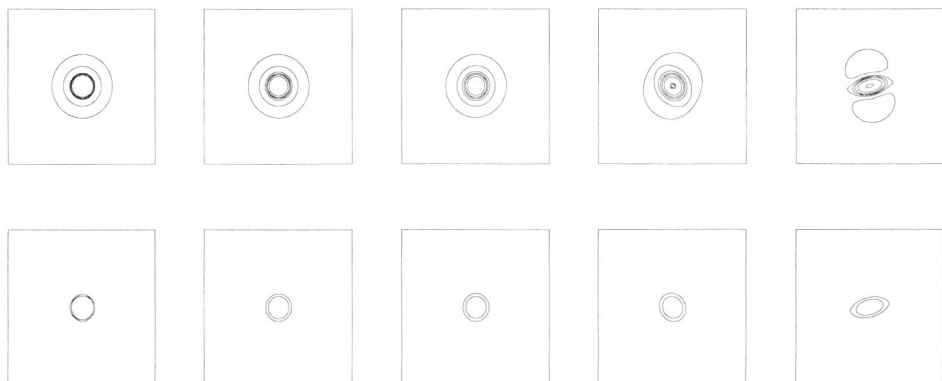

Figure 2. Time-evolution of potential vorticity maps showing the formation of a surface intensified tripole ($Ro = 0.25$, $Bu = 1$, $\kappa = 0.5$, time interval= 200, contour interval=0.075).

controlled by the Courant-Friedrichs-Lewy stability criterion. A biharmonic viscosity removes small-scale noise. The perturbation is elliptical (angular mode $l = 2$).

When $\alpha \geq 0$ and whatever the value of Burger and Rossby numbers, the initial vortex turns into a surface-intensified tripole (Figure 2). The core and the external ring in the upper layer have opposite-sign vorticity. The shear of the perturbation breaks the external ring which generates two satellites, rotating around the core vortex which becomes elliptical in both layers. In the lower layer, the major axis of the core is aligned with that of the upper core.

When $\alpha < 0$ and for large values of Burger number ($Bu = 1$), the vortex most often breaks into a baroclinic dipole (Figure 3): The upper and lower cores break into two poles which combine with part of the original ring to move outward. When the Burger number becomes small, baroclinic dipolar

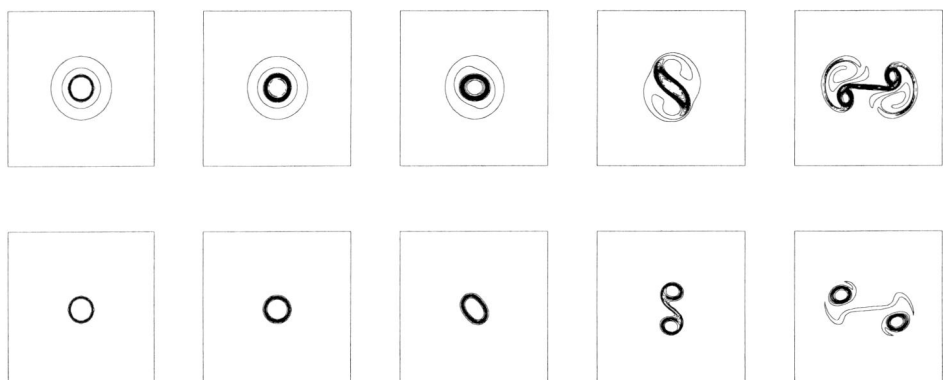

Figure 3. Time-evolution of potential vorticity maps showing the formation of a baroclinic dipolar breaking ($Ro = 0.25$, $Bu = 1$, $\kappa = -0.5$, time interval= 200, contour interval=0.02).

breaking occurs only for large baroclinicity ($\alpha = -1$). Again for small Bu but for moderate baroclinicity ($\alpha = -0.5$), the initial vortex stabilizes nonlinearly into a baroclinic tripole.

These results show that nonlinear stabilization of linearly unstable circular vortices is easier for barotropic than for baroclinic instability. Moreover, similar results are obtained for triangular perturbations ($l = 3$).

4. Stationarity and stability of tripoles

The surface intensified tripole is stationary in a frame of reference rotating at its rotation rate Ω. Then we can combine the dynamical equations to get the relation $J(Q_j, B_j) = 0$ where J() is the Jacobian operator, Q_j is the potential vorticity and B_j is the Bernoulli function

$$B_j = \frac{p_j}{\rho_j} + \frac{1}{2}(u_j{}^2 + (v_j - r.\Omega)^2) - \frac{1}{2}r^2\Omega(f + \Omega)$$

Figure 4 compares the potential vorticity and Bernoulli function at large times for the surface intensified tripole. The angular velocity is determined experimentally. The Bernoulli function and the potential vorticity have identical isolines; thus they are univoquely related. A (Q_j, B_j) scatter plot also exhibits little dispersion (not shown). This confirms that these tripoles are stationary.

Finally, the stability of these stationary tripoles is assessed:
- first, white noise is added initially to their initial velocity fields. A relative

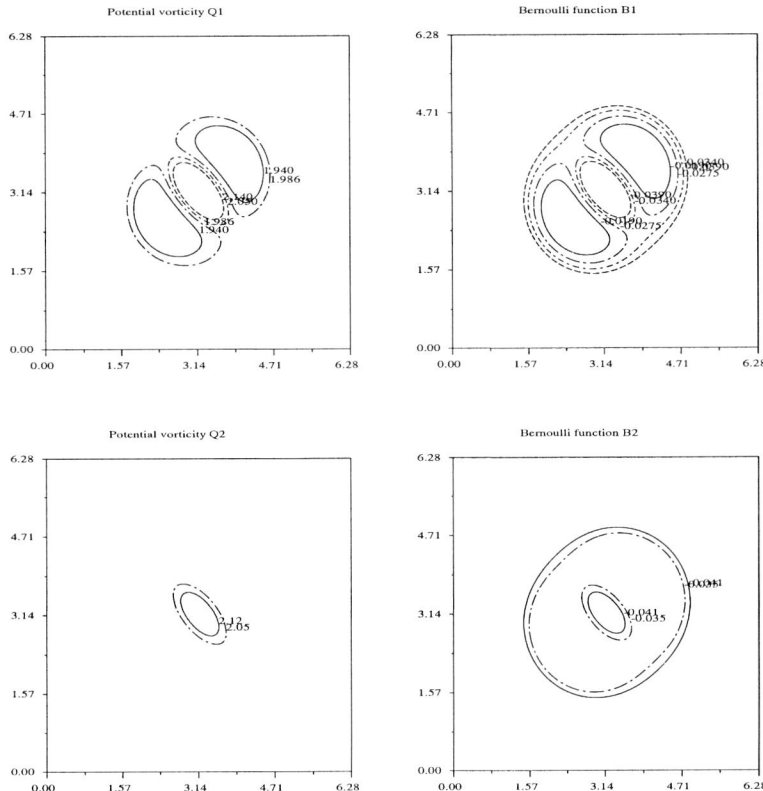

Figure 4. Comparison between potential vorticity and Bernoulli function in a framework rotating with angular velocity $\Omega = 0.0075$ at t=1600 model time units for the surface intensified tripole ($Ro = 0.25$, $Bu = 1$, $\kappa = 0.5$). The upper (lower) graphs corespond to the upper (lower) layer.

noise level close to unity is necessary to affect the tripole. Only when the noise level is equal to 5 is the tripole severely perturbed: in the lower layer, a tripole replaces the initial elliptical vortex. This indicates that the large structures of the upper layer have organized noise in the lower layer to form satellites around the core.

β	5.10^{-4}	10^{-3}	5.10^{-3}	10^{-2}	5.10^{-2}
T_{life} for SITP	1τ	0.5τ	0.25τ	0.25τ	0

Table 1. Life time (T_{life}) of the surface intensified tripole (SITP) when evolving on the β-plane. τ is the period of rotation of the initial vortex.

- secondly, the stationary surface-intensified tripole shown hereabove is used as an initial condition for new simulations with β-effect (see Table 1). The life time of the tripole is then small even for values of β comparable with those on Earth at mid-latitudes.

5. Conclusion

This study confirms, with piecewise-constant vortices, previous findings for unstable circular vortices, with continuous potential vorticity, in stratified rotating shallow-water flows (Carton and Baey, 2000): the elliptical mode of perturbation is the most unstable at moderate values of Ro and Bu; barotropic instability can more easily stabilize at finite-amplitude than the baroclinic one; tripoles with a complex vertical structure exist as in quasi-geostrophic flows; they are robust to initial perturbations but do not resist sustained anisotropy (beta-effect). If observed in the ocean, they are therefore likely to disappear within a few weeks. More complex multipoles such as quadrupoles can be formed from perturbed circular vortices (e.g. with wavenumber $l = 3$), but they are only transient vortex aggregates in two-layer rotating shallow-water flows.

References

Benilov, E. S., D. Broutman and E.P. Kuznetsova (1998) On the stability of large-amplitude vortices in a continuously stratified fluid on the f-plane. *J. Fluid Mech.*, **Vol. no. 355**, pp.139-162.

Benilov, E. S. (2000) Baroclinic instability of vortices in a two-layer ocean with thin upper layer. Part 1: Quasigeostrophic vortices. *Submitted to J. Fluid Mech.*

Carton, X.J. and Baey, J.M. (2000) Tripolar vortices in a two-layer shallow-water model. Proceeding of the *5th International Symposium on Stratified Flows*, **Vol. no. 1**, G.A. Lawrence Ed., Vancouver, pp.25-30.

Carton, X.J. and Legras, B. (1994) The life-cycle of tripoles in two-dimensional incompressible flows, *J. Fluid Mech.*, **Vol. no. 267**, pp.53–82.

Corréard, S.M. and X.J. Carton (1999) Formation and stability of tripolar vortices in stratified geostrophic flows. *Il Nuovo Cimento, C*, **Vol. no. 22**, 6, pp.767-777.

Dewar, W.K. and P.D. Killworth (1995) On the stability of oceanic rings. *J. Phys. Oceanogr.*, **Vol. no. 25**, pp.1467-1487.

Stegner, A. and D.G. Dritschel (2000) A numerical investigation of the stability of isolated shallow-water vortices. To appear in *J. Phys. Oceanogr.*.

A DYNAMICAL STABILIZER IN THE CLIMATE SYSTEM: A MECHANISM SUGGESTED BY A SIMPLE MODEL AND SUPPORTED BY GCM EXPERIMENTS AND AN OBSERVATIONAL DATA STUDY

J. R. BATES and V.A. ALEXEEV
Danish Center for Earth System Science, Niels Bohr Institute for Astronomy, Physics and Geophysics, University of Copenhagen, Denmark.

1. Introduction

A fundamental problem in climate research is that of explaining how the Earth's climate remains stable on very long time scales. Positive feedback mechanisms such as the ice-albedo feedback and the lower tropospheric water vapour/infrared radiative feedback on sea surface temperature perturbations are known to exist which could, in principle, drive the climate system far from its observed mean state even in the absence of any external forcings. Extreme scenarios that have been envisaged are a completely ice-covered earth on the one hand and a runaway greenhouse such as appears to have occurred on Venus on the other.

There is at present no generally accepted explanation for the stability of the Earth's climate in the face of these positive feedbacks. The most obvious candidate as a stabilizing mechanism, the basic radiative Stefan-Boltzmann feedback, guarantees only that the emission temperature of the planet will adjust so that the time-averaged outgoing longwave radiation balances the absorbed solar radiation. It places little constraint on the surface temperature. The latter is determined by the strength of the greenhouse effect, which depends not only on the concentration of greenhouse gases and their radiative properties but also on the dynamics of the atmosphere and oceans. A number of possible stabilizing mechanisms have been proposed, based on various processes such as the radiative effects of clouds and upper tropospheric water vapour. All of the proposed stabilizing mechanisms are controversial and their relative importance a matter of debate.

A new stabilizing mechanism has recently been proposed (Bates, 1999). It is based on an empirically discovered linear relationship between the poleward transport of atmospheric angular momentum (AM) across 30° and the difference between the mean heights of the 500hPa surface in the tropics and the extratropics. This relationship has been incorporated into a simple two-zone atmosphere-ocean model. Assuming that the atmosphere is in a state of dynamic balance on climatic timescales, the mean zonal surface winds, and hence the evaporation, are determined by the AM transport. The model includes the solar radiation and the net infrared radiation emitted from the surface. It also includes the poleward heat transport by ocean currents. Using energy equations for the tropical and extratropical ocean basins, a mean climate state for the model is determined and the stability of small perturbations about the mean state is examined. It is found that the evaporative heat losses determined by the AM transport act as a negative feedback on sea surface temperature (SST) perturbations. The feedback is of sufficient strength to overcome the water vapour/infrared radiative feedback and hold the model climate stable. Supporting evidence for the stabilizing mechanism is provided by experiments with a GCM in which the SST is perturbed from its equilibrium (Alexeev and Bates, 1999) and supporting evidence for the validity of the parameterizations used in the simple model is provided by an observational study using a 40-year atmospheric data set (Alexeev and Bates, 2001). The present paper provides a summary of the results of the three papers referred to above.

2. The Model

Adopting the notation of Bates (1999) that subscripts 1 and 2 refer in all cases to Zones 1 (0°-30°) and 2 (30°-90°), respectively, the ocean energy equations are

$$c_{O1}\frac{dT_{S1}}{dt} = S_1 - [\,(F_L)_1 + (F_H)_1 + (F_I)_1\,] - F_{OH} \quad (1)$$

$$c_{O2}\frac{dT_{S2}}{dt} = S_2 - [\,(F_L)_2 + (F_H)_2 + (F_I)_2\,] + F_{OH} \quad (2)$$

where c_{O1} and c_{O2} are the ocean heat capacities, S_1 and S_2 are the fluxes of solar energy absorbed at the surface, and ($(F_L)_1$, $(F_L)_2$), ($(F_H)_1$, $(F_H)_2$) and ($(F_I)_1$, $(F_I)_2$) are the surface energy losses due to latent heat flux, sensible heat flux and net (upward minus downward) IR radiation, respectively.

2.1 THE MODEL'S EQUILIBRIUM CLIMATE

The latent, sensible and IR radiative heat fluxes corresponding to the model's equilibrium climate are calculated using a combination of empirical input and theoretical calculation. The equilibrium value of the ocean heat transport is taken from observational estimates as $\overline{F}_{OH} = 2.4$ PW. The latent and sensible heat fluxes from the ocean surface are calculated from the bulk aerodynamic formulae, with wind speeds determined from AM/torque balance (the mean AM transport between the zones being taken from observation as 30 Hadleys) and with low level relative humidity and air-sea temperature difference fixed at their mean observed values of 80% and 1 K, respectively. The net infrared fluxes at the surface are taken from the radiative model calculations of Hartmann and Michelsen ((1993), with mean SSTs taken from observation as $\overline{T}_{S1} = 300$ K and $\overline{T}_{S2} = 278$ K. The solar fluxes S_1 and S_2 are then determined from the requirement that the equilibrium climate satisfy the steady-state version of eqns. (1) and (2), i.e.,

$$(\overline{F}_L)_1 + (\overline{F}_H)_1 + (\overline{F}_I)_1 + \overline{F}_{OH} = S_1 \tag{3}$$

$$(\overline{F}_L)_2 + (\overline{F}_H)_2 + (\overline{F}_I)_2 - \overline{F}_{OH} = S_2 \tag{4}$$

where the overbars denote equilibrium quantities.

It is found that $(\overline{F}_L)_1 = 16$ PW, $(\overline{F}_L)_2 = 5$ PW, $(\overline{F}_H)_1 = 1$ PW, $(\overline{F}_H)_2 = 1.4$ PW, $(\overline{F}_I)_1 = 5.6$ PW and $(\overline{F}_I)_2 = 11.5$ PW. Eqs. (3) and (4) then give $S_1 = 25$ PW and $S_2 = 15.5$ PW. We see that the dominant term balancing the solar radiation absorbed at the surface in Zone 1 is evaporation, while in Zone 2 it is net IR radiation. Both of these terms are much larger than the oceanic heat flux $\overline{F}_{OH}$. There is good agreement between the calculated equilibrium fluxes and the best available estimates obtained using observations, providing support for the premise that the model represents a reasonable first-order model of the climate system.

2.2 STABILITY OF THE MODEL CLIMATE

Having determined the model's equilibrium climate, its stability with respect to small perturbations is studied. The governing equations for the perturbations are the linearized perturbation form of the ocean energy equations (1) and (2). Setting $T_{S1} = \overline{T}_{S1} + T_{S1}'(t)$, etc., and regarding (S_1, S_2) as fixed at their equilibrium values, we have the perturbation equations

$$c_{O1} \frac{dT_{S1}'}{dt} = - [(F_L)_1' + (F_H)_1' + (F_I)_1'] - F_{OH}' \tag{5}$$

$$c_{O2} \frac{dT_{S2}'}{dt} = - [(F_L)_2' + (F_H)_2' + (F_I)_2'] + F_{OH}' \tag{6}$$

where the flux quantities on the r.h.s. are calculated as linearized perturbations about the model's equilibrium climate. They involve the partial derivatives of each flux quantity with respect to variations in T_{S1} and T_{S2}. These, along with their calculated values, are given in Table 1.

Using the above, eqs. (5) and (6) become

$$\frac{dT_{S1}'}{dt} = -\beta_1 T_{S1}' - \beta_2 T_{S2}' \tag{7}$$

$$\frac{dT_{S2}'}{dt} = -\beta_3 T_{S1}' - \beta_4 T_{S2}' \tag{8}$$

where $(\beta_1, \beta_2, \beta_3, \beta_4) = (\alpha_1/c_{O1}, \alpha_2/c_{O1}, \alpha_3/c_{O2}, \alpha_4/c_{O2})$, with

$$\alpha_1 = \gamma_{L11} + \gamma_{H11} + \gamma_{O1} + \gamma_{I1} \tag{9}$$

$$\alpha_2 = \gamma_{L12} + \gamma_{H12} + \gamma_{O2} \tag{10}$$

$$\alpha_3 = \gamma_{L21} + \gamma_{H21} - \gamma_{O1} \tag{11}$$

$$\alpha_4 = \gamma_{L22} + \gamma_{H22} - \gamma_{O2} + \gamma_{I2} \tag{12}$$

Table 1. Partial derivatives of the latent, sensible, radiative and oceanic heat fluxes

Symbol	Definition	Value (PW K^{-1})
γ_{L11}	$\partial(F_L)_1/\partial T_{S1}$	1.66
γ_{L12}	$\partial(F_L)_1/\partial T_{S2}$	-0.44
γ_{L21}	$\partial(F_L)_2/\partial T_{S1}$	0.22
γ_{L22}	$\partial(F_L)_2/\partial T_{S2}$	0.21
γ_{H11}	$\partial(F_H)_1/\partial T_{S1}$	0.04
γ_{H12}	$\partial(F_H)_1/\partial T_{S2}$	-0.03
γ_{H21}	$\partial(F_H)_2/\partial T_{S1}$	0.06
γ_{H22}	$\partial(F_H)_2/\partial T_{S2}$	-0.04
γ_{I1}	$\partial(F_I)_1/\partial T_{S1}$	-0.37
γ_{I2}	$\partial(F_I)_2/\partial T_{S2}$	-0.04
γ_{O1}	$\partial F_{OH}/\partial T_{S1}$	0.37
γ_{O2}	$\partial F_{OH}/\partial T_{S2}$	-0.23

The solution of the coupled equations (7) and (8) is composed of two normal modes, which can be written

$$(T_{S1}', T_{S2}') = A \left(1, \frac{\beta_3}{\eta+\xi}\right) \exp\left[-\left(\frac{\beta_1+\beta_4}{2} + \eta\right) t\right] \tag{13}$$

$$(T_{S1}', T_{S2}') = B \left(-\frac{\beta_2}{\eta+\xi}, 1\right) \exp\left[-\left(\frac{\beta_1+\beta_4}{2} - \eta\right) t\right] \tag{14}$$

where $\xi = (\beta_1 - \beta_4)/2$, $\eta^2 = \xi^2 + \beta_2 \beta_3$ and A and B are arbitrary constants.

The conditions for the stability of the system are that both normal modes be exponentially decaying, i.e.,

$$\frac{\beta_1 + \beta_2}{2} + \eta > 0 \qquad (15)$$

$$\frac{\beta_1 + \beta_2}{2} - \eta > 0 \qquad (16)$$

Using the values of the parameters in Table 1, it is found that both these criteria are satisfied, this being the case because of the influence of the latent heat flux terms. If the latent heat flux sensitivities are set to zero, the model climate is unstable because of the positive water vapour/infrared radiative feedback inherent in the radiative terms.

3. The GCM experiments

The general circulation model (GCM) experiments (Alexeev and Bates, 1999) use aquaplanet boundary conditions with a prescribed latitudinally-varying SST distribution that approximates that observed. The surface albedo is first varied until an approximate equilibrium climate is found. (An equilibrium climate is defined as one in which the long term mean surface energy fluxes integrate globally to zero.) In the approximate equilibrium climate of the GCM, the solar radiation absorbed over a hemisphere is 46.13 PW while the infrared radiative and turbulent heat losses are 46.46 PW. A uniform perturbation of 2 K is then added to the equilibrium SST and the model is integrated until it reaches a (non-equilibrium) steady state, in which the integrated surface fluxes have become unbalanced. The imbalances in the surface energy fluxes in each zone then indicate how the model attempts to return to equilibrium. The imbalances and their component parts are found to be in good agreement with what is expected on the basis of the simple model. The total imbalance for a hemisphere is found to be -1.6 PW, the negative sign indicating a tendency for the perturbed state to return to equilibrium (a negative feedback). A resolution of the imbalance into its components shows that the net radiative flux change at the surface gives a positive feedback, the longwave part being a positive feedback of $+0.90$ PW. This corresponds to the positive water vapor/infrared radiative feedback in the simple model. The principal negative feedback countering this is the evaporative imbalance, which amounts to -1.74 PW. The eddy AM transport was found to increase by 10% as a result of the SST perturbation, in the manner predicted by the simple model. The stronger AM transport led to a strengthening of the surface winds. A simple analysis showed that, for the particular form of SST perturbation used, the wind and humidity components of the evaporative flux imbalance have comparable magnitudes in the tropics while the humidity component dominates in the extratropics.

The relative contributions of water vapour and clouds to the infrared and solar radiative changes at the surface were also estimated. It was found that for the solar radiation the cloud and water vapour influences were of about equal importance, while for the infrared radiation the effect of water vapour was dominant, as assumed in the simple model.

4. The observational study

The observational study (Alexeev and Bates, 2001) uses data from the NCEP/NCAR Reanalysis (Kalnay et al., 1996). The validity of the following features of the simple model was verified from the data study: (a) the use of the bulk aerodynamic formulae to calculate long term mean evaporation averaged over the sea points of the model zones, (b) the relation between AM transport and the difference between the mean heights of the 500hPa surface in the tropics and extratropics , (c) the relationship between variations in SST and the mean heights of the 500hPa surfaces. The importance of the wind factor relative to the humidity factor in evaporation was also found to be in accord with that indicated by the simple model.

5. References

Alexeev, V. and J.R. Bates (1999) GCM experiments to test a proposed dynamical stabilizing mechanism in the climate system. Tellus, **51A**, 630-651.

Alexeev, V.A.and J.R. Bates (2001) A study of the basic parameterizations underlying a proposed dynamical stabilizing mechanism in the climate system. (Submitted for publication)

Bates, J.R. (1999) A dynamical stabilizer in the climate system: A mechanism suggested by a simple model. Tellus, **51A**, 349-372.

Hartmann, D.L. and Michelsen, M.L. (1993) Large scale effects on the regulation of tropical sea surface temperature. *J.Climate* , **6**, 2049-2062.

Kalnay, E. et. al. (1996) The NCEP/NCAR 40-year Reanalysis Project. *Bull. Amer. Met. Soc.* **77**, 437-471.

WAVES ON THE BETA-PLANE OVER TOPOGRAPHY

E.S. BENILOV
University of Limerick
Limerick, Ireland

We examine linear waves on the beta-plane over topography which consists of isolated radially-symmetric irregularities. We assume that the radii of those are much smaller than the characteristic distance between neighbouring topographic features, and that the latter parameter is much smaller than the wavelength.

1. Introduction

Bottom irregularities exist everywhere in the ocean and play an important role in the dynamics of oceanic waves. Even *small-amplitude, short-horizontal-scale* topography can strongly affect planetary motions and give rise to the so-called topographic wave modes. The latter type of waves was discovered by Rhines and Bretherton (1973) for the case of one-dimensional topography (i.e. such that the isobaths are straight lines). Their results have been later generalised for the two-layer case (Samelson 1992, Reznik and Tsybaneva 1999), linearly-stratified case (Bobrovich and Reznik 1999), and for waves in zonal currents (Benilov 2000a,c).

However, generalisation of Rhines's and Bretherton's (1973) results for *two*-dimensional topography turned out to be much more difficult. So far, the two-dimensional case has been only examined under the assumption that topography is sparse, i.e. consists of isolated, radially symmetric features separated by large distances (Vanneste 2000, Benilov 2000b).

The present paper examines two-dimensional short-scale topography. Our objective is to clarify the difficulties of the two-dimensional case and outline ways of bypassing them.

2. Governing equations

The standard linear equation which describes quasigeostrophic flows on the beta-plane over topography is:

$$\nabla^2 \psi_t + J(\psi, d) + \beta \psi_x = 0, \qquad (1)$$

where (x, y) and t are the Cartesian coordinates and time, $\psi(x, y, t)$ is the streamfunction, β is the meridional gradient of the Coriolis parameter, and $d(x, y)$ describes the deviation of the ocean depth $H(x, y)$ from its mean value H_0:

$$d = \frac{f}{H_0}(H - H_0)$$

(f is the Coriolis parameter). We are concerned with oscillations that are harmonic in t, but not in x and y. However, as a matter of convenience, we shall separate the harmonic dependence on the spatial variables from the dependence "induced" by topography:

$$\psi(x, y, t) = \phi(x, y) \exp(ikx + ily - i\omega t), \qquad (2)$$

where ω and (k, l) are the frequency and wavevector, and $\phi(x, y)$ describes the effect of bottom irregularities (if $d = 0$, then $\phi = const$). Substituting (2) into (1), we obtain

$$-i\omega \left[\nabla^2 \phi + 2i(k\phi_x + l\phi_y) - (k^2 + l^2)\phi\right] \\ + \phi_x d_y - \phi_y d_x + i(kd_y - ld_x)\phi + \beta(\phi_x + ik\phi) = 0. \qquad (3)$$

We shall assume that the horizontal spatial scale L_D of topography is much smaller than the wavelength, i.e.

$$L_D \ll (k^2 + l^2)^{-1/2},$$

which correspond mathematically to a change of variables:

$$x_* = \frac{x}{\varepsilon}, \qquad y_* = \frac{y}{\varepsilon},$$

where ε is a small parameter. Equation (3) becomes (asterisks omitted)

$$\begin{aligned}\omega \nabla^2 \phi &+ i\left(d_y \phi_x - d_x \phi_y\right) \\ &= \varepsilon\left[(kd_y - ld_x)\phi - 2i\omega\left(k\phi_x + l\phi_y\right) - i\beta\phi_x\right] \\ &\quad + \varepsilon^2\left[\beta k + \omega\left(k^2 + l^2\right)\right]\phi. \quad (4)\end{aligned}$$

3. Asymptotic analysis

We expand the eigenfunction and eigenvalue of equation (4) in powers of ε:

$$\phi = \phi^{(0)} + \varepsilon \phi^{(1)} + \varepsilon^2 \phi^{(2)} + ..., \qquad \omega = \omega^{(0)} + \varepsilon \omega^{(1)} + \varepsilon^2 \omega^{(2)} + ...$$

In the zeroth order, we can assume that the wave mode is not affected by topography:

$$\phi^{(0)} = 1. \qquad (5)$$

In the next order, we obtain the following equation for $\phi^{(1)}$:

$$\omega^{(0)} \nabla^2 \phi^{(1)} + i\left(d_y \phi_x^{(1)} - d_x \phi_y^{(1)}\right) = kd_y - ld_x. \qquad (6)$$

If the isobaths are straight lines, i.e. if

$$d = d(\xi), \qquad \xi = const_1\, x + const_2\, y,$$

the solution to equation (6) can be sought in the form

$$\phi^{(1)}(x,y) = \phi^{(1)}(\xi).$$

The Jacobian in (6) disappears, and the resulting equation can be readily solved (see Rhines and Bretherton 1973).

In the general case, where $d(x,y)$ is an arbitrary function, (6) has no "easy" solutions and thus becomes the main stumbling block of the two-dimensional problem. We shall postpone its discussion until the next section, and proceed to the next order of perturbation expansion. Taking into account (5), we obtain

$$\omega^{(0)}\nabla^2\phi^{(2)} + \omega^{(1)}\nabla^2\phi^{(1)} + i\left(d_y\phi_x^{(2)} - d_x\phi_y^{(2)}\right)$$
$$= (kd_y - ld_x)\phi^{(1)} - 2i\omega^{(0)}\left(k\phi_x^{(1)} + l\phi_y^{(1)}\right)$$
$$- i\beta\phi_x^{(1)} + \beta k + \omega^{(0)}\left(k^2 + l^2\right). \quad (7)$$

As usual, the solution $\phi^{(2)}$ of equation (7) is not necessarily bounded at infinity. To demonstrate this, average (7) over the (x,y)-plane, i.e. consider

$$\lim_{S\to\infty} \frac{1}{4S^2} \int_{-S}^{S}\int_{-S}^{S} (7)\, dx\, dy.$$

Integrating by parts and assuming that $\phi^{(1)}$ and $\phi^{(2)}$ are bounded as $x, y \to \infty$, one can see that the left-hand side of (7) vanishes when $S \to \infty$, and so do some of the terms on the right-hand side. Eventually, we obtain

$$\left\langle (kd_y - ld_x)\phi^{(1)}\right\rangle + \beta k + \omega^{(0)}\left(k^2 + l^2\right) = 0, \quad (8)$$

where $\langle ... \rangle$ denotes spatial averaging. Dispersion relation (8) and the "definition" of $\phi^{(1)}$ [equation (6)], determine the frequency of the wave modes (if any).

4. The approximation of sparse topography

Equation (6) is a linear PDE with variable coefficients and it does not admit analytical solution in the general case [at least, no-one was able to obtain such since 1973, when Rhines and Bretherton derived this equation]. The only tractable case found so far is the case of *sparse topography* (Vanneste 2000, Benilov 2000b).

Assume that topography "consists" of an infinite number of isolated radially symmetric irregularities. We shall also assume for simplicity that the irregularities are localised within circles of various radii, R_n (n is the "number" of irregularity). Introducing the distances R_{nm} between the centres of n-th and m-th irregularities, we assume that those are rare and far between, i.e.

$$\overline{R}_n \ll \overline{R}_{nm},$$

where $\overline{R}_n$ and $\overline{R}_{nm}$ are the typical values of R_n and R_{nm}, respectively. In this case, *the solution of (6) can be approximated by a sum of the contributions of individual irregularities.*

This approach has been applied to the simplest particular case of *cylindrical* irregularities of various heights h and radii R. We considered the case where all irregularities have the same height, and the case where heights are distributed randomly between $-h_0$ and h_0. The following results have been obtained (see Figure 1):

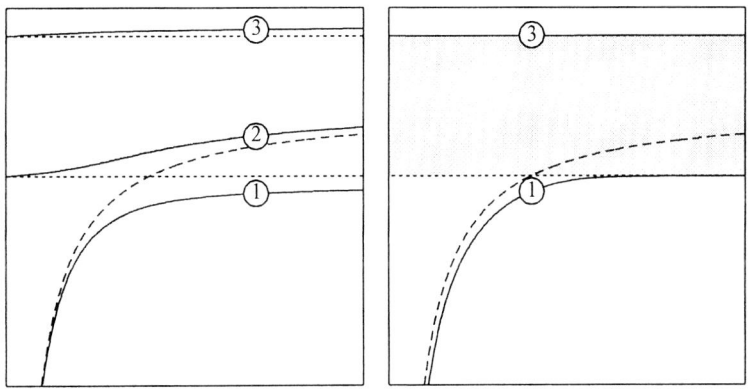

Figure 1: Typical dispersion curves, ω vs. k, for the case of cylindrical irregularities of identical heights (left panel), and for the case where the heights of irregularities are distributed randomly (right panel).

Curves 1 and 2 correspond to mixed topographic-Rossby modes, curve 3 corresponds to a purely topographic mode.

Dashed line shows the unperturbed Rossby-wave dispersion curve, the dotted lines show the "natural" topographic frequency ω_{topo}.

- If all irregularities are of the same height h, there are three wave modes: a topographic mode and two mixed, topographic-Rossby modes, separated by horizontal lines at $\omega_{topo} = \pm \frac{fh}{2H_0}$. Physically, ω_{topo} is the "natural" frequency of an oscillation trapped by an individual irregularity, which explains why free waves cannot have this frequency (they get captured by topographic features).

- If the height of cylindrical irregularities is randomly distributed between $-h_0$ and h_0, the frequency of the eigenmodes cannot be within the band $\left(-\frac{fh_0}{2H_0}, \frac{fh_0}{2H_0}\right)$ (for the same reason as that in the case of irregularities of identical height).

References

Benilov, E.S., (2000a) The stability of zonal jets in a rough-bottomed ocean on the barotropic beta-plane, *J. Phys. Oceanogr.* **30**, 733-740.

Benilov, E.S. (2000b) Barotropic Rossby waves over two-dimensional topography, *J. Fluid Mech.* **423**, 263-273.

Benilov, E.S. (2000c) Baroclinic instability of two-layer flows over bottom topography, *J. Phys. Oceanogr.* (in press).

Bobrovich, A.V., and Reznik, G.M. (1999) Planetary waves in a stratified ocean of variable depth. Part 2. Continuously stratified ocean, *J. Fluid Mech.* **388**, 147-169.

Reznik, G.M. and Tsybaneva, T.B. (1999) Planetary waves in a stratified ocean of variable depth. Part 1. Two-layer model, *J. Fluid Mech.* **388**, 115-145.

Rhines, P.B. and Bretherton, F (1973) Topographic Rossby waves in a rough-bottomed ocean, *J. Fluid Mech.* **61**, 583-607.

Samelson, R.M. (1992) Surface-intensified Rossby waves over rough topography, *J. Mar. Res.* **50**, 367-384.

Vanneste, J. (2000) Rossby-wave frequency change induced by small-scale topography, *J. Phys. Oceanogr.* **30**, 1820-1826.

ASYMPTOTIC MODELS AND APPLICATION TO VORTEX DYNAMICS

M. BEN JELLOUL

L.M.D./E.N.S.

24 rue Lhomond, 75005 Paris , France

AND

X. J. CARTON

S.H.O.M.

Établissement Principal B.P. 426, 29275 Brest Cedex, France

Abstract

We study the evolution of initially axisymmetric, large scale (Bu $\ll$ 1), geostrophic (Ro $\ll$ 1), barotropic vortices. In such a parameter regime, asymptotic expansion of rotating shallow water (RSW) leads to the frontal geostrophic (FG) model which is numerically implemented. Both in FG and in RSW, anticyclones travel westward faster than cyclones. In FG, vortices of both signs show initially a secondary meridional northward drift which we detail. However this last feature is not present in RSW numerical simulations. Vortices behave like quasi-geostrophic (QG) vortices i.e. showing an initial meridional drift that depends on vortex sign.

1. Introduction

Coherent vortices are widely observed in the ocean (Olson, 1991) and the planetary atmospheres (Marcus , 1993). Similar coherent vortical structures also appear in numerical simulations of geophysical flows or two-dimensional turbulence (Polvani et al. , 1994). The present study of geophysical vortices focuses on freely evolving large-scale oceanic rings (e.g. when they have left the current which has generated them): Agulhas rings southwest of Africa (Olson et al., 1986) are a standard example of such isolated structures (with rapidly decreasing velocity at finite distance from the core, hence with zero circulation).

Geophysical flow regimes can be characterized by two dimensionless numbers: the Rossby number Ro which measures the distance to geostrophy and the Burger number Bu which is the squared ratio of the Rossby radius of deformation R_d to the length scale L of the flow. Asymptotic models can be derived from the shallow-water equations depending on the values of these two parameters (Filatoff et al., 1997; Stegner et al., 1995). The best known model is quasi-geostrophy (QG, with $\text{Ro} \ll 1, \text{Bu} = O(1)$), but large vortices are described by frontal geostrophic dynamics (FG, with $\text{Ro} \ll 1, \text{Bu} \ll 1$).

We study numerically the behavior of vortices both in a one-layer FG model by means of a pseudo-spectral code. First we recall the one layer FG equations, then we study the evolution of both cyclonic and anticyclonic vortices on the beta-plane (both are stable on the f-plane).

2. One layer frontal geostrophic dynamics

2.1. ANALYTICAL RESULTS

We briefly recall the derivation of the FG equations. Using dimensionless velocity $\vec{u}$ and elevation η the rotating shallow water equations are

$$\partial_t \vec{u} + \text{Ro}\vec{u} \cdot \vec{\nabla}\vec{u} + (1 + \tilde{\beta}y)\vec{k} \wedge \vec{u} + \frac{\lambda \text{Bu}}{\text{Ro}}\vec{\nabla}\eta = 0$$
$$\partial_t \eta + \text{Ro}\vec{\nabla} \cdot [\vec{u}(\lambda^{-1} + \eta)] = 0. \qquad (1)$$

with the layer thickness $h = 1 + \lambda\eta$ and

$$\text{Ro} = \frac{U}{f_0 L}, \quad \text{Bu} = \frac{g'H}{f_0^2 L^2} = \frac{R_d^2}{L^2}, \quad \lambda = \frac{\Delta H}{H}, \quad \tilde{\beta} = \frac{\beta}{f_0 L} \qquad (2)$$

where $U, L, H, \Delta H$ are the horizontal velocity and extent, vertical scale of the flow and interface deviation, g' is reduced gravity, f_0 is the mean Coriolis parameter (which scales time). Here the Rossby number is small and the flow is in geostrophic balance at first order. The following relation holds between the dimensionless parameters: $\lambda\text{Bu} \simeq \text{Ro}$.

The FG equation results from a multiple time scale expansion in the case $\text{Bu}, \text{Ro} \ll 1, \lambda = O(1)$ (large vortices, surface elevation on the order of layer depth):

$$\partial_t h = J(h, h\nabla^2 h + \frac{|\nabla h|^2}{2}) + \beta h \partial_x h \qquad (3)$$

with the Jacobian operator $J(A, B) = \partial_x A \partial_y B - \partial_x B \partial_y A$. Here time has been rescaled by a factor Ro^2. Finally note that the inviscid FG equation possesses two classes of invariants [1] which are every functional of h and the

---
[1] These invariants can easily be related to the Hamiltonian formulation of the FG equation.

kinetic energy

$$C \equiv \int\int dxdy \, \Phi(h), \quad KE \equiv \frac{1}{2}\int dxdy \, h|\nabla h|^2 + \beta y h^2. \quad (4)$$

where Φ is any function of a real variable.

2.2. NUMERICAL RESULTS

On the f-plane, axisymmetric vortices with monotonic $h(r)$ profiles, both cyclonic and anticyclonic, are proved to be stable (Ben Jelloul et al., 1999). On the β-plane, the evolution of Gaussian vortices ($h = 1 + \lambda e^{-\frac{r^2}{2}}$) is now studied numerically. The model is pseudo-spectral in space and has a mixed Euler-leapfrog time stepping scheme; its resolution is 128^2 and bi-Laplacian viscosity dissipate small-scale noise. We seek to understand how the different terms in the FG equation act

$$\partial_t \eta - \beta \partial_x \eta = \beta \lambda \eta \partial_x \eta + \lambda J(\eta, \nabla^2 \eta) + \lambda^2 J(\eta, \eta \nabla^2 \eta + \frac{|\nabla \eta|^2}{2}). \quad (5)$$

The advection term $\beta \partial_x \eta$ produces the well known westward drift also observed in QG models. The "shock" term $\beta \lambda \eta \partial_x \eta$ is expected to produce a finite time singularity, when all other terms are assumed null. These gradients could be smoothed by dissipation (as for the Burgers equation) but here they are prevented from diverging by the advective terms (fig. 1). First note that the kinetic energy KE for a smooth and compact initial condition is finite. Since KE is conserved in time and unless a non zero elevation occurs at infinite y (which is highly unlikely to happen in finite time) infinite gradients are prohibited. Secondly, when sharp gradients of η appear on the zonal axis of the vortex (east of cyclones, west of anticyclones), wave breaking is inhibited by enhanced differential advection. Indeed, in the absence of β, eq. (5) can be considered as the advection of η with the following azimuthal velocity

$$v_\theta = v_1 + v_2 = \lambda \partial_r (\nabla^2 \eta) + \lambda^2 \partial_r \left(\frac{|\nabla \eta|^2}{2}\right), \quad (6)$$

where v_1 is symmetric with respect to the vortex sign and v_2 is antisymmetric. On the zonal axis, these azimuthal velocity profiles are plotted hereafter for cyclones and anticyclones (see fig.2). Since the isopycnic isolines are curved, the gradients are prevented from growing by the shear which advects them azimuthally. Therefore gradients are no more aligned with the zonal axis and the term $\beta \lambda \eta \partial_x \eta$ stops growing[2]. Note also that

---

[2] Note that the curvature of the vortex is essential. With parallel isolines $h(x)$, the meridional advection would not prevent wave breaking.

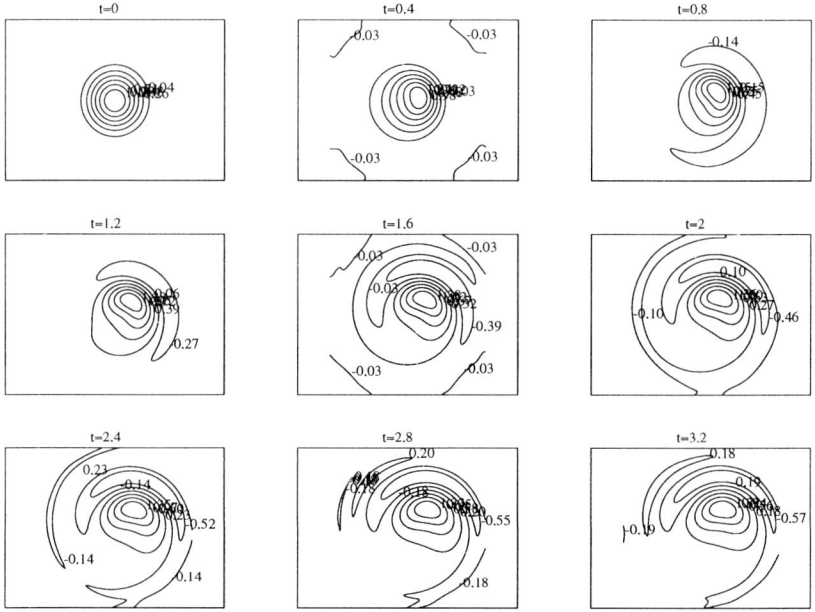

Figure 1. Time sequence of vorticity maps for a cyclonic vortex with $\lambda = -0.7$ and $\beta = 1$. The vortex is plotted in a reference frame translating with speed $-\beta$ (which coincides with the Rossby velocity in dimensional notation).

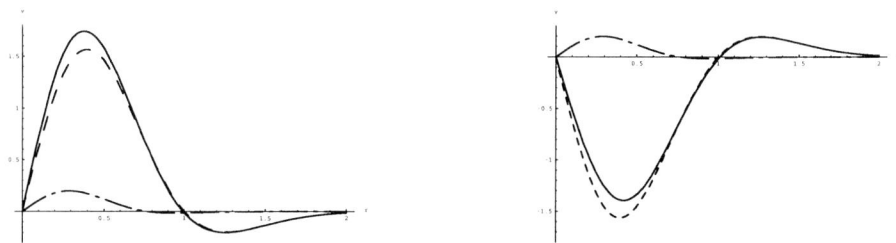

Figure 2. Radial profile of the azimuthal velocity (solid line) of anticyclonic (left graph) and cyclonic (right graph) vortices for a steepening surface in the case $\lambda = .1$. The two components v_1 (dashed) and v_2 (dot-dashed) are displayed.

the cyclone-anticyclone asymmetry present in the FG model for substantial λ results in a more vigorous shear for anticyclones than for cyclones as can be seen in fig. 3. As the nonlinear Rossby wave term produce sharp gradients west (resp. east) of the anticyclonic (resp. cyclonic) vortex core, these

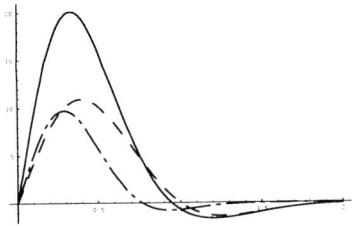 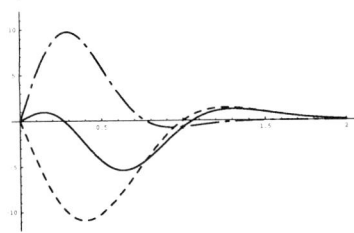

Figure 3. Same as figure 2 but for $\lambda = .7$.

opposite-sign vorticity anomalies induce a global northward advection of the vortex, irrespective of its polarity contrary to the QG model.

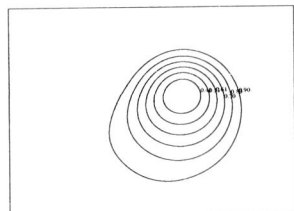 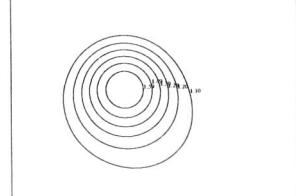

Figure 4. Isolines of the height field h at time $t = 4$ starting from initially Gaussian vortices verifying $h = 1 + \lambda \exp(-r^2/2)$ with $\lambda = \pm.7$. Vortices are plotted in the reference frame translating with speed $-\beta$. The cyclone (left) and the anticyclone (right) both drift northward.

3. Comparison with rotating shallow water vortices

We also perform numerical simulations with a RSW model, for a set of parameters in the FG regime. The results are plotted on fig. 5. The zonal drift velocity is enhanced for anticyclonic vortices and decreased for cyclones as in the FG regime. However, the secondary meridional drift is similar to the QG results and depends on the vortex sign. This discrepancy suggest that the FG model equations are not a convergent expansion of the RSW model.

4. Conclusions

In one-layer FG dynamics on the β-plane, vortices experience finite gradient intensification due to azimuthal advection. Their westward motion is

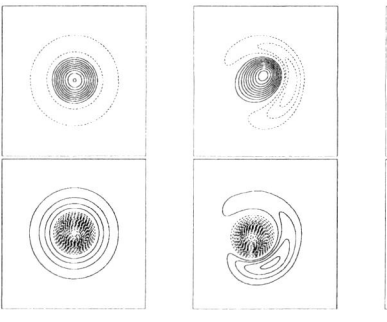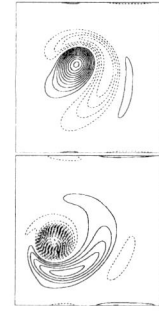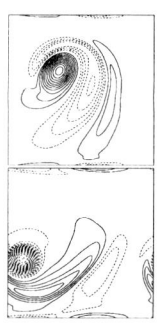

Figure 5. Time sequence of vorticity maps for a cyclonic ($\lambda = -0.7$, upper plot) and anticyclonic ($\lambda = 0.7$, lower plot) vortices (Static reference frame).

slower for cyclones than for anticyclones as in RSW dynamics (Nycander *et al.*, 1992; Benilov, 1996), but they undergo an initial northward drift (irrespective of their polarity), which is not confirmed by RSW simulations. Caution must therefore be exerted in using truncated expansions of RSW dynamics.

References

Benilov E. S. Beta-induced translation of strong isolated eddies. *J. Phys. Oceanogr.*, 26:2223–2229, 1996.

Ben Jelloul M. and V. Zeitlin. Remarks on the stability of the rotating shallow water vortices in the frontal dynamics regime. *Nuovo Cim.*, 22:931–941, 1999.

Cushman-Roisin B. , G. G. Sutyrin, and Benyang Tang. Two-layer geostrophic dynamics. part one: Governing equations. *J. Phys. Oceanogr.*, 22:117–127, 1992.

Filatoff N., X. Carton, and S. Pous. Intermediate model based on geostrophic dynamics. Technical Report 006/97, S.H.O.M., Brest, 1997.

V. M. Kamenkovich. Theory of quasigeostrophic motions in a two-layer atmosphere with scale greater than rossby scale. *Oceanology*, 29(2):133–138, 1989.

Marcus P. S. Jupiter's great red spot and other vortices. *Annu. Rev. Astron. Astrophys.*, 31:523–573, 1993.

Nycander J. and G. G. Sutyrin Steadily translating anticyclones on the beta plane. *Dyn. Atmos. Oceans*, 16:473–498, 1992.

Olson D. B. Rings in the ocean. *Annu. Rev. Earth Planet. Sci.*, 19:283–311, 1991.

Olson D. B. and R. H. Evans. Rings of the agulhas current. *Deep-Sea Research*, 33(1):27–42, 1986.

Pedlosky J. *Geophysical fluid dynamics.* Springer-Verlag, 2 edition, 1987.

Polvani L. M., J. C. McWilliams, M. A. Spall, and R. Ford. The coherent structures of shallow-water turbulence: Deformation-radius effects, cyclone/anticyclone asymmetry and gravity-wave generation. *Chaos*, 4(2):177–186, 1994.

Stegner A. and Zeitlin V. What can asymptotic expansions tell us about large-scale quasi-geostrophic anticyclonic vortices. *Nonlin. Proc. Geophys.*, 2:186–193, 1995.

Swaters G. E. On the baroclinic dynamics, Hamiltonian formulation and general stability characteristics of density-driven surface currents and fronts over a sloping continental shelf. *Phil. Trans. R. Soc. Lond. A*, (345):295–325, 1993.

OROGRAPHICALLY FORCED VARIABILITY IN THE COASTAL MARINE ATMOSPHERIC BOUNDARY LAYER

STEPHEN D. BURK, TRACY HAACK, AND RICHARD M. HODUR
Naval Research Laboratory
Marine Meteorology Division
Monterey, California, USA

1. Introduction

Coastal orography can impose distinct imprints upon the structure and dynamic behavior of the marine atmospheric boundary layer (MABL). Abrupt changes in MABL depth, wind speed, and MABL inversion strength have been noted in several over ocean field experiments adjacent to mountainous coastlines. These variations, which occur over much smaller spatial scales than that of the synoptic forcing, tend to be particularly pronounced in the vicinity of topographic points and capes. Regions of large stress, and curl of stress, frequently appear in the lee of such coastal protrusions. Mesoscale mountain gravity waves, diurnally varying cross-coast baroclinity and supercritical flow effects can all contribute to these pronounced, localized gradients.

Supercritical flow occurs when the Froude number (Fr), a dimensionless quantity given by the ratio of fluid speed to the phase speed of the fastest internal wave mode, is greater than unity. Atmospheric supercritical flow interacting with orography contains dynamic similarities to supercritical shallow water flow in a variable width channel (Ippen 1951; Samelson 1992; Rogerson 1999). However, the Froude number may be defined much more precisely in simple layered fluids having discrete density jumps than in continuously stratified flows such as the atmosphere. Expansion fans within supercritical flow around a convex bend create localized regions of pronounced stress and stress curl favorable for upwelling (Samelson 1992). Enriquez and Friehe (1995) provided observational and modeling evidence from several field experiments along the California coast during the 1980's that these supercritical flow effects occur in the atmosphere. Here we use the Naval Research Laboratory's Coupled Ocean/Atmosphere Mesoscale Prediction System (COAMPS) to produce forecasts along the Oregon/California coast during the Coastal Waves 1996 (CW96; Rogers *et al.* 1998) field experiment. During several days of CW96, we investigate supercritical MABL flow around Cape Blanco, OR and Cape Mendocino, CA, as well as the flow interaction occurring between these two capes. To generalize the results obtained from these CW96 case studies, in section 3 we examine COAMPS monthly averaged reanalysis fields. For the full year of 1999, COAMPS was run in an on-going data assimilation cycle in which observations were assimilated and a model initialization performed every 12 h at 00 and 12 UTC. Monthly averaged statistics are computed based on fields saved hourly from 24 h forecasts beginning at 00 UTC.

The atmospheric portion of COAMPS is a semi-implicit, compressible, nonhydrostatic model that uses a terrain-following coordinate system. The atmospheric model is an operational mesoscale NWP system, having a suite of advanced physical parameterizations and utilizing an intermittent data assimilation procedure, while the ocean model portion is in its final development stage. COAMPS may be run with multiple grid nests, with each inner mesh having a factor of 3 better resolution than its

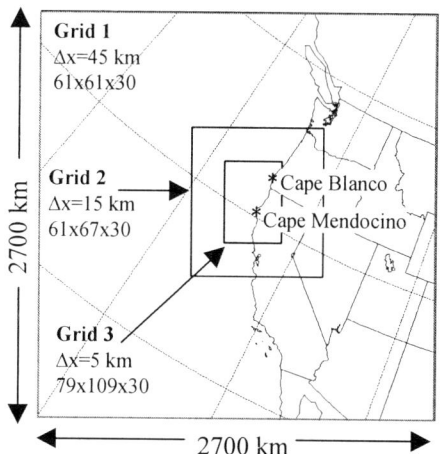

Figure 1. COAMPS nested grid domains for June/July 1996 case studies.

immediate predecessor. The CW96 simulations presented here are run with three nested grids having horizontal resolutions of 45, 15 and 5 km. Figure 1 depicts this nested grid structure and labels geographical locations on the innermost grid. There are 30 vertical levels in which more than half the levels are placed in the lowest kilometer, for an average MABL vertical spacing of ~50 m. Further description of COAMPS appears in Hodur (1997).

2. Supercritical Flow Interaction Between Capes

The COAMPS forecast fields for 7 June, 12 June, and 1 July 1996, when CW96 research aircraft flights were conducted in the Cape Blanco-Cape Mendocino vicinity, have been extensively analyzed (Haack *et al.* 2000) and some of those results are discussed in this section. On each of these days, expansion fans develop south of the two Capes while a compression jump forms between the Capes. Figure 2 displays the grid-3 10-m wind speed (m s^{-1}) and the surface stress field, τ, valid 00 UTC 2 July. [Here $\tau = \rho C_d U^2$, where ρ is atmospheric density, C_d the stability-dependent drag coefficient, and U the 10-m wind speed]. Northerly flow of ~10 m s^{-1} rounds Cape Blanco and accelerates to a maximum of ~16 m s^{-1} (Fig. 2a) creating local τ values greater than 0.4 Pa (Fig. 2b), while the marine boundary layer depth (not shown) gradually decreases by ~200 m within the expansion fan. The areal extent of this expansion fan varies diurnally, reaching a maximum size at night and contracting during the day.

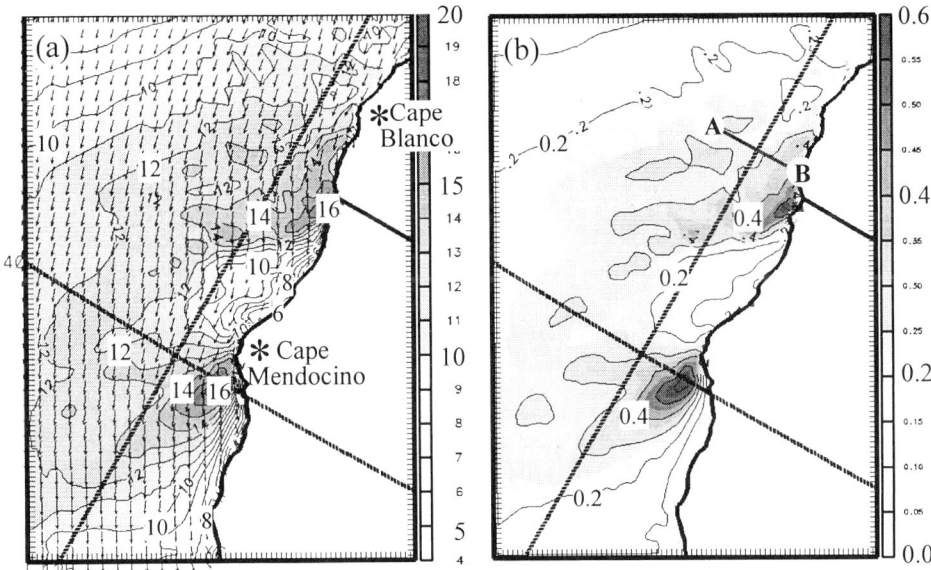

Figure 2. 12-h forecast valid 00 UTC 2 July 1996 of (a) 10-m wind speed (m s^{-1}) and; (b) surface stress (Pa) on grid 3 (Δx=5 km)

Flow south of Cape Blanco is strongly blocked as the coastline turns into the flow and forms Cape Mendocino (Fig. 2a). Similar to idealized, triple bend results presented in

Figs. 11-14 of Burk et al. 1999, a sharp drop in wind speed and change in flow direction north of Cape Mendocino is characteristic of a compression jump. Boundary layer depth rapidly increases when crossing this compression jump from north to south. A second expansion fan, again creating large stress values, forms in the lee of Cape Mendocino.

Comparison of COAMPS fields with available aircraft measurements during CW96 generally has yielded favorable agreement (Haack et al. 2000; Burk and Haack 2000; Dorman et al. 1999; Rogers et al. 1998). As an example, Fig. 3 presents results of an aircraft transect normal to the California coast during CW96 together with COAMPS cross sections along flight path A-B of Fig. 2b. Shown are observed and 9h forecast fields of potential temperature (K), relative humidity (%), and wind speed (m s^{-1}) valid 21-22 UTC 1 July 1996. As evident from Fig. 2, A-B lies in the entrance region of the expansion fan around Cape Blanco. The observed and forecast potential temperature and relative humidity cross-sections show the thermally well-mixed MABL sloping downward toward the coast (Fig. 3). There is high relative humidity within the MABL, with a sharp gradient to drier air at the capping inversion. The sloping baroclinic zone, coupled with expansion fan dynamics, produces a strong, localized low-level jet that is quite well forecast by COAMPS.

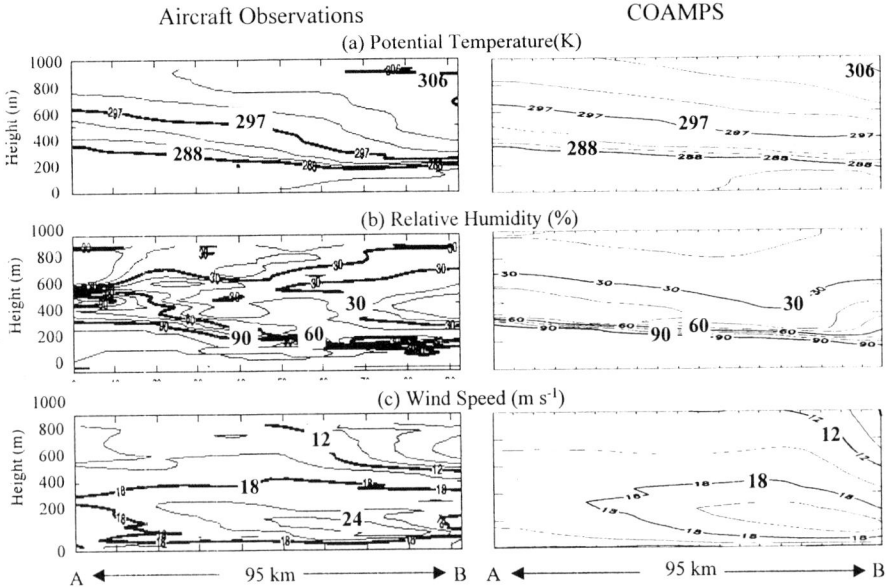

Figure 3. Cross-sections of CW96 aircraft observations in column 1 and COAMPS grid 3 9-h forecast fields in column 2. The cross-section location is indicated by A-B in Fig. 2b.

Observed and modeled diurnal variations in wind speed and direction over the coastal waters in this region occur due to thermal forcing over land. However, because the flow is largely blocked and channeled by the coastal orography, the manifestation of such thermal forcing is more subtle than in classic sea/land breezes. A turning of the nearly coast parallel flow slightly onshore during the afternoon and offshore in the evening is characteristic. For each CW96 day studied here, the modeled compression jump at night and in the early morning forms an oblique shock that is attached to the windward face of Cape Mendocino. During the late morning and afternoon, the flow turns more onshore reducing the curvature of the flow around Cape Blanco. This causes the Cape Blanco expansion fan to contract and also reduces the maximum Fr attained. As the Froude number diminishes, the oblique shock detaches from Cape Mendocino and propagates northward. However, as the flow curvature around Cape Blanco lessens in the afternoon, the flow on the windward side of Cape Mendocino is more strongly blocked and deflected, resulting in greater flow curvature around this cape. In a numerical experiment with the ground temperature fixed at its early morning value, the Cape Blanco expansion fan does not contract during the afternoon, nor does the compression jump propagate northward. Hence, this study demonstrates that numerous complex supercritical flow responses about closely spaced coastal bends cannot be analyzed by independently modeling the effect upon the MABL of each bend separately.

3. Monthly Average Surface Stress and Heat Flux

Localized regions of enhanced surface stress in supercritical expansion fans connected to points and capes form all along the California coast during the summer. This pattern in the stress field shows up not only on select days (e.g., Fig. 2b), but is also very evident in monthly average fields. The surface stress and heat flux in COAMPS are computed from stability-dependent, bulk aerodynamic relationships (Louis 1979). Although monthly average 10-m wind speed plots such as presented by Dorman *et al.* 2000 provide a qualitative estimate of the surface stress pattern, they cannot be used quantitatively to compute stress. Because stress is a nonlinear function, its monthly average cannot be computed from monthly-averaged drag coefficient and wind speed values, i.e., $<\tau> \neq <\rho><C_d><U>^2$, where the angular brackets represent a monthly average.

The sea surface temperature (SST) field within the coastal waters of California is quite inhomogeneous, containing localized regions of cold, upwelled water that appear to be strongly correlated with regions of large atmospheric stress and curl of stress. These SST inhomogeneities substantially impact the surface stability, and hence the drag and heat transfer coefficients. Figure 4 displays the July 1999 surface stress and sensible heat flux fields from the COAMPS reanalysis on a 9 km horizontal grid. The monthly average surface stress along the coast in Fig. 4a contains local maxima that are more than double the τ value well off the coast. Also, the local coastal minima where flow is blocked are approximately half of the open ocean value. Figure 4a indicates the alternating max/min values of over water stress all along the coast. The sensible heat flux (Fig. 4b) is computed as $H = \rho c_p C_h U \Delta \Theta$, where c_p is the specific heat at constant pressure, C_h is the stability-dependent heat transfer coefficient, and $\Delta \Theta$ is the sea-air temperature difference (i.e., $\Delta \Theta = \Theta_{sea} - \Theta_{air}$). If the sea-air temperature difference, $\Delta \Theta$, were constant everywhere in Fig. 4, we would expect the patterns of τ and H to be rather similar, with maxima and minima in the same locations. Evident from Figs. 4a-b, however, are substantial differences in the patterns of stress and heat flux. The local maxima/minima in surface stress (Fig. 4a) do not correlate well with local maxima/minima in surface heat flux.

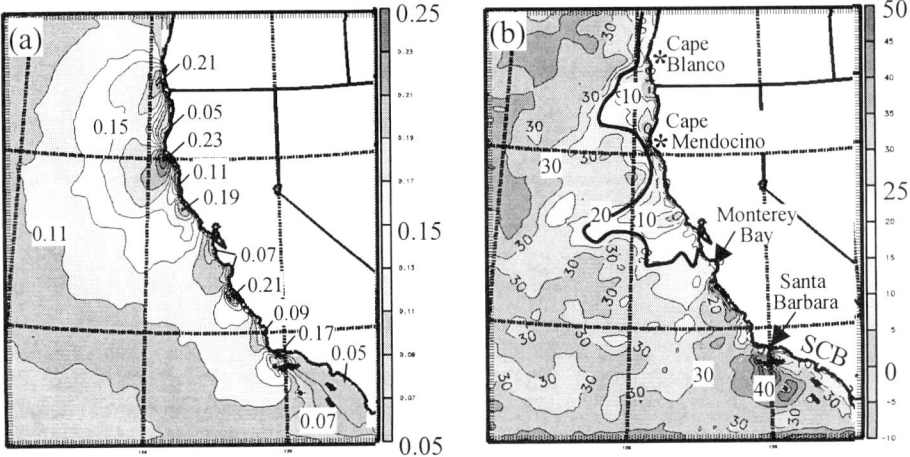

Figure 4. (a) COAMPS surface stress (N m^{-2}) averaged for July 1999 (Δx=9km); (b) as in (a) except for sensible heat flux (W m^{-2}). SCB=Southern California Bight.

Much of the structure evident in the monthly average surface heat flux (Fig. 4b) is also evident in the monthly average SST field (Fig. 5). The 287.5 K SST isotherm (bold line Fig. 5) follows rather closely the 20 W m^{-2} contour (bold line in Fig. 4b) of H that begins near Monterey Bay and meanders northward. Cold, upwelled water and weaker surface heat fluxes are found coastward of this contour. A different situation exists in the Southern California Bight. With the exception of the Santa Barbara Channel, the stress is weak (because the winds are light) in most of the Bight. However, the surface heat fluxes in the Bight are not particularly small because the SST is substantially warmer than values further north along the coast. The SST field has considerable more small-scale variability than either the 10-m wind speed or air temperature, and this is manifest, through $\Delta\Theta$, in the pattern of H.

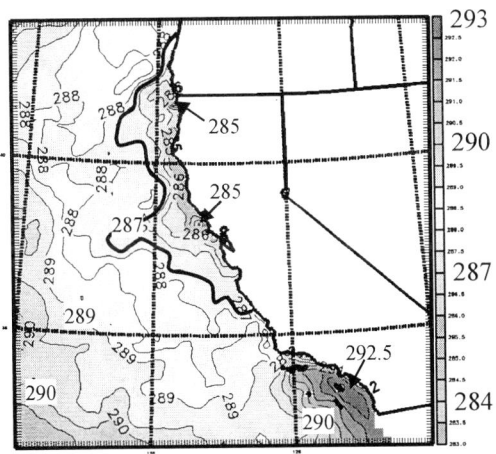

Figure 5. Monthly averaged COAMPS sea surface temperature [SST(K)] for July 1999.

The correlation between observed coastal cold SST 'pools' and modeled regions of large stress and stress curl suggests that pronounced air-sea interaction processes are driving oceanic upwelling. The extent of this interaction may be investigated further when the fully coupled COAMPS model becomes available. We hypothesize that the stabilization of the atmospheric surface layer associated with cold SST values will tend to reduce the sensible heat flux (Fig. 4b) as well as the surface stress, thereby producing a negative feedback in the coupled system.

Acknowledgments. This work was supported by the Office of Naval Research, Program Elements 0601153N, 0602315N and 0602435N.

References

Burk, S.D. and T. Haack, 2000: The dynamics of wave clouds upwind of coastal orography. *Mon. Wea. Rev.*, **128**, 1438-1455.

Burk, S.D., T. Haack, and R.M. Samelson, 1999: Mesoscale simulation of supercritical, subcritical, and transcritical flow along coastal topography. *J. Atmos. Sci.*, **56**, 2780-2795.

Dorman, C.E., T. Holt, D.P. Rogers, and K. Edwards, 2000: Large-scale structure of the June-June 1996 marine boundary layer along California and Oregon. *Mon. Wea. Rev.*, **128**, 1632-1652.

Dorman, C.E., D.P. Rogers, W. Nuss, and W.T. Thompson, 1999: Adjustment of the summer marine boundary layer around Point Sur, California. *Mon. Wea. Rev.*, **127**, 2143-2159.

Enriquez, A.G., and C.A. Friehe, 1995: Effects of wind stress and wind stress curl variability on coastal upwelling. *J. Phys. Oceanography*, **25**, 1651-1671.

Haack, T., S.D. Burk, C.E. Dorman, D.P. Rogers, 2000: Supercritical flow interaction between the Cape Blanco-Cape Mendocino orographic complex. To appear in *Mon. Wea. Rev.*

Hodur, R.M., 1997: The Naval Research Laboratory's coupled ocean/atmosphere mesoscale prediction system (COAMPS). *Mon. Wea. Rev.*, **125**, 1414-1430.

Ippen, A.T., 1951: Mechanics of supercritical flow. *Trans. Amer. Soc. Civil Eng.*, **116**, 268-295.

Louis, J.-F., 1979: A parametric model of vertical eddy fluxes in the atmosphere. *Bound.-Layer Meteor.*, **17**, 187-202.

Rogers, D.P. *et al.*, 1998: Highlights of Coastal Waves 1996. *Bull. Amer. Meteor. Soc.*, **79**, 1307-1326.

Rogerson, A.M., 1999: Transcritical flows in the coastal marine atmospheric boundary layer. *J. Atmos. Sci.*, **56**, 2761-2779.

Samelson, R.M., 1992: Supercritical marine-layer flow along a smoothly varying coastline. *J. Atmos. Sci.*, **49**, 1571-1584.

THE INFLUENCE OF THERMOCLINE TOPOGRAPHY ON THE OCEANIC RESPONSE TO FLUCTUATING WINDS: A CASE STUDY IN THE TROPICAL NORTH PACIFIC

A. CAPOTONDI, M. A. ALEXANDER
NOAA-CIRES Climate Diagnostics Center
325 Broadway, R/CDC1, Boulder, CO 80305-3328

1. Introduction

Thermocline processes likely play an important role in climate variability at decadal timescales. Surface anomalies subducting at midlatitudes propagate in the main thermocline toward the tropics with a timescale close to decadal. Thermocline variability is also associated with baroclinic Rossby waves, a fundamental agent in the adjustment of the ocean circulation, whose propagation time can introduce a delay of several years in the oceanic feedback to atmospheric forcing.

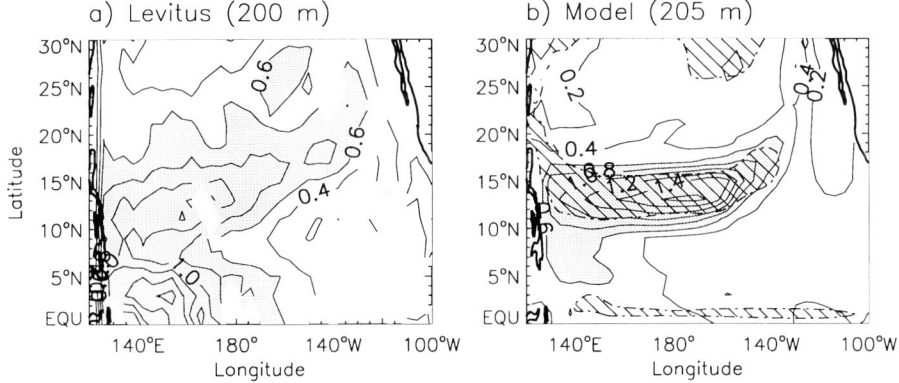

Figure 1. Temperature standard deviation at 200m depth in the Pacific using data from Levitus (a), and from a numerical model simulation performed with the NCAR ocean model forced with observed surface forcing over the period 1958-1997 (b). Contour interval is 0.2°C. Values larger than 0.6°C are shaded. In (b) the hatching indicates areas where the meridional gradients of the depth of the 15°C isotherm exceed 1.2×10^{-4}.

In the Pacific the thermocline exhibits three centers of variability, as indicated by temperature variance at 200m depth, one in the Kuroshio extension around 40°N, the second between 10°N and 15°N, and the third around 10°S. In this paper we will focus on the variability in the 10°–15°N latitude band (Figure 1). What is special about this region? Is

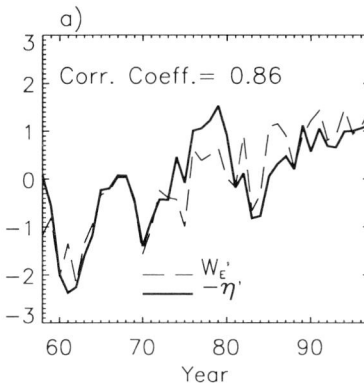 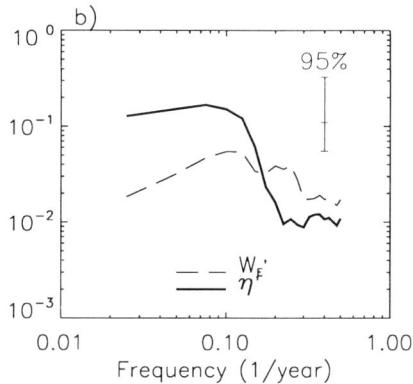

Figure 2. a) Time series of annual anomalies of Ekman pumping (dashed line) averaged over the area 179°E-140°W, 11°–14°N, and time series of the annual depth anomalies of the 25.5σ_θ isopycnal, with the sign reversed (thick solid line). The two time series are normalized to unit standard deviation. b) Spectra of the time series in a).

the enhanced variance associated with a stronger surface forcing, or is the ocean response more pronounced in this area? Capotondi and Alexander (2000, CA hereafter) have analyzed the variability at 10°–15°N using the results from the National Center for Atmospheric Research (NCAR) ocean General Circulation Model (GCM) (Gent et al. 1998) referred to in Figure 1b. CA have shown that the band of enhanced variance around 13°N is associated with Rossby waves forced by anomalous Ekman pumping east of approximately 180°, while west of this longitude the waves appear to be freely propagating. Although several aspects of the waves can be explained in terms of characteristics of the surface forcing, the reason for the particularly large variance in the 10°-15°N latitude band remains unclear. In this paper we examine the possibility that spatial changes in mean thermocline depth may be responsible for the amplitude of the oceanic response to variable wind forcing. We will see that meridional movements of the thermocline can account for a large fraction of the temperature variability at 13°N.

The paper is organized as follows: in section 2 we summarize the major findings of CA concerning the nature of the variability at 13°N, while in section 3 we examine the importance of the thermocline "topography" for determining the amplitude of the variability. We conclude in section 4.

2. Variability at 10°-15°N

2.1 THE OCEAN SIGNAL

Zonal sections of temperature and salinity anomalies along 13°N (CA) show that the largest amplitudes are found in the thermocline, and are associated with large changes in isopycnal depth, suggesting that the variability in this latitude band is primarily due to adiabatic movements of the thermocline. The time evolution of the 25.5 σ_θ depth anomalies is dominated by the low-frequencies (periods longer than ~ 7 years) as shown in Fig-

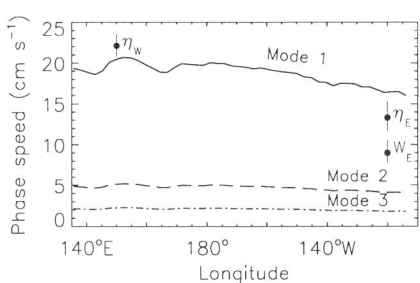
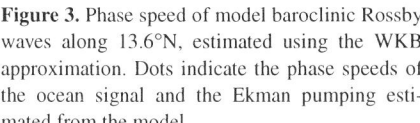
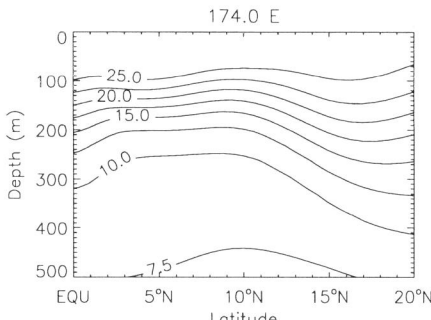

Figure 3. Phase speed of model baroclinic Rossby waves along 13.6°N, estimated using the WKB approximation. Dots indicate the phase speeds of the ocean signal and the Ekman pumping estimated from the model.

Figure 4. Meridional section of the time averaged temperature field in the model at 174°E. Contour interval is 2.5°C.

ure 2a, b, where the changes in thermocline depth are compared with the local Ekman pumping in a box centered at 160°W, 13°N. In addition to an upward trend in both time series, the ocean model has low-frequency variability, with a timescale close to 10 years (Figure 2a). The oceanic spectrum (Figure 2b) is redder than the spectrum of the forcing, with enhanced power at periods longer than ~7 years.

The propagation characteristics have been studied in CA using cross-spectral analysis. In the decadal range (7-10 years), the thermocline displacements along 13°N remain coherent, at the 90% significance level, with the thermocline displacements in the eastern end of the basin all the way to the western boundary. Significant coherence is also found along part of the western boundary, both to the north and south of 13°N, indicating that after reaching the boundary the incoming signal propagates along the boundary. The southern branch reaches the equator, and produces thermocline depth anomalies of 5-10m along the equator, with the potential of modulating ENSO on decadal timescales (Meehl et al. 2000, Fedorov and Philander 2000).

In the 7-10 years spectral band, phase lags increase monotonically from east to west. Phase speeds can be estimated by relating the phase lags between signals at different points to the distance between those points. It is found that in the eastern half of the basin the phase speed is ~13 cm s^{-1}, while west of the dateline the phase speed increases to ~22 cm s^{-1}, a value in agreement with previous observational estimates (Kessler 1990).

These results can be understood in the framework of Rossby wave dynamics. The phase speeds of the first three baroclinic Rossby wave modes have been estimated for the model using the WKB approximation (Morse and Feshbach 1953). The result along 13°N is shown in Figure 3. For the first baroclinic mode, the phase speed predicted by linear theory increases from about 15 cm s^{-1} close to the eastern boundary to ~20 cm s^{-1} in the western half of the basin, in agreement with the observational study of Chelton et al. (1998). Thus, the oceanic signal is slower than the first baroclinic Rossby wave mode east of the dateline (η_E in Figure 3), while west of the dateline it is slightly faster (η_W in Figure 3). The different propagation characteristics in the eastern and western halves of the basin can be understood by examining the nature of the forcing.

2.2 FORCING

Ekman pumping is the component of the surface forcing which can be expected to produce adiabatic thermocline displacements. For this model simulation, Ekman pumping is derived from the NCEP/NCAR reanalyses. Local correlations as large as 0.8 are found between Ekman pumping and changes in thermocline depth east of ~180°, while west of the dateline correlations drop to values smaller than 0.5. A direct comparison between the Ekman pumping and thermocline displacements time series is shown in Figure 2a, the correlation coefficient being 0.86. Cross-spectral analysis shows that east of the dateline Ekman pumping anomalies in the 7-10 year spectral band propagate westward with a phase speed of ~9 cm s^{-1} (indicated by W_E in Figure 3). Thus, we can interpret the ocean model signal east of the dateline as forced Rossby waves which propagate westward closely tracking the Ekman pumping anomalies. The westward propagation of "decadal" Ekman pumping anomalies with a phase speed which is very close to the phase speed of free Rossby waves (Figure 3) can explain the "decadal" timescale of the waves as a preferred response to the forcing in that frequency band. West of the dateline, on the other hand, thermocline variability mainly consists of free, first-mode baroclinic Rossby waves.

While the spectrum of the ocean signal can be explained in terms of the characteristics of the forcing, the amplitude of the ocean response around 13°N seems to be independent of propagating Ekman pumping anomalies. A preliminary analysis of a coupled integration performed with the NCAR climate model shows the presence of a similar band of enhanced thermal variance at 13°N, although the Ekman pumping characteristics in the coupled model appear to be different. Thus, the origin of the enhanced variance in that latitude band may be associated with the ocean structure.

3. Effects of thermocline topography

Figure 1b shows the presence of large meridional gradients in the depth of the 15°C isotherm (a proxi for thermocline depth comparable to the 25.5σ_θ isopycnal) which closely overlap the band of enhanced thermal variance at 10°–15°N. At these latitudes, the thermocline deepens poleward (Figure 4), so that an equatorward displacement of the thermocline will result in a positive temperature anomaly for an observer sitting at a fixed point. The isotherms are almost parallel with a maximum slope achieved at ~13°N (Figure 4). The instantaneous position of the maximum meridional gradient can be used as an indicator of meridional displacements of the thermocline. An example of the evolution of that position is shown in Figure 5a for 174°W. The time series of temperature anomalies at 174°W, 13°N, 230m depth (the approximate depth of the 15°C isotherm at this point) is also shown on the same axis for comparison. Poleward displacements are associated with negative temperature anomalies, and vice versa, and the two time series are nearly opposite, with a correlation coefficient of -0.97.

What is the mechanism responsible for the meridional displacements, and how much of

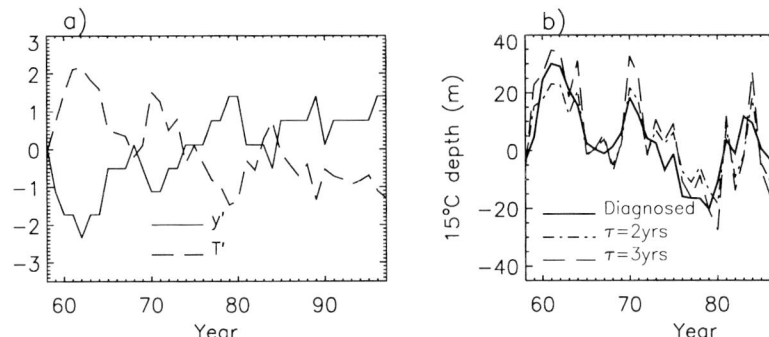

Figure 5. a) Time series of the position of the maximum meridional thermocline slope (y', solid line), and time series of temperature anomalies at 174°W, 13.6°N, 230 m depth (T', dashed line). Both time series are normalized to unit standard deviation. b) Time series of the depth anomalies of the 15°C isotherm diagnosed from the model in the box 180°–140°W, 9°–15°N (thick solid line), and computed from equation (4) in the same box for τ=2 years (dot-dashed line), and τ=3 years (dashed line). The correlation coefficient is 0.88 in both cases.

the signal can it account for? The evolution of the thermocline depth at a fixed point is governed by the equation

$$\frac{\partial \eta}{\partial t} = w - u\frac{\partial \eta}{\partial x} - v\frac{\partial \eta}{\partial y} - r\eta \tag{1}$$

where η is the thermocline depth, u, v, and w are the zonal, meridional, and vertical components of the velocities, respectively, and r is the friction coefficient. Equation (1) says that at a fixed spatial point η changes can be due to vertical displacements of the isotherm, or to horizontal advection of depth gradients. At the long timescales we are considering (periods longer than ~7 years) dissipation can be important, and for simplicity we include it in the form of Rayleigh friction. Each variable can be decomposed in time mean and deviation components ($a=\bar{a}+a'$, where a is any of the variables, the overbar indicates the time mean, and the prime indicates the anomaly). After introducing this decomposition in Equation (1) and linearizing, we examine the relevance of the balance

$$\frac{\partial \eta'}{\partial t} \sim -v'\frac{\partial \bar{\eta}}{\partial y} - r\eta' \tag{2}$$

for which changes in thermocline depth are primarily due to anomalous advection of the mean meridional gradient, where the anomalous velocity v' may be associated with the Rossby waves. If we assume an oscillatory time dependence for v' ($=v_0 e^{i\omega t}$) the solution of Equation (2) will also be oscillatory ($\eta=\eta_0 e^{i\omega t}$) with

$$\eta_o = -v_o \frac{\partial \bar{\eta}}{\partial y}\frac{(r-i\omega)}{(r^2+\omega^2)} \tag{3}$$

For periods much longer than the frictional timescale τ ($=1/r$), we have $\omega \ll r$, so that

$$\eta_o \sim -\tau v_o \frac{\partial \bar{\eta}}{\partial y} \tag{4}$$

When friction is important, the amplitude of the response is in opposition of phase with the meridional advection, and increases with decreasing friction. Equation (4) is tested in the box 180°–140°W, 9°–15°N. The results are shown in Figure 5b, where the changes in the depth of the 15°C isotherm are compared with those predicted by (4) using the anomalous velocities obtained directly from the ocean model at the depth of the mean 15°C isotherm in the box. Frictional timescales of 2 and 3 years are considered. As a reference, the timescale estimated from the isopycnal eddy diffusion coefficient used in the model simulation (0.8×10^3 m^2 s^{-1}) and from the model zonal grid size (2.4°) is 2.7 years. The result shows that for reasonable values of the frictional timescale the anomalous meridional advection of mean thermocline slope can account for a large fraction of the thermocline depth variability in that box.

4. Conclusions

We have examined the effect of spatial variations of mean thermocline depth on the oceanic response to fluctuating winds. We have focused our analysis on a relatively narrow latitude band in the tropical North Pacific (~10°–15°N), where enhanced variability is found in the thermocline. The variability is associated with baroclinic Rossby waves forced by anomalous Ekman pumping in the eastern half of the basin.

The 10°–15°N band coincides with an area of large meridional gradients in mean thermocline depth. We have shown that anomalous meridional advection of those gradients is primarily responsible for the enhanced variability in that region. Thus, the vertical velocities directly induced by the Ekman pumping are not necessarily the dominant factor in determining the amplitude of the wave field, but the meridional advection of mean thermocline gradients by the anomalous velocities associated with the Rossby waves may play a fundamental role.

5. References

Capotondi, A. and Alexander, M. A. (2000) Rossby waves in the tropical North Pacific and their role in decadal thermocline variability, *J. Phys. Oceanogr.*, submitted.

Chelton, D. B., DeSzoeke, R. A., and Schlax, M. G. (1998) Geographical variability of the first baroclinic Rossby radius of deformation, *J. Phys. Oceanogr.*, **28**, 433-460.

Fedorov, A. V. and Philander, S. G. (2000) Is El Niño changing?, *Science*, **288**, 1997-2002.

Gent, P. R., Bryan, F. O., Danabasoglu, G., Doney, S. C., Holland, W. R., Large, W. G., McWilliams, J. C. (1998) The NCAR Climate System Model global ocean component, *J. Climate*, **11**, 1287-1306.

Kessler, W. S. (1990) Observations of long Rossby waves in the northern tropical Pacific, *J. Geophys. Res.*, **95**, 5183-5217.

Meehl, G. A., Gent, P. R., Arblaster, J. M., Otto-Bliesner, B. L., Brady, E. C., and Craig, A. (2000) Factors that affect the amplitude of El Niño in global climate models, *Climate Dyn.*, in press.

Morse, P. M., Feshbach, H. (1953) *Methods of theoretical physics*, McGraw-Hill.

MODELLING THE DYNAMICS OF ABYSSAL EQUATOR-CROSSING CURRENTS

P.F. CHOBOTER AND G.E. SWATERS
Applied Mathematics Institute
Department of Mathematical Sciences
and
Institute for Geophysical Research
University of Alberta
Edmonton, Alberta, Canada

1. Introduction

Abyssal flows, as part of the global thermohaline circulation, make a significant contribution to the flux of heat over the earth, and therefore affect the planet's climate. In the Atlantic, the deepest flow consists of Antarctic Bottom Water, which originates in the Weddell Sea near Antarctica and flows northward along the western boundary of the Atlantic ocean. While part of this flow recirculates within the Brazil Basin, remaining in the southern hemisphere, part of the flow is observed to cross the equator into the northern hemisphere (DeMadron & Weatherly, 1994; Friedrichs & Hall, 1993).

Potential vorticity is conserved following the flow if friction effects are neglected. However, the fluid is relatively quiescent before and after crossing the equator, that is, planetary vorticity dominates relative vorticity. Therefore, since the planetary vorticity changes sign over the path of the flow, the potential vorticity of the fluid has also changed sign, and so is certainly not conserved! This violation of potential vorticity conservation in cross-equatorial flows and the breakdown of the geostrophic approximation at the equator constitute two significant challenges in modelling these flows.

We present a simplified model of large-scale flow across the equator. Edwards & Pedlosky (1998a, b) show that the presence of friction in the dynamics is necessary for potential vorticity modification, and thus for cross-equatorial flow to exist. Additionally, Nof & Borisov (1998; see also

Borisov & Nof, 1998) find that the geometry of the bottom topography plays a crucial role in the equator-crossing process. Accordingly, the model we study retains frictional and topographic effects. We compare the simplified model to the more sophisticated shallow water theory to identify to what extent the model captures the essential physics of the problem.

2. Frictional geostrophic model

One simple model which is geostrophic to leading order away from the equator, yet predicts well-defined velocities at the equator, may be written

$$-fv = -g'\frac{\partial(h+h_B)}{\partial x} - ru, \qquad (1)$$

$$fu = -g'\frac{\partial(h+h_B)}{\partial y} - rv, \qquad (2)$$

$$\frac{\partial h}{\partial t} + \nabla \cdot (h\boldsymbol{u}) = 0, \qquad (3)$$

where $\boldsymbol{u} = (u,v)$ is the horizontal velocity, h is the height of the fluid layer, h_B is the bottom topography elevation, g' is the reduced gravity, f is the Coriolis parameter, and r is a small damping coefficient to be specified. Note that this model does, in fact, retain the effects of an arbitrary bottom topography and parameterizes the effects of friction.

Models in which the momentum equations have been reduced to the geostrophic relations with the addition of a linear term representing the effects of friction have been used recently to study large-scale motions by several authors (see Stephens & Marshall 2000; Edwards, Willmott & Killworth 1998; Samelson 1998; Samelson & Vallis 1997; and further references therein). In particular, Stephens & Marshall (2000) numerically integrate a similar model over realistic bottom topography out to steady state in order to model the path of Antarctic Bottom Water across the equator. The resulting steady flow is found to be broadly consistent with observations.

In this model, the velocities may be solved for in a diagnostic relation in terms of the pressure gradients,

$$u = g'\frac{-fp_y - rp_x}{f^2 + r^2}, \quad v = g'\frac{fp_x - rp_y}{f^2 + r^2}, \qquad (4)$$

where $p = h + h_B$ and subscripts denote partial derivatives. Thus, the model contains a geostrophic component (terms proportional to f in the numerator), and a down-pressure-gradient component (terms proportional to r in the numerator). In the limit as $f \to 0$, the motion is that of a potential flow.

By substituting the velocity relations (4) into the conservation of mass equation (3), a single evolution equation for the height field may be written

$$h_t + J\left(g'(h+h_B), \frac{hf}{f^2+r^2}\right) = r\nabla \cdot \left[\frac{g'h\nabla(h+h_B)}{f^2+r^2}\right], \qquad (5)$$

where $J(A,B) = A_x B_y - A_y B_x$. In this form, the model is clearly nonlinear and diffusive, with the amount of diffusion controlled by the parameter r.

The potential vorticity equation of this model is

$$\frac{\partial}{\partial t}\left(\frac{f}{h}\right) + \boldsymbol{u} \cdot \nabla \left(\frac{f}{h}\right) = -\frac{r}{h}\zeta, \qquad (6)$$

where $\zeta = v_x - u_y$ is the relative vorticity. This model, then, effectively neglects relative vorticity in favour of planetary vorticity, and has the feature that it simulates the dissipation of potential vorticity by Ekman friction.

The major disadvantage of this model is its oversimplification of the dynamics. In particular, fluid inertia has been neglected. Since the fluid must always move down the pressure gradient, a mass of fluid flowing down one side of a valley does not have the momentum to flow back up the other side.

3. Frictional geostrophic versus shallow water

3.1. SCALINGS

We numerically integrate forward in time the reduced-gravity shallow water model and the frictional geostrophic model in order to compare the two models. The shallow water model may be written in non-dimensional form as

$$\frac{\partial \boldsymbol{u}}{\partial t} + \boldsymbol{u} \cdot \nabla \boldsymbol{u} + \frac{f}{Ro}\mathbf{k} \times \boldsymbol{u} = -\frac{1}{Ro}\nabla(h+h_B) + \boldsymbol{F_{fric}}, \qquad (7)$$

$$\frac{\partial h}{\partial t} + \nabla \cdot (\boldsymbol{u}h) = 0, \qquad (8)$$

where $\boldsymbol{u}$ is the horizontal velocity vector, $\boldsymbol{F_{fric}}$ represents the friction term, $Ro = U/f_0 L$ is the Rossby number, and U, L, f_0, and h_0 are typical scales for the velocity, length, Coriolis parameter and fluid depth. It has been assumed that the time variable is scaled advectively, $T = L/U$, for a time scale T, and that the scale slope for the bottom topography is the same as the scale slope of the fluid height, h_0/L. We have also employed the geostrophic scaling $U^2/(g'h_0) = Ro$. Since f passes through zero in the domain of interest, f_0 is taken to be the maximum dimensional value of f in the domain. It is assumed that the flow is geostrophic at that latitude.

The numerical methods used are based upon the methods of Hallberg & Rhines (1996). For brevity, the details are not reported here, but may be found in Choboter & Swaters (2000).

The simple model and the shallow water model are compared for flow over simplified bottom topography. The topography takes the shape of a meridional channel. Simulations were performed with the fluid initially south of the equator, flowing northward along the western half of the channel, in the form of an eddy, i.e. the height field initially has compact support in the domain. These initial conditions were chosen, in part, to simulate the Antarctic Bottom Water flow, which flows northward along the western slope toward the equator.

The bottom topography and functional form of Coriolis parameter are chosen in a particular way to provide a clean testing ground for the comparison of the two models. In particular, we are interested in diagnosing how well the propagation speed of the eddy agrees with the Nof (1983) speed, $g's/f$, where s is the bottom slope. A nearly constant bottom slope and Coriolis parameter away from the channel bottom and equator facilitates computing this diagnostic. Therefore, the bottom topography is chosen to be a simplified meridional channel of hyperbolic cross section, $h_B = \sqrt{x^2 + 1}$, which has a slope approaching ± 1 away from $x = 0$, and the Coriolis parameter is chosen to be $f = \tanh(\beta_0 L y / f_0)$, which tends to a non-dimensional f-plane value of unity away from $y = 0$, and has a slope at $y = 0$ of $\beta_0 L / f_0$. For simulations reported here, $\beta_0 L / f_0 = 1$, which, for f_0 evaluated at 5° latitude, corresponds to choosing a horizontal length scale of $L = 500$ km.

3.2. RESULTS

Several simulations of an isolated abyssal dome of fluid approaching the equator from the south have been carried out varying only the damping parameter r in the case of the simple model, or the Rossby number Ro in the case of the shallow water model. In figure 1, we show snapshots from a typical simulation employing the shallow water equations. The eddy is observed to propagate along the shelf without losing much height until almost at the equator, when fluid starts to accelerate downhill. Part of the fluid is located slightly north of the equator while flowing downhill. The fluid rises up the other side of the channel, and ultimately splits into two eddies, one flowing north and one flowing south. This is qualitatively consistent with the simulations of Borisov & Nof (1998), who investigated eddies crossing the equator in a meridional channel.

Figure 2 displays the simulation of the motion of the same initial eddy, but as predicted by the simple model. The eddy is seen to initially travel

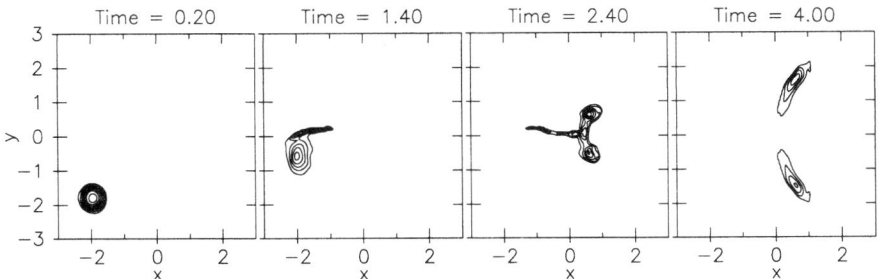

Figure 1. The results of a shallow water simulation, $Ro = 0.02$ The contour spacing is 0.02.

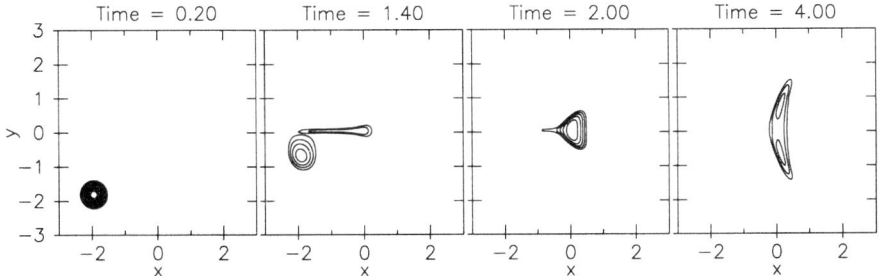

Figure 2. The results of a frictional geostrophic simulation, $r = 0.02$. The contour spacing is 0.02.

along the slope, as in the shallow water simulation, but upon reaching the equator, flows directly downhill, with very little fluid found north of the equator as it does so. The fluid pools at the bottom of the channel at the equator, and then *proceeds to split into two parts*, one flowing north, and the other recirculating back south. Despite the simplicity of the model, it captures the characteristic splitting of the fluid into northward and southward flowing parts seen in the shallow water simulation. The lack of inertia in the model is seen in both the sharp turn from along-slope flow to downhill flow and the immediate deceleration from fast downhill flow to nearly stationary fluid pooling at the equatorial channel bottom. Thus the net result of the lack of inertia in the model is that the north-south splitting of the flow is very symmetric, and that the final flow is very near the bottom of the channel.

Further analysis of this model (Choboter & Swaters 2000) shows that the simple model simulations capture well the along-shelf Nof (1983) speed. For all the simple model runs, the fluid does not flow as high onto the opposite bank as for the shallow water runs, and a very symmetric north-south splitting of the fluid is predicted. This further points to the lack of fluid inertia in the simple model.

4. Concluding remarks

The model studied here parameterizes frictional and other ageostrophic effects into a simple Rayleigh damping term. We have investigated the viability of this model by comparing its predictions to the predictions of shallow water theory. Despite the simplicity of the model, it broadly captures certain aspects of shallow water flow quite well, such as the Nof (1983) along-slope eddy speed and the north-south splitting of the fluid. However, the model neglects the inertia of the fluid, which restricts the motion.

The bottom topography of the Atlantic Ocean is certainly more complicated than a meridional channel. It remains for future research to compare the predictions of these models over more realistic topography.

Preparation of this paper was supported by a Killam Postgraduate Scholarship awarded to P.F.C. and a research grant awarded to G.E.S. by the Natural Sciences and Engineering Research Council.

References

Borisov, S. & Nof, D. (1998) Deep, cross-equatorial eddies. *Geophys. Astrophys. Fluid Dyn.* **87**, 273–310.

Choboter, P. F. & Swaters, G. E. (2000) Modelling equator-crossing currents on the ocean bottom. Submitted to *Can. App. Math. Quart.*

DeMadron, X. D. & Weatherly, G. (1994) Circulation, transport and bottom boundary layers of the deep currents in the Brazil Basin. *J. Mar. Res.* **52**, 583–638.

Edwards, N. R., Willmott, A. J. & Killworth, P. D. (1998) On the role of topography and wind stress on the stability of the thermohaline circulation. *J. Phys. Oceanogr.* **28**, 756–778.

Edwards, C. A. & Pedlosky, J. (1998a) Dynamics of nonlinear cross-equatorial flow. Part I: Potential vorticity transformation. *J. Phys. Oceanogr.* **28**, 2382–2406.

Edwards, C. A. & Pedlosky, J. (1998b) Dynamics of nonlinear cross-equatorial flow. Part II: The tropically enhanced instability of the western boundary current. *J. Phys. Oceanogr.* **28**, 2407–2417.

Friedrichs, M. A. M. & Hall, M. M. (1993) Deep circulation in the tropical North Atlantic. *J. Mar. Res.* **51**, 697–736.

Hallberg, R. & Rhines, P. (1996) Buoyancy-driven circulation in an ocean basin with isopycnals intersecting the sloping boundary. *J. Phys. Oceanogr.* **26**, 913–940.

Nof, D. (1983) The translation of isolated cold eddies on a sloping bottom. *Deep-Sea Res.* **30**, 171–182.

Nof, D. & Borisov, S. (1998) Inter-hemispheric oceanic exchange. *Q. J. R. Meteorol. Soc.* **124**, 2829–2866.

Samelson, R. M. (1998) Large-scale circulation with locally enhanced vertical mixing. *J. Phys. Oceanogr.* **28**, 712–726.

Samelson, R. M. & Vallis, G. K. (1997) A simple friction and diffusion scheme for planetary geostrophic basin models. *J. Phys. Oceanogr.* **27**, 186–194.

Stephens, J. C. & Marshall, D. P. (2000) Dynamical pathways of Antarctic Bottom Water in the Atlantic. *J. Phys. Oceanogr.* **30**, 622–640.

Toward Accurate Coastal Ocean Prediction

Peter C. Chu
Naval Postgraduate School, Monterey, CA 93943, USA

November 28, 2000

1 Introduction

Several major problems, namely, uncertain surface forcing function, unknown open boundary conditions (OBC), and pressure gradient error using the σ-coordinate, affect the accuracy of coastal ocean prediction. At open lateral boundaries where the numerical grid ends, the fluid motion should be unrestricted. Ideal open boundaries are transparent to motions. The most popular and successful scheme is the adjoint method. The disadvantages that may restrict its use are ocean-model dependency and difficulty in deriving the adjoint equation when the model contains rapid (discontinuous) processes, such as change of ocean mixed layer from entrainment to shallowing regime. Development of a ocean-model independent algorithm for determining the OBC becomes urgent.

Reduction of horizontal pressure gradient error is another key issue of using σ-coordinate ocean models, especially of using coastal models. The error is caused by the splitting of the horizontal pressure gradient term into two parts and the subsequent incomplete cancellation of the truncation errors of those parts. As advances of the computer technology, use of highly accurate schemes for ocean models becomes feasible.

2 Uncertainty of Surface Forcing

To investigate the uncertainty of surface wind forcing and its effect on the coastal prediction, the Princeton Ocean Model (POM, Blumberg and Mellor, 1987) was used with 20 km horizontal resolution and 23 sigma levels conforming to a realistic bottom topography during the life time of tropical cyclone Ernie 1996 over the South China Sea (SCS). A study (Chu, et al, 1999) shows that the root-mean-square (RMS) difference of each component (zonal or latitudinal) between the two wind data (NCEP and NSCAT) over the whole SCS during November 1996 fluctuated between 2.7 m/s to 6.5 m/s. The uncertainty of the whole SCS response to the two wind data sets were 4.4 cm for surface elevation, 0.16 m/s for surface current velocity, and 0.5°C for near-surface temperature, respectively.

3 Jacobian Matrix Method for Determining OBCs

Improvement of coastal prediction largely depends on determination of lateral OBCs, vector, $\mathbf{B} = (b_1, b_2, ..., b_n)$. The observation forms an m-dimensional vector (observation vector) $\mathbf{O} = (O_1, O_2, ... O_m)$, located at the interior. If $\mathbf{B}$ is given, we can solve the dynamic system and obtain the solution S. At the observational points, the solutions form a solution vector $\mathbf{S} = (S_1, S_2, ... S_m)$, which depends on $\mathbf{B}$ (Fig. 1).

It is reasonable to determine $\mathbf{B}$ with the given $\mathbf{O}$ by minimize the RMS error

$$I = \sqrt{\frac{1}{m}\sum_{j=1}^{m}(S_j - O_j)^2}. \tag{1}$$

which leads to a set of n equations implicitly solvable for $b_1, b_2, ... b_n$,

$$\sum_{j=1}^{m}(S_j - O_j)R_{ij} = 0, \quad i = 1, 2, ..., n \tag{2}$$

where

$$R_{ij} \equiv \frac{\partial S_j}{\partial b_i}, \quad i = 1, 2, ..., n;\ j = 1, 2, ...m. \tag{3}$$

are components of a $n \times m$ Jacobian matrix $R = \{R_{ij}\}$.

From a first guess boundary vector $\mathbf{B}^*$, a solution vector $\mathbf{S}^*$ is obtained by solving the numerical ocean model. The RMS between $\mathbf{S}^*$ and $\mathbf{O}$ might not be minimal. We update the boundary parameter vector components by increments $\{\delta b_i |\ i = 1, 2, ..., n\}$, and therefore components of the solution vector become

$$S_j = S_j^* + \sum_{i=1}^{n} R_{ij}\delta b_i + \text{ high order terms} \tag{4}$$

Substituting (4) into (2) and neglecting higher order terms leads to a set of n linear algebraic equations for $\{\delta b_i\}$,

$$\sum_{l=1}^{n} P_{il}\delta b_l = d_i, \quad i = 1, 2, ..., n \tag{5}$$

where

$$P_{il} \equiv \sum_{j=1}^{m} R_{lj}R_{ij}, \quad d_i \equiv \sum_{j=1}^{m} R_{ij}(O_j - S_j^*); \quad i = 1, 2, ..., n;\ l = 1, 2, ...n. \tag{6}$$

Both O_j and S_j^* are known quantities. Therefore, the linear algebraic equations (5) have definite solutions when the Jacobian matrix $\{R_{ij}\}$ is determined and

$$\det\{P_{il}\} \neq 0 \tag{7}$$

This method was verified by a flat bay centered at 35°N and bounded by three rigid boundaries (Fig. 1). This bay expands 1000 km in both the north-south and east-west directions. The northern, southern, and western boundaries are rigid, and the eastern boundary is open. Using the optimization method, the temporally varying OBC, $\mathbf{B}(t)$, is determined. After 10 day's of integration, the magnitude of relative error

$$E^{(O)} \equiv \frac{\sum |O_j - S_j|}{\sum |O_j|} \tag{8}$$

is on the order of 10^{-4}-10^{-5} (Fig. 2), which is almost in the noise level. The Jacobian matrix method performs well even when random noises are added to the 'observational' points. This indicates that we can use real-time data to invert for the unknown open boundary values.

4 High-Order Difference Schemes

4.1 A Hidden Problem

Improvement of the prediction also partially depends on the selection of the discretization schemes. Most coastal models use second-order difference schemes (such as second-order staggered C-grid scheme) to approximate first-order derivative

$$\left(\frac{\partial p}{\partial x}\right)_i \simeq \frac{p_{i+1/2} - p_{i-1/2}}{\Delta} - \frac{1}{24}\left(\frac{\partial^3 p}{\partial x^3}\right)_i \Delta^2, \tag{9}$$

where p, Δ represent pressure and grid spacing. This scheme uses the local Lagrangian Polynomials whose derivatives are discontinuous.

4.2 Combined Compact Scheme

Recently, Chu and Fan (1997, 1998, 1999, 2000) proposed a new three-point combined compact difference (CCD) scheme,

$$\left(\frac{\delta f}{\delta x}\right)_i + \alpha_1\left(\left(\frac{\delta f}{\delta x}\right)_{i+1} + \left(\frac{\delta f}{\delta x}\right)_{i-1}\right) + \beta_1 h\left(\left(\frac{\delta^2 f}{\delta x^2}\right)_{i+1} - \left(\frac{\delta^2 f}{\delta x^2}\right)_{i-1}\right) + ...$$
$$= \frac{a_1}{2h}(f_{i+1} - f_{i-1})$$

$$\left(\frac{\delta^2 f}{\delta x^2}\right)_i + \alpha_2\left(\left(\frac{\delta^2 f}{\delta x^2}\right)_{i+1} + \left(\frac{\delta^2 f}{\delta x^2}\right)_{i-1}\right) + \beta_2 \frac{1}{2h}\left(\left(\frac{\delta f}{\delta x}\right)_{i+1} - \left(\frac{\delta f}{\delta x}\right)_{i-1}\right) + ... \tag{10}$$
$$= \frac{a_2}{h^2}(f_{i+1} - 2f_i + f_{i-1})$$

to compute f'_i, f''_i, ...$f^{(k)}_i$ by means of the values and derivatives at the two neighboring points. Moving from the one boundary to the other, CCD forms a global algorithm to compute various derivatives at all grid points, and guarantees continuity of all derivatives at each grid point.

4.3 Seamount Test Case

4.3.1 Model Description

Suppose a seamount (Fig. 3) located inside a periodic f-plane ($f_0 = 10^{-4}\text{s}^{-1}$) channel with two solid, free-slip boundaries along constant y. Unforced flow over seamount in the presence of resting, level isopycnals is an idea test case for the assessment of pressure gradient errors in simulating stratified flow over topography. The flow is assumed to be reentrant (periodic) in the along channel coordinate (i.e., x-axis). We use this seamount case of the Semi-spectral Primitive Equation Model (SPEM) version 3.9 (Haidvogel et al., 1991) to test the new difference scheme.

4.4 Temporal Variations of Peak Error Velocity

Owing to a very large number of calculations performed, we discuss the results exclusively in terms of the maximum absolute value the spurious velocity (called peak error velocity) generated by the pressure gradient errors. Fig. 4 shows the time evolution of the peak error velocity for the first 20 days of integration with the second-, fourth-, and sixth-order ordinary schemes. The peak error velocity fluctuates rapidly during the first few days integration. After the 5 days of integration, the peak error velocity show the decaying inertial oscillation superimposed into asymptotic values. The asymptotic value is around 0.19 cm/s for the ordinary scheme and 0.15 cm/s for the compact scheme. For the sixth order difference the asymptotic value is near 0.04 cm/s for the ordinary scheme and 0.02 cm/s for the compact scheme.

5 Conclusions

(1) The surface forcing function contains uncertainty. The difference between commonly used NSCAT and NCEP surface wind data is not negligible. The response of the South China Sea to the uncertain surface forcing is also evident. Therefore, it is quite urgent to study the model sensitivity to surface boundary conditions.

(2) The Jacobian matrix method provides a useful scheme to obtain unknown open boundary values from known interior values. The optimization method performs well even when random noises are added to the 'observational' points. This indicates that we can use real-time data to invert for the unknown open boundary values.

(3) The σ-coordinate, pressure gradient error depends on the choice of difference schemes. Fourier analysis shows that the fourth-order scheme may reduce

the truncation errors by 1-2 order of magnitude compared to the second-order scheme, and the sixth-order scheme may reduce the truncation errors further by 1-2 order of magnitude compared to the fourth-order scheme. Within the same order of the difference the combined compact scheme leads to a minimum truncation error. The compact scheme may reduce near 55% error, and the combined compact scheme may reduce near 84% error compared to ordinary sixth-order difference scheme.

6 ACKNOWLEDGMENTS

This research is sponsored by the Office of Naval Research, Naval Oceanographic Office, and Naval Postgraduate School.

References

[1] Blumberg, A.F., and Mellor, G.L. (1987) A description of a three-dimensional coastal ocean circulation model. *Three Dimensional Coastal Ocean Models*, edited by N.S. Heaper, American Geophysical Union, 1-16.

[2] Chu, P.C., and Fan, C. (1997) Sixth-order difference scheme for sigma coordinate ocean models. *Journal of Physical Oceanography*, **27**, 2064-2071.

[3] Chu, P.C., and Fan, C. (1998) A three-point combined compact difference scheme. *Journal of Computational Physics*, **140**, 1-30.

[4] Chu, P.C., and Fan, C. (1999) A non-uniform three-point combined compact difference scheme, *Journal of Computational Physics*, **148**, 663-674.

[5] Chu, P.C., and Fan, C. (2000) A staggered three-point combined compact difference scheme, *Mathematical and Computer Modeling*, **32**, 323-340.

[6] Chu, P.C., Chen, Y.C. and Lu, S.H. (1998) On Haney-type surface thermal boundary conditions for ocean circulation models. *Journal of Physical Oceanography*, **28**, 890-901.

[7] Chu, P.C., Fan, C. and Ehret, L. (1997) Determination of open boundary conditions with an optimization method. *Journal of Atmospheric and Oceanic Technology*, **14**, 723-734.

[8] Chu, P.C., Lu, S.H. and Liu, W.T. (1999) Uncertainty of the South China Sea prediction using NSCAT and NCEP winds during tropical storm Ernie 1996. *Journal of Geophysical Research*, 104, 11273-11289.

[9] Haidvogel, D.B., Wikin, J.L., and Young, R. (1991) A semi-spectral primitive equation model using vertical sigma and orthogonal curvilinear coordinates. *Journal of Computational Physics*, **94**, 151-185.

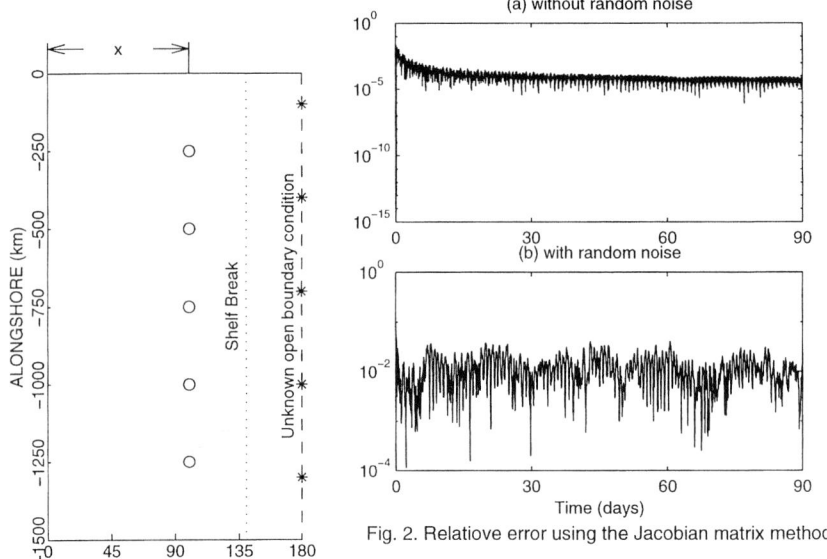

Fig. 1. Determination of open boundary condition

Fig. 2. Relatiove error using the Jacobian matrix method.

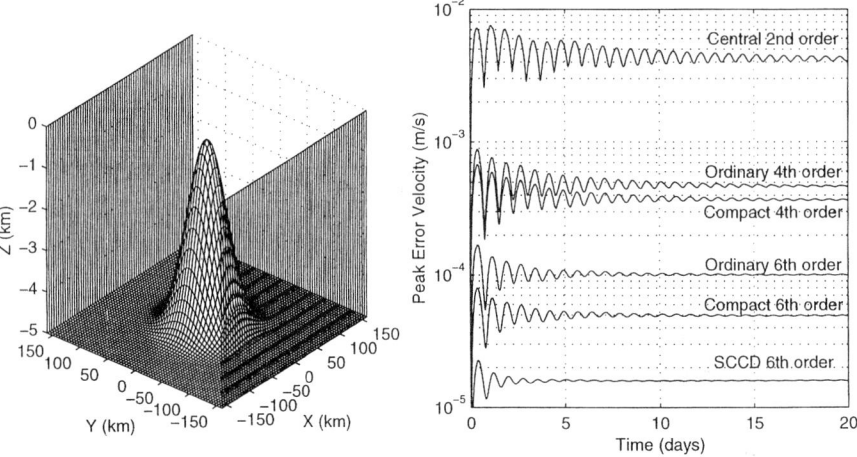

Fig. 3. Seamount geometry

Fig. 4. Peak error velocity for different order schemes

An Air-Ocean Coupled Nowcast/Forecast System for the East Asian Marginal Seas

Peter C. Chu, Shihua Lu, and Chenwu Fan
Naval Postgarduate School, Monterey, California

September 12, 2000

1 Introduction

The South China Sea (SCS), Yellow/East China Sea (YES), and Japan/East Sea (JES) are major east Asian marginal seas (EAMS). The complex topography includes the broad shallows of the Sunda Shelf in the south/southwest of SCS; the continental shelf of the Asian landmass in the north, extending from the Gulf of Tonkin to the YES; a deep, elliptical shaped SCS and JES basins, and numerous reef islands and underwater plateaus scattered throughout (Fig. 1a). The shelf that extends from the Gulf of Tonkin to the YES is consistently near 70 m deep, and averages 150 km in width.

The EAMS is subjected to a seasonal monsoon system. From April to August, the weaker southwesterly summer monsoon winds result in a wind stress of just over 0.1 N/m^2. From November to March, the stronger northeasterly winter monsoon winds corresponds to a maximum wind stress of nearly 0.3 N/m^2. Recent observational studies show that the EAMS is energetic and has multi-eddy structure. For example, the SCS synoptic eddy structure was identified in May 1995 using the airborne expendable bathythermograph (AXBT) data (Chu et al., 1998a), the eddy spatiotemporal scales in the YES were identified using the Navy's Master Oceanographic Observational Data Set (MOODS) during 1929-1991 (Chu et al., 1997a,b), and the seasonal JES multi-eddy structure from a composite analysis on the U.S. National Centers for Environmental Prediction (NCEP) monthly SST fields during 1981-1994 (Chu et al. 1998b).

The fundamental scientific issues are: What are the dynamical balances controlling mesoscale eddy variations in the EAMS, including the effect of wind and thermohaline forcing, wave processes, current instabilities, coastline geometry and topographic slope? What is the role of coastal air-ocean coupling on oceanic and atmospheric dynamics of mesoscale eddies? To study these problems, a coastal atmosphere-ocean coupled system (CAOCS) was developed at the Naval Postgraduate School. The model domain (67°-142°E, 7°S-55°N) covers the whole EAMS and surrounding land and islands. The surface fluxes of water, heat (excluding solar radiation), and momentum are applied synchronously with opposite signs in the atmosphere and ocean. Flux adjustments are not used.

2 CAOCS for the EAMS Nowcast/Forecast

2.1 Ocean Component

2.1.1 Numerics

The CAOCS ocean component (i.e., the POM) is a time dependent, primitive equation circulation model on a three dimensional grid that includes realistic topography and a free surface. From a series of numerical experiments (Chu et al., 1998a; 1999a,b; 2000b), the qualitative and quantitative effects of non-linearity, wind forcing and lateral boundary transport on the EAMS are analyzed, yielding considerable insight into the external factors affecting the regional oceanography. We use a rectilinear grid with horizontal spacing of 0.25° by 0.25° and 23 nonuniform vertical σ levels. The model uses realistic bathymetry data from the Naval Oceanographic Office Digital Bathymetry Data Base with 5 minute resolution (DBDB5).

2.1.2 Open Boundary Conditions

Closed lateral boundaries, i.e., the modeled ocean bordered by land, were defined using a free slip condition for velocity and a zero gradient condition for temperature and salinity. No advective or diffusive heat, salt or velocity fluxes occur through these boundaries.

At open boundaries, we use monthly varying volume transports from a global inverse model (Chu and Fan, 2000) for the open boundaries (Fig. 1a). When the water flows into the model domain, temperature and salinity at the open boundary are likewise prescribed from the climatological data. When water flows out of the domain, the radiation condition was applied,

$$\frac{\partial}{\partial t}(\theta, S) + U_n \frac{\partial}{\partial n}(\theta, S) = 0 \tag{1}$$

where the subscript n is the direction normal to the boundary.

2.1.3 Initial Conditions and Model Initialization

Before coupling to the atmospheric model that is MM5, the POM model was integrated for four years and four months from zero velocity, and January climatological temperature and salinity fields and forced by monthly mean surface wind stress from the Comprehensive Ocean and Atmosphere Data Set (COADS) and by the restoring-type surface heat and salinity fluxes which are relaxed to the surface monthly values. The final states of temperature, salinity, and velocity were taken as the initial ocean conditions for 1 April 1998.

2.2 Atmospheric Component

2.2.1 Model Description

We take Version-3 of the Fifth-Generation Penn State University/NCAR Mesocale Model (MM5) as the atmospheric component for CAOCS. The MM5 is a limited-area, nonhydrostatic, terrain-following sigma-coordinate model designed to simulate or predict mesoscale and regional-scale atmospheric circulation. It has been developed at Penn State and NCAR as a community mesoscale model and is continuously being improved by contributions from users at several universities and government laboratories.

2.2.2 Land Surface Parameterization

The horizontal resolution is 60 km. Sixteen pressure levels were used with the top at 10 mb. A soil water availability function was used for computing the soil water content by using 10 specified vegetation types. Dominant deciduous and coniferous forests cover the most land areas. There are small areas of tropical forest in Taiwan, and Indo-China Peninsula.

2.2.3 Initial and Lateral Boundary Conditions

The initial and horizontal lateral boundary conditions for wind, temperature, water vapor, and surface pressure are interpolated from analyses of observations from European Center for Medium-Range Weather Forecasts (ECMWF). For the present study, they were projected on a spectral T42 grid, and a vertical resolution of 14 pressure levels. The initial conditions were the fields on 00Z April 1, 1998. The lateral boundary conditions were provided via a relaxation method and updated every 12 hours.

2.3 Ocean-Atmosphere Coupling

Atmospheric and oceanic surface fluxes of water, heat (excluding solar radiation), and momentum are of opposite sign and are applied synchronously. Flux adjustments are not used. We use the same flux parameterization as the MM5 for CAOCS. The only difference between the two is the use of sea surface temperature (SST) data. In stand-alone MM5 model, SST is prescribed as a given parameter. However, in the CAOCS, SST is predicted by the POM model at each time step for the atmospheric component (MM5).

3 Predicted EAMS Circulation

We integrate the CAOCS model for four months from 1 April 1998. Figures 1b,c,d show the predicted temperature and velocity vector fields at three different depths (surface, 50m, and 200 m) in June 1998. The North Equatorial

Current (NEC) was well predicted in all depths. After encountering the western boundary along the Philippine coast, the NEC bifurcates into the northward flowing Kuroshio and the southward flowing Mindanao Current (MC).

The MC and the simulated New Guinea Coastal Undercurrent flow equatorward, feeding both the North Equatorial Counter Current (NECC) and the flow from the Pacific to the Indian Ocean, that is, the Indonesian throughflow. The MC is an essential part of the shallow, wind and buoyancy flux driven subtropical circulation cell that carries the subtropical waters to the equator. It is typically colder and fresher than the South Equatorial Current (SEC) and the New Guinea Coastal Current over the thermocline, due to ventilation in the North Pacific subtropical gyre. The northward flowing Kuroshio is the major western boundary current in the Pacific Ocean. The Kuroshio Water intrudes into the SCS through the southern Luzon Strait (the strait between Taiwn and Phillippines), loops around anticyclonically, bifurcates southwest of Taiwan into the northeast flowing South China Sea Warm Current (SCSWC) and southeast flowing current which exits the SCS through the northern Luzon Strait. In the SCS, a multi-eddy structure was predicted with central SCS warm anticylonic eddy with several surrounding cyclonic eddies. The continuation of the SCSWC becomes the Taiwan Warm Current after passing the Taiwan Strait. Major current system in the EAMS were realistically predicted using the CAOCS. The flow pattern is consistent with recent observational studies (e.g., Chu and Li 2000) and earlier studies. Interested readers are referred to a recent review paper on the EAMS (Su 1998).

The CAOCS model was assessed using the data collected from the intensive observational period (IOP) of the international South China Sea Monsoon Experiment (SCSMEX) and from the Navy's AXBT measurements in April-August 1998.

4 CONCLUSIONS

The CAOCS has capability to predict the current system and thermal structure of the EAMS. Comparing to the SCSMEX and AXBT measurements, the CAOCS successfully predicted the South China Sea multi-eddy structure during the SCSMEX IOP in 1998. These warm-core and cool-core eddies have radii varying from 100 km to 300 km and maximum tangential velocities ranging from 10 cm/s to 20 cm/s. The cool-core eddies are cyclonic and the warm-core eddies are anticyclonic.

5 Acknowledgments

The authors wish to thank George Mellor and Tal Ezer of the Princeton University for most kindly providing them with a copy of the POM code, and to appreciate the Pennsylvania State University/ National Center for Atmospheric Research (PSU/NCAR) mesoscale modeling group for allowing them to use the

MM5. This work was funded by the Office of Naval Research NOMP Program, the Naval Oceanographic Office, and the Naval Postgraduate School.

References

[1] Blumberg, A., and Mellor, G. (1987) A description of a three dimensional coastal ocean circulation model, Three-Dimensional Coastal Ocean Models, edited by N.S. Heaps, American Geophysics Union, Washington D.C., 1-16.

[2] Chu, P.C., Tseng, H.C., Chang, C.P., and Chen, J.M. (1997a) South China Sea warm pool detected in spring from the Navy's Master Oceanographic Observational Data Set (MOODS). *J. Geophys. Res.*, **102,** 15761-15771.

[3] Chu, P.C., Lu, S.H., and Chen, Y.C. (1997b) Temporal and spatial variabilities of the South China Sea surface temperature anomaly. *J. Geophys. Res.*, **102,** 20937-20955.

[4] Chu, P.C., Chen, Y.C., and Lu, S.H. (1998a) Wind-driven South China Sea deep basin warm-core/cool-core eddies. *J. Oceanogr.*, **54,** 347-360.

[5] Chu, P.C., Fan, C.W., Lozano, and Kerling, J.L. (1998b) An airborne expandable bathythermograph survey of the South China Sea, May 1995. *J. Geophys. Res.*, **103,** 21637-21652.

[6] Chu, P.C., Edmons, N.L., and Fan, C.W. (1999a) Dynamical mechanisms for the South China Sea seasonal circulation and thermohaline variabilities. *J. Phys.Oceanogr.*, **29**, 2971-2989.

[7] Chu, P.C., Lu, S.H., and Liu, W.T. (1999b) Uncertainty of the South China Sea prediction using NSCAT and NCEP winds during tropical storm Ernie 1996. *J. Geophys. Res.*, **104**, 11273-11289.

[8] Chu, P.C., and Li, R.F. (2000): South China Sea isopycnal surface circulations. *J. Phys. Oceanogr.*, **30**, 2419-2438.

[9] Chu, P.C., Lan, J., and Fan, C.W. (2000a) Japan/East Sea (JES) circulation and thermohaline structure, Part 1 Climatology. *J. Phys. Oceanogr.*, in press.

[10] Chu, P.C., Veneziano, J. M., and Fan, C.W. (2000b) Response of the South China Sea to tropical cyclone Ernie 1996. *J. Geophys. Res.*, **105**, 13991-14009.

[11] SCSMEX Science Working Group (1995) *The South China Sea Monsoon Experiment (SCSMEX) Science Plan.* NASA/Goddard Space Flight Center, Greenbelt, 65pp.

[12] Su, J. (1998) Circulation dynamics of the China seas north of 18°N. *The Sea*, **11**, 483-505.

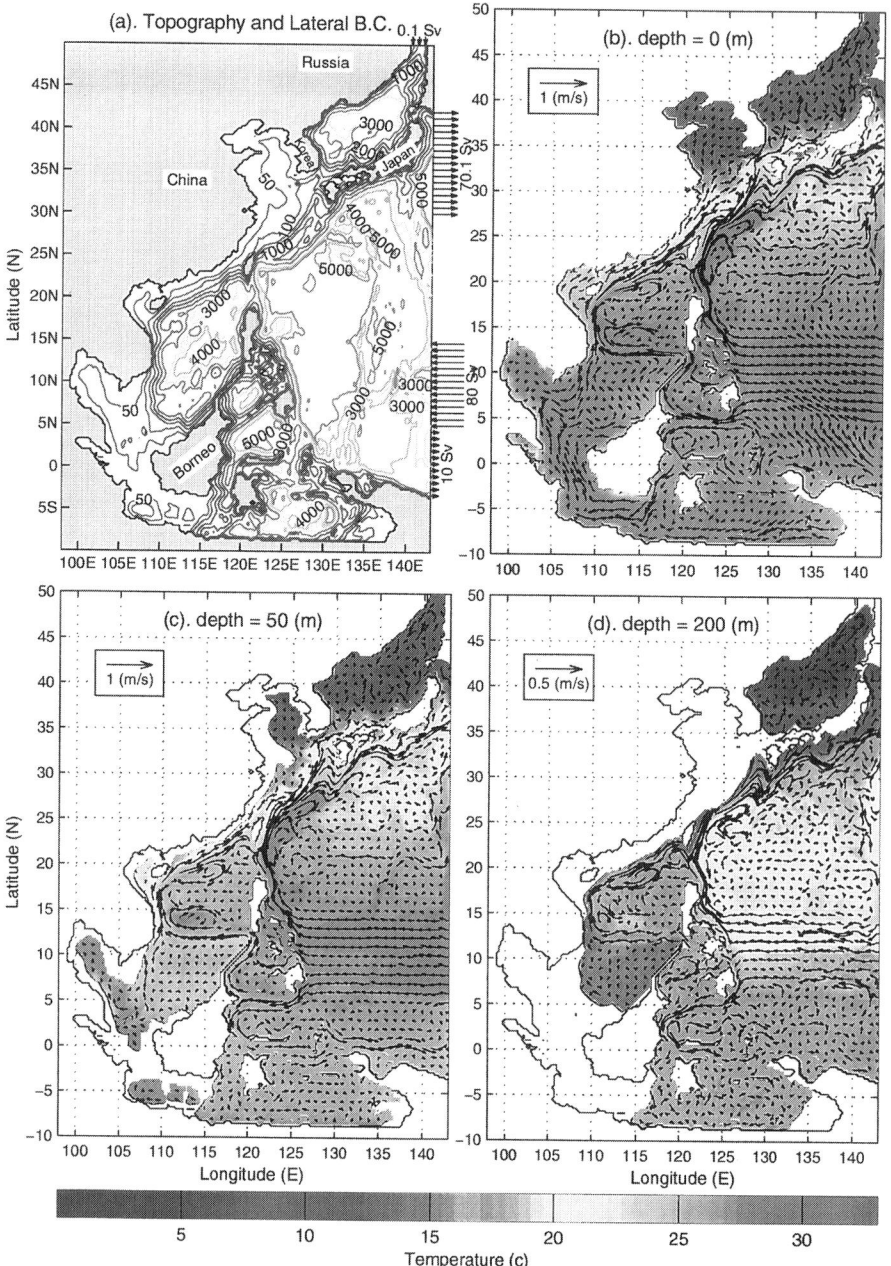

Fig. 1. Predicted temperature and velocity fields in June 1998

DEVELOPMENT AND SIMULATION OF ATLANTIC STORMS DURING FASTEX

KLARA FINKELE (klara.finkele@met.ie) and PETER LYNCH
Met Eireann, Glasnevin Hill, Dublin 9

1. Introduction

FASTEX (Fronts and Atlantic Storm Track Experiment) took place in January and February of 1997. The main purpose of the experiment was to observe in detail the evolution of weather systems crossing the Atlantic using airborne Doppler radar and dropsonde observations as well as an enhanced coverage of conventional and ship observations (Joly et al., 1997).

2. Model Description

HIRLAM (HIgh Resolution Limited Area Model) is an operational forecast model for short range prediction (Källén, 1996). It has been developed in collaboration with several European countries (Denmark, Finland, France, Iceland, Ireland, Norway, Spain, Sweden and The Netherlands).

It is a hydrostatic atmospheric model on a hybrid coordinate system with a choice of Eulerian and semi-Langrangian advection schemes. In most countries the model is run operationally with 31 vertical levels. For this study the vertical levels have been increased to 45 to obtain a higher vertical resolution in the troposphere.

The new turbulent kinetic energy parameterisation scheme has been used Cuxart et al. (2000). The predicted TKE is used to parameterise a trigger for cumulus convection in the convection scheme of Kain and Fritsch (1990) (Colin Jones, pers. communication). The new condensation scheme is the Rasch and Kristjánsson (1998). The reference convection and condensation scheme in HIRLAM is STRACO, a modified Kuo and Sundqvist scheme to allow for the soft transition of convection (Sass et al., 1999)

3. Simulation Results

The selected storm is the intensive observation period (IOP16) on the 17 February 1997. The storm is a fast moving and rapidly deepening cyclone and has been used for model intercomparisons. The model simulations start on the 17th at 00Z and finish at 12Z. The model domain covers the whole northern Atlantic on a rotated grid of 162x109 grid points with a horizontal resolution of 24km and 45 vertical levels.

3.1. INITIAL CONDITIONS

Tow model intercomparison studies have used IOP16 as a case study and two sets of initial and boundary conditions are available. The first model intercomparison is part of the FASTEX Cloud System Study. The initial and boundary fields have been provided by 4DVAR of the ARPEGE model from Meteo France on a 0.4° grid. The second model intercomparison is undertaken by GEWEX working group 3. The initial and boundary fields are provided by the UKMO Unified Model on a 0.21° grid. Partaking in both intercomparison studies gives the opportunity to run parallel tests from different initial conditions. The rapid development of the secondary low is quite sensitive to the initial conditions which are slightly different in the two analysis. The ARPEGE analysis is somewhat smoother in the region where the secondary low starts to develop.

The secondary low at the end of the simulation at 12Z is 963mbar with the ARPEGE analysis and 966mbar with the UKMO analysis. However, the location is quite different with the UKMO analysis being further south east than the ARPEGE analysis. The cloud top temperature is diagnosed from the model using a threshold for cloud water. This diagnostic can be used as a surrogate and compared to the METEOSAT Infra Red. The cloud top temperature using the UKMO analysis compares better to the satellite image at 12Z. The cloud head has greater separation from the polar frontal cloud band with the ARPEGE analysis and thus the different location of the center of the secondary low. The main precipitation occurs in the polar frontal cloud band with the ARPEGE analysis whereas with the UKMO analysis it is along the track of development of the secondary low.

3.2. CONVECTION AND CONDENSATION SCHEME

Two different convection and condensation schemes have been used for a 12 hour forecast using the UKMO analysis and boundary fields. The first convection and condensation scheme is STRACO (Sass et al., 1999) which is the reference scheme in HIRLAM. The second scheme is the newly implemented Kain and Fritsch (1990) and Rasch and Kristjánsson (1998), abbreviated by KFRK.

The depth of the secondary low at 12Z is somewhat lower in the KFRK simulation (966.3mbar) than in the STRACO simulation (966.5mbar). The location of the lows are similar, however, the trough extends further southwest in the STRACO simulation.

The accumulated precipitation over the 12 hour period shows two main rain bands. One associated with the track of the development of the secondary low and another further south associated with a trough underneath the polar frontal cloud band. The separation of the rain bands is larger for STRACO than KFRK. The maxima in the main rain band are similar. The main difference lies in the split of total precipitation into convective and large scale precipitation. In KFRK almost all precipitation is large scale. The split in STRACO with most precipitation being classified as convective seems quite unrealistic for this cyclone where most precipitation would be large scale driven.

The cloud top temperature is compared to the METEOSAT infra red satellite images at 6Z and 12Z. At 6Z the cloud head emerges from underneath the polar frontal cloud band and is fully developed at 12Z. The STRACO simulation shows overall too high cloud top temperatures and therefore too much cloud water compared to the satellite image. The KFRK simulation shows a more distinguished cloud head. STRACO shows little structure in the polar frontal cloud band whereas KFRK compares better to the satellite images.

A cross section at 6Z of the cross frontal circulation has been illustrated by Forbes et al. (2000) using drop sonde and radar data from aircraft. It shows a multi stacked cloud head and the slantwise ascent and descent circulation of two cloud heads. The KFRK simulation shows the slantwise cross frontal circulation of the first cloud head and second cloud head. Whereas in the STRACO simulation the slantwise updraft is reduced and the slantwise downdraft is increased in the first cloud head. The increased slantwise downdraft is most likely due to the empirical parameterisation of the sublimation process in STRACO.

4. Conclusion

The FASTEX storm IOP 16 has been successfully simulated using HIRLAM and compared to satellite images and cross sections derived from aircraft data.

The strong sensitivity to initial conditions has been shown by using the 4DVAR analysis of ARPEGE and the UKMO Unified Model analysis. The central pressure of the secondary low is 3mbar lower with the ARPEGE analysis. Moreover the location of the low is quite different as is the cloud top temperature and the precipitation field.

The newly implemented condensation (Rasch and Kristjánsson, 1998) and convection scheme (Kain and Fritsch, 1990) has been shown to be an improvement compared to the reference scheme (STRACO). Both schemes predict similar depth of the secondary low and location. However, the cloud field is better predicted by KFRK as seen in the cloud top temperature. The cloud head and polar frontal cloud band compare in structure and height better to the satellite images. The slantwise cross frontal circulation which is the cause for the emergence of the cloud head is slightly stronger and better defined in the KFRK simulation.

References

Cuxart, J., Bougeault, P., and J.-L. Redelsperger: 2000, 'A Turbulence Scheme allowing for Mesoscale and Large-scale Simulations'. *Quart. J. Roy. Meteor. Soc.* **126**, 1–30.

Forbes, R. M., Lean, H. W., Roberts, N. M., and P. A. Clark: 2000, 'Implications for Mesoscale Modeling from a Study of the FASTEX IOP 16 Mid-latitude Cyclone'. *JCMM Internal Report* **107**, U.K. Met. Office.

Joly, A., Jorgensen, D., Shapiro, M. A., Thorpe, A. J., Bessemoulin, P., Browning, K. A., Cammas. J.-P., Chalon, J.-P., Clough, S. A., Emanuel, K. A., Eymanrd, L., Gall, R., Hildebrand, P. H., Langland, R. H., Lemaitre, Y., Lynch, P., Moore, J. A., Persson, P. O. G., Snyder, C., and R. M. Wakimoto: 1997, 'The Fronts and Atlantic Storm-Track Experiment (FASTEX): Scientific Objectives and Experimental Design'. *Bull. Amer. Meteor. Soc.* **78**, 1917–1940.

Källén, E., (ed.): 1996, *HIRLAM Documentation Manual Level 2-5*. Swedish Meterological and Hydrological Institute.

Kain, J. S., and J. M. Fritsch: 1990, 'A One-Dimensional Entraining/Detraining Plume Model and Its Applications in Convective Parameterization'. *J. Atmos. Sci.* **47**, 2784–2802.

Rasch, P.J., and J. E. Kristjánsson: 1998, 'A Comparison of the CCM3 Model Climate Using Diagnosed and Predicted Condensate Parameterizations'. *J. Climate* **11**, 1587–1614.

Sass, B. H., Nielsen, N. W., Jøergensen, J. U., and B. Amstrup: 1999, 'The Operational DMI-HIRLAM System'. *DMI Tech. Rep.* **99-21**, Danish Meteorological Institute.

LOW-ORDER MODELS OF ATMOSPHERIC DYNAMICS WITH PHYSICALLY SOUND BEHAVIOR

ALEXANDER GLUHOVSKY AND CHRISTOPHER TONG
Purdue University
West Lafayette, Indiana 47907, USA

1. Introduction

Low-order models (LOMs) are commonly developed by employing the Galerkin technique. Unfortunately, along with a number of highly attractive features, the method does not provide criteria for selecting modes, nor a guarantee that a model based on a particular set of modes will behave anything like the original system. Moreover, fundamental conservation properties of the fluid dynamical equations are sometimes violated in LOMs, and they may exhibit unphysical behavior (throughout the paper conservation is assumed in the absence of forcing and dissipation).

One important example is provided by the famous Howard – Krishnamurti (1986) model of convection with shear that was the first to demonstrate tilting of convection cells giving rise to large scale horizontal motions. Howard and Krishnamurti themselves noticed in the model trajectories going to infinity that they rightly attributed to deficiencies of the Galerkin truncation.

In this paper we address the problem of mode selection in the Galerkin procedure by constructing LOMs in the form of coupled 3-mode systems known in mechanics as Volterra gyrostats, introduced in section 2. Such models also arise as Galerkin approximations in problems of GFD, but they retain the conservation properties of the original equations (Gluhovsky 1982, Gluhovsky and Agee 1997, Gluhovsky and Tong 1999).

The Howard – Krishnamurti (1986) model cannot be transformed into a system of coupled gyrostats, which indicates the lack of energy conservation. To retain energy conservation, Thiffeault and Horton (1996) added a term to the Galerkin temperature expansion. We demonstrate, in section 3, that this improvement permits to transform the thus modified original model into coupled gyrostats form, thereby ensuring energy conservation and boundedness of trajectories. Hermiz et al. (1995) added a term into the stream function expansion of the Howard – Krishnamurti (1986) model, to ensure total vorticity conservation. Their model, however, still lacks energy conservation. Thiffeault and Horton

(1996) suggested that energy conservation here may be restored again by adding a term to the temperature expansion. Modifying accordingly the Howard – Krishnamurti (1986) model, we obtain a system of coupled gyrostats that conserves both energy and total vorticity. Moreover, we identify a coupled gyrostats subsystem that also describes tilting and conserves both quantities.

In Section 4, a LOM for the quasi-geostrophic barotropic potential vorticity equation is presented. In both models, an increase in the order of approximation results in adding certain gyrostats to the model, while incorporating new effects results in adding certain terms to existing gyrostats. This suggests a *modular approach* to developing LOMs whereby simple well understood components (gyrostats) are used to construct models of complex systems. We employed the modular approach to develop a cascade system of coupled gyrostats that exhibits Kolmogorov spectral behavior (Gluhovsky and Tong, 1999).

2. The Volterra Gyrostat and the Lorenz Model

The Volterra (1899) gyrostat (also Wittenburg 1977)

$$\dot{v}_1 = pv_2v_3 + bv_3 - cv_2,$$
$$\dot{v}_2 = qv_3v_1 + cv_1 - av_3, \quad (1)$$
$$\dot{v}_3 = rv_1v_2 + av_2 - bv_1,$$

where p, q, r, a, b, c are constants, $p+q+r=0$, can be thought of as a rigid body containing a rotor revolving with a constant angular velocity about an axis fixed in the carrier. The *nonlinear terms* in Eqs. (1) compose the Euler rigid body. Like the Euler gyroscope, gyrostat (1) has two quadratic invariants: the kinetic energy and the square of the angular momentum. The *linear terms* are caused by the relative motion of the rotor. Unlike ordinary viscous terms, they do not affect the conservation of energy or phase space volume. These linear gyrostatic terms occur in LOMs due to various factors peculiar to GFD: stratification, rotation, and topography. One example is provided by the celebrated Lorenz (1963) model of Rayleigh – Bénard convection

$$\dot{x} = \sigma(y-x), \quad \dot{y} = -xz + Rx - y, \quad \dot{z} = xy - bz, \quad (2)$$

that after a linear change of variables becomes (Gluhovsky 1982) the simplest gyrostat in a forced regime (we shall call it the *Lorenz gyrostat*):

$$\dot{x}_1 = -x_2x_3 \qquad -\alpha_1 x_1 + f,$$
$$\dot{x}_2 = x_3x_1 - x_3 - \alpha_2 x_2, \quad (3)$$
$$\dot{x}_3 = \qquad x_2 - \alpha_3 x_3,$$

with the parameter correspondence: $\alpha_2 = 1/b$, $\alpha_3 = \sigma/b$, $f = 1 + \sigma R/b^2$.

3. The Modified Howard – Krishnamurti (1986) Model

The dimensionless Boussinesq equations for Rayleigh – Bénard convection are

$$\frac{\partial}{\partial t}\nabla^2\Psi = \sigma\nabla^4\Psi + \sigma\frac{\partial\theta}{\partial x} + \frac{\partial\Psi}{\partial x}\frac{\partial\nabla^2\Psi}{\partial z} - \frac{\partial\Psi}{\partial z}\frac{\partial\nabla^2\Psi}{\partial x},$$

$$\frac{\partial\theta}{\partial x} = \nabla^2\theta + Ra\frac{\partial\Psi}{\partial x} + \frac{\partial\Psi}{\partial x}\frac{\partial\theta}{\partial z} - \frac{\partial\Psi}{\partial z}\frac{\partial\theta}{\partial x},$$

(4)

where Ψ is the stream function, θ is the deviation from the equilibrium vertical temperature profile, σ is the Prandtl number, and Ra is the Rayleigh number. To arrive at the modified Howard – Krishnamurti model that possesses appropriate conservation properties, we used the following Galerkin expansions:

$$\Psi = A(t)\sin\alpha x\sin z + B(t)\sin z + C(t)\cos\alpha x\sin 2z + G(t)\sin 3z,$$ (5)

$$\theta = D(t)\cos\alpha x\sin z + E(t)\sin 2z + F(t)\sin\alpha x\sin 2z + H(t)\sin 4z.$$

Substituting expansions (5) into Eqs. (4) results in a LOM that may be transformed (by a linear change of variables) into the following system of six coupled gyrostats:

$$\begin{aligned}
\dot{x}_1 &= -\alpha_1 x_1 + f_1 - x_2 x_3 \\
\dot{x}_2 &= -\alpha_2 x_2 + x_3 x_1 - x_3 \qquad\qquad -cx_4 x_6 \qquad\qquad\qquad +(c/3)x_8 x_6 \\
\dot{x}_3 &= -\alpha_3 x_3 \qquad\quad x_2 + px_4 x_5 \qquad\qquad\qquad\qquad +Px_4 x_8 \\
\dot{x}_4 &= -\alpha_4 x_4 \qquad\qquad\quad +qx_5 x_3 \qquad\qquad\qquad +x_6 +Qx_8 x_3 \\
\dot{x}_5 &= -\alpha_5 x_5 \qquad\qquad\quad +rx_3 x_4 \\
\dot{x}_6 &= -\alpha_6 x_6 \qquad\qquad\qquad\qquad +cx_4 x_2 +dx_4 x_7 - x_4 \quad -(c/3)x_8 x_2 \\
\dot{x}_7 &= -\alpha_7 x_7 + f_2 \qquad\qquad\qquad\qquad\qquad\quad -dx_6 x_4 \\
\dot{x}_8 &= -\alpha_8 x_8 \qquad\qquad\qquad\qquad\qquad\qquad\qquad\qquad\quad +Rx_3 x_4
\end{aligned}$$

(6)

$$\quad\text{I}\qquad\quad\text{II}\qquad\quad\text{III}\qquad\quad\text{IV}\qquad\quad\text{V}\qquad\quad\text{VI}$$

In Eqs. (6), $p+q+r = P+Q+R = 0$, variables $x_1 - x_8$ correspond to E, D, A, C, B, F in expansions (7) (composing the Howard – Krishnamurti (1986) model) and H, G, respectively. The expressions for the coefficients in Eqs. (6) as well as Eqs. (8) below will be given in a forthcoming paper. The gyrostats are particular cases of system (1): gyrostats I and IV are the Lorenz gyrostats (3), while gyrostats II and V are the Euler gyroscopes, and gyrostats III and VI are the degenerative Euler gyroscopes. Taken alone (i.e. with $\dot{x}_4 = \dot{x}_8 = 0$), the latter are just simple linear oscillators. But coupled here with other gyrostats possessing time varying modes x_4 and x_8, they are nonlinear subsystems.

It may be easily checked that in any system of coupled gyrostats the sum of squares of all modes, representing an energy quantity (see discussion in Gluhovsky and Tong (1999)), is conserved. Thus, the failure to convert a LOM into coupled gyrostats may signal violation of energy conservation. This is the case with the Howard – Krishnamurti (1986) model. Its modification by Thiffeault and Horton (1996), who added the $H(t)\sin 4z$ term in Eqs. (5), can be converted (Gluhovsky and Tong, 1999) into a system of coupled gyrostats (within dashed rectangle in Eqs. (6)). This model lacks, however, total vorticity conservation due to an insufficient number of shearing modes. Hermiz et al. (1995) added the second shearing mode $G(t)\sin 3z$ to the Howard – Krishnamurti (1986) model to provide total vorticity conservation, but their model still lacks energy conservation like the original model. Model (6), however, possesses tilting (Figure 1) and both conservation properties. Total vorticity conservation here is ensured by the linear integral $I = Rx_5 - rx_8$.

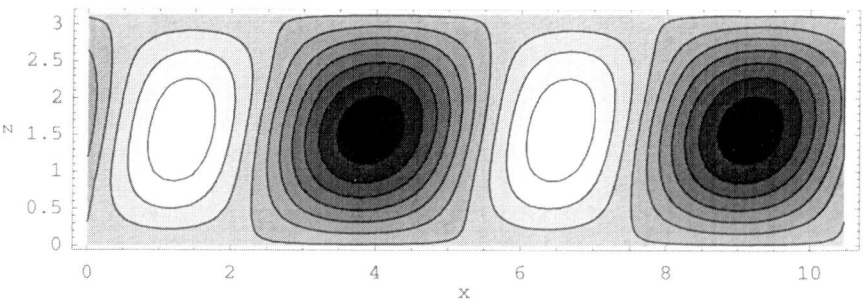

Figure 1. Contour plot of the stream function showing the tilting of convection cells for LOM (6) for a steady state solution at $Ra/Ra_{cr} = 8$, $\sigma = 3$, $\alpha = 1.2$.

Presenting a system in a gyrostatic form may also help to identify a subsystem still possessing desired properties. It has to contain enough modes to exhibit the effect of interest, while maintaining a coupled gyrostats structure to have conservation properties. With this in view, we reduced system (6) to the subsystem of three gyrostats composed of modes $x_i = X_i, i = 1, ..., 5;\ x_8 = X_6$.

$$\begin{aligned}
\dot{X}_1 &= -\gamma_1 X_1 + f_1 - X_2 X_3 \\
\dot{X}_2 &= -\gamma_2 X_2 + X_3 X_1 - X_3 \\
\dot{X}_3 &= -\gamma_3 X_3 + X_2 + p X_4 X_5 + P X_4 X_6 \\
\dot{X}_4 &= -\gamma_4 X_4 + q X_5 X_3 + Q X_6 X_3 \\
\dot{X}_5 &= -\gamma_5 X_5 + r X_3 X_4 \\
\dot{X}_6 &= -\gamma_6 X_6 + R X_3 X_4
\end{aligned} \qquad (7)$$

In Eqs. (7), $\gamma_i = \alpha_i, i = 1, ..., 5;\ \gamma_6 = \alpha_7$, other coefficients are from Eqs. (6). Figure 2 demonstrates that tilting is preserved in LOM (7).

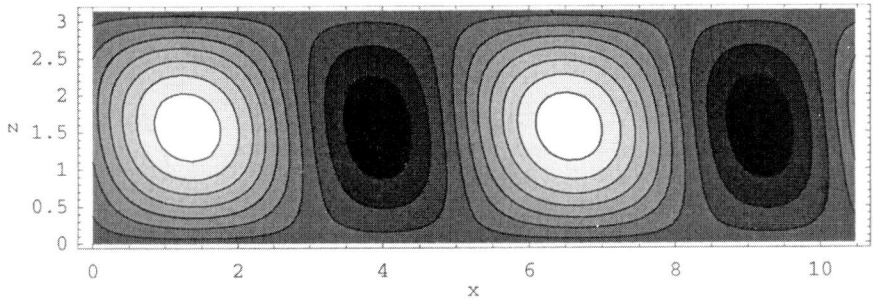

Figure 2. Contour plot of the stream function showing the tilting of convection cells for LOM (7) for a steady state solution at $Ra/Ra_{cr} = 30$, $\sigma = 3$, $\alpha = 1.2$.

4. A LOM for the Quasi – Geostrophic Potential Vorticity Equation

Charney and DeVore (1979) studied the combined effect of topographic forcing and rotation. We converted a LOM for this problem obtained by De Swart (1988) into a system of coupled gyrostats.

The 3-mode approximation (a gyrostat within the rectangle in Eqs. (8)) describes the interaction of an externally forced zonal flow mode v_1 and a single Rossby wave represented by two other modes v_2 and v_3. The gyrostat has two pairs of gyrostatic terms: those with coefficient a are due to topography, those with coefficient b are due to rotation. Thus, each mechanism gives rise to a pair of gyrostatic terms in the model.

The 6-mode approximation (Eqs. (8)) describes the evolution of two zonal flow modes v_1 and v_4 and two Rossby waves (modes v_2, v_3 and v_5, v_6), which results in three more gyrostats: a degenerative gyrostat, a gyrostat with three nonlinear terms, and the Euler gyroscope.

$$
\begin{aligned}
\dot{v}_1 &= F_1 - Cv_1 + b_1 v_3 \\
\dot{v}_2 &= -Cv_2 \; q_1 v_3 v_1 - a_1 v_3 \; + q_3 v_4 v_6 \\
\dot{v}_3 &= -Cv_3 \; -q_1 v_1 v_2 + a_1 v_2 - b_1 v_1 \; - q_3 v_4 v_5 \\
\dot{v}_4 &= F_4 - Cv_4 \; + p_3 v_6 v_2 + b_3 v_6 \; - p_3 v_5 v_3 \\
\dot{v}_5 &= -Cv_5 \; q_2 v_6 v_1 - a_2 v_6 \; - r_3 v_3 v_4 \\
\dot{v}_6 &= -Cv_6 \; -q_2 v_1 v_5 + a_2 v_5 \; + r_3 v_2 v_4 - b_3 v_4
\end{aligned} \quad (8)
$$

Both approximations were studied by Charney and DeVore (1979); see also Yoden (1985). In the 10-mode approximation, De Swart (1988) discovered chaotic behavior interpreted as vacillations between high-index and low-index regimes. As in section 3, we were able to find a smaller, 8-mode subsystem of the De Swart model that also exhibits these vacillations. It includes the third Rossby wave and adds to Eqs. (10) the same familiar combinations of gyrostats.

In conclusion, restricting LOMs to have a gyrostatic form

1) ensures energy conservation, thus preventing unphysical behavior often observed in LOMs based on *ad hoc* truncations;
2) helps to design a system of optimum size to describe the effects of interest.
3) allows a modular implementation of the Galerkin technique using gyrostats as elementary building blocks;

5. Acknowledgments

The authors are grateful to E. M. Agee for useful discussions. This work was supported by NSF grant ATM-9909009 and by Purdue Research Foundation.

6. References

Charney, J. G. and DeVore, J. G. (1979) Multiple flow equilibria in the atmosphere and blocking, *J. Atmos. Sci.* **36**, 1205-1216.

De Swart, H. E. (1988) Low-order spectral models of the atmospheric circulation: a survey, *Acta Appl. Math.* **11**, 49 – 96.

Gluhovsky, A. (1982) Nonlinear systems that are superpositions of gyrostats, *Sov. Phys. Doklady* **27**, 823 – 825.

Gluhovsky, A. and Agee, E. M. (1997) An interpretation of atmospheric low-order models, *J. Atmos. Sci.* **54**, 768 – 773.

Gluhovsky, A. and Tong, C. (1999) The structure of energy conserving low-order models, *Phys. Fluids* **11**, 334 – 343.

Hermiz, K. B., Guzdar, P. N., and Finn, J. M. (1995) Improved low-order model for shear flow driven by Rayleigh-Bénard convection, *Phys. Rev. E* **51**, 325 (1995).

Howard, L. N., and Krishnamurti, R. (1986) Large-scale flow in turbulent convection: a mathematical model, *J. Fluid Mech.* **170**, 385-410.

Lorenz, E. N. (1963) Deterministic nonperiodic flow, *J. Atmos. Sci.* **20**, 130-141.

Thiffeault, J.-L., and Horton, W. (1996) Energy-conserving truncations for convection with shear flow, *Phys. Fluids* **8**, 1715-1719

Volterra, V. (1899) Sur la théorie des variations des latitudes, *Acta Math.*, **22**, 201-358.

Wittenburg, J. (1977) *Dynamics of Systems of Rigid Bodies*, B. G. Teubner, Stuttgart.

Yoden, S. (1985) Bifurcation properties of a quasi-geostrophic, barotropic, low-order model with topography, *J. Meteor. Soc. Japan* **63**, 535-546.

MODELS FOR INSTABILITY IN GEOPHYSICAL FLOWS

Roger Grimshaw
Department of Mathematical Sciences
Loughborough University, UK
R.H.J.Grimshaw@lboro.ac.uk

Georg Gottwald
INLS-CNRS, Sophia-Antipolis, France
georg.gottwald@inln.cnrs.fr

Abstract The generation of instability in inviscid, non-diffusive geophysical flows is generically caused by a resonance between two wave modes. The weakly nonlinear unfolding of this situation is described in the long-wave regime, using a particular two-layer quasi-geostrophic model as an illustrative example. The outcome is a system of two coupled Korteweg-de Vries equations. This system contains a very rich solution set, consisting typically of solitary wave interactions. We will describe some numerical solutions of the coupled Korteweg-de Vries equations, supplemented by perturbation analyses. We also report on some preliminary analogous numerical simulations of the full two-layer quasi-geostrophic system.

1. INTRODUCTION

In inviscid fluid flows it is well-known that instability generally arises due to a resonance between two waves. Thus, as an appropriate external parameter is varied, the phase speeds of two waves coincide for some critical parameter value. The generic unfolding of this resonance yields either a stable "kissing" configuration, or a "bubble" of instability in the space of the external parameter. There are many examples of this situation (see, for instance, the monograph by Craik (1985) for shear flows, or Baines and Mitsudera (1994) for a discussion of the physical processes involved).

Our concern here is with the unfolding of this resonance in the limit of long waves. Recent work has established that there are two generic canonical models (see, for instance, Grimshaw (2000)). For the case when the wave modes coincide at criticality, the canonical model is the Boussinesq equation

$$A_{tt} - \Delta^2 A_{xx} + \frac{1}{2}\mu(A^2)_{xx} + \lambda A_{xxxx} = \delta A_{xx}, \qquad (1.1)$$

Here Δ is the aforementioned external parameter, δ is an unfolding parameter, while μ and λ are the nonlinear and dispersive coefficients respectively. In the linear, long-wave limit the dispersion relation for waves of speed c is just $c^2 = \Delta^2 + \delta$. Resonance occurs for $\Delta = \delta = 0$, and the flow is linearly stable, or unstable, according as $\delta > 0$, or < 0. Equations of this form have been derived by Hickernell (1938a, b) for Kelvin-Helmholtz instability, and by Helfrich and Pedlosky (1993) and Mitsudera (1994) for certain quasi-geostrophic flows.

However, our main interest here is with the alternative scenario, when the wave modes remain distinct at criticality. In this case a suitable canonical model consists of the coupled Korteweg-de-Vries (KdV) equations

$$\begin{aligned} A_t + \Delta_1 A_x + \mu_1 A A_x + \lambda_1 A_{xxx} + \kappa_1 B_x &= 0, \\ B_t + \Delta_2 B_x + \mu_2 B B_x + \lambda_2 B_{xxx} + \kappa_2 A_x &= 0, \end{aligned} \qquad (1.2)$$

Here $\Delta_1 - \Delta_2$ is the detuning parameter, κ_1, κ_2 are unfolding parameters, while μ_1, μ_2 and λ_1, λ_2 are nonlinear and dispersive coefficients respectively. In the linear long-wave limit the dispersion relation for waves of speed c is

$$(c - \Delta_1)(c - \Delta_2) = \kappa_1 \kappa_2. \qquad (1.3)$$

This is equivalent to that for the Boussinesq equation (1.1) if we put $\Delta = \frac{1}{2}(\Delta_1 - \Delta_2)$ and $\delta = \kappa_1 \kappa_2$. Hence there is instability if $\kappa_1 \kappa_2 < 0$, and stability if $\kappa_1 \kappa_2 > 0$. Equations of the form (1.2) have been derived by Mitsudera (1994) and Gottwald and Grimshaw (1999) for certain quasi-geostrophic flows, and by Grimshaw (2000) for a certain three-layer stratified shear flow. The coupled KdV sytem is Hamiltonian, and so conserves "energy"; the system also conserves two Casimirs (the integrals of A and B) as well as the "momentum" invariant,

$$P = \int_{-\infty}^{\infty} \{\kappa_1 A_2^2 + \kappa_2 A_1^2\} \, dx. \qquad (1.4)$$

Note that P is sign-definite when the system is linearly stable, but sign-indefinite if the system is linearly unstable.

Here we shall briefly review the derivation of the coupled KdV system (1.2) for a two-layer quasi-geostrophic flow, and some of the more relevant solutions of (1.2). More details can be found in the work of Grimshaw and Gottwald (1999) (hereafter denoted by GG). We shall also describe the relationship between these solutions of the coupled KdV system and the corresponding solutions of the full quasi-geostrophic equations, based on some preliminary numerical studies of the latter system.

2. TWO-LAYER QUASI-GEOSTROPHIC FLOW

In order to illustrate the basic concepts in a relatively simple geophysical fluid dynamics context, we introduce the familiar two-layer quasi-geostrophic model on a β-plan, in the Boussinesq approximation when the layer densities are almost equal. We shall use a nondimensional coordinate system, based on a typical horizontal lengthscale L_0, typical vertical scales for each layer D_1, D_2 with $H_0 = D_1 + D_2$, and a typical Coriolis parameter f_0. A typical velocity U is taken to be the maximum of the mean current velocity and the timescale is given by U/L_0. If we separate the meanflow U_1 and U_2 from the perturbation pressure fields p_1 and p_2, we obtain the following equations (Pedlosky (1987)) for the nondimensional perturbation pressure fields

$$\left(\frac{\partial}{\partial t} + U_n \frac{\partial}{\partial x}\right) q_n + \psi_{nx} Q_{ny} + J(\psi_n, q_n) = 0, \tag{2.1}$$

where $n = 1, 2$ respectively, and

$$\begin{aligned} q_n &= \nabla^2 \psi_n \pm F_n (\psi_2 - \psi_1), \\ Q_{ny} &= \beta - U_{nyy} \mp F_n (U_2 - U_1). \end{aligned} \tag{2.2}$$

with the Jacobian defined by $J(a,b) = a_x b_y - a_y b_x$. Here the alternate signs refer to n=1,2 respectively. The boundary conditions are $\psi_{1,2} = $ const at $y = -L, 0$. Here the pressure fields are scaled by $\rho_0 f_0 U_0 L_0$, where ρ_0 is a reference density, and, in this quasigeostrophic approximation, also serve as streamfunctions for the velocity fields in each layer. The subscripts 1 and 2 are associated with the upper and lower layers, respectively. We have introduced the nondimensional meridional gradient of planetary vorticity β, and the Froude-numbers $F_n = (L_0/R_i)^2$, where R_i is the internal Rossby radius of deformation for each layer, i.e $R_i = f_0^{-1}\sqrt{gD_n(\Delta\rho/\rho_o)}$, where $\Delta\rho$ is the density difference across the interface.

Our concern here is with weakly nonlinear long waves in a parameter regime where two wave modes have nearly coincident phase speeds.

Hence we introduce the following scaled variables,

$$X = \delta x, \quad T = \delta^3 t,$$
$$\psi_n = \delta^2 \psi_n^{(0)} + \delta^4 \psi_n^{(1)} + \cdots,$$
$$U_n = U_n^{(0)} + \delta^2 U_n^{(1)} + \cdots, \quad (2.3)$$

where δ is a small parameter, the inverse of which measures the horizontal scale of the waves. This scaling is typical for KdV systems. Next, we rescale the parameters

$$F_n \to \delta^2 F_n, \quad \beta \to \delta^2 \beta. \quad (2.4)$$

As we shall see the scaling of the Froude numbers is to ensure that the desired resonance between two waves can be realised. It implies that our model is valid for situations where the internal Rossby radius of deformation is of the order of the long horizontal scale. Further, the scaling of β implies that $Q_{ny} \approx -U_{nyy}$ at the lowest order. Substituting this scaling into equation (2.1), we obtain at the lowest order, $\mathcal{O}(\delta^3)$

$$U_n^{(0)} \psi_{nXyy}^{(0)} - U_{nyy}^{(0)} \psi_{nX}^{(0)} = 0, \quad (2.5)$$

from which we conclude that

$$\psi_1^{(0)} = A(X,T) U_1^{(0)}(y) \quad \text{and} \quad \psi_2^{(0)} = B(X,T) U_2^{(0)}(y). \quad (2.6)$$

Hence the meridional structure of ψ_i is entirely determined by the mean currents at the leading order. Note that two independent amplitudes $A(X,T), B(X,T)$ appear, indicating that there is indeed a resonance between two waves, whose phase speeds are identically zero in this present case.

The $\mathcal{O}(\delta^5)$ terms give us two evolution equations for the amplitudes A, B for each layer. We reiterate that the reason for the occurrence of two coupled equations is the scaling of the Froude numbers (2.4), which implies the existence of two independent modes at leading order. We obtain

$$U_n^{(0)} \psi_{nXyy}^{(1)} - U_{nyy}^{(0)} \psi_{nX}^{(1)} + G_n = 0, \quad (2.7)$$

where the inhomogeneous terms G_n contain the terms A_T, A_X, B_X, A_{XXX} and AA_X, or B_T, B_X, A_X, B_{XXX} and BB_X respectively, (for details see GG). Applying a compatibility condition to each equation then yields a coupled KdV system, which is just (1.2) on reverting to the unscaled coordinates. Here the coefficients are give by,

$$I_n = \left[U_{ny}^{(0)} \right]_{-L}^{0}, \quad I_n \lambda_n = \int_{-L}^{0} U_n^{(0)2} dy, \quad I_n \mu_n = - \left[U_{ny}^{(0)2} \right]_{-L}^{0},$$

$$I_n\Delta_n = \int_{-L}^{0} (\beta - F_n U_m^{(0)})U_n^{(0)}dy + \left[U_n^{(1)}U_{ny}^{(0)}\right]_{-L}^{0},$$

$$I_n\kappa_n = F_n \int_{-L}^{0} U_1^{(0)}U_2^{(0)}dy. \tag{2.8}$$

Here $n = 1, 2$ and $m = 2, 1$. Note that the linear instability criterion $\kappa_1\kappa_2 < 0$ is here equivalent to $I_1 I_2 < 0$ which is just the familiar condition for baroclinic instability in a two-layer system with the present scaling. It should also be noted that in order for the nonlinear coefficients μ_n to be non-zero, the basic flow $U_n^{(0)}$ should be assymetric.

3. INTERACTING SOLITARY WAVES

The coupled KdV system (1.2) can support a rich variety of solutions. Here we shall describe just one scenario of interest. For more details of this and other dynamics, see GG. In the absence of any coupling between the component equations in (1.2), the system reduces to two KdV equations, each of which can support a solitary wave. In the presence of coupling, we then expect these to interact with each other. This process can be described by seeking asymptotic solutions of the form of solitary wave solutions of the KdV equation, but with an amplitude and speed which are allowed to evolve in time, that is,

$$A_0(B_0) = a_n(t)\text{sech}^2[w_n(t)x - \Phi_n(t)],$$
$$\text{where} \quad a_n = 12\frac{\lambda}{\mu}w_n^2. \tag{3.1}$$

The time-evolution of the amplitudes a_n (n=1,2) and the relative position $\Phi = \Phi_2 - \Phi_1$ are determined by the following set of three ordinary differential equations,

$$\frac{da_n}{dt} = F_n(a_1, a_2, \Phi), \quad \frac{d\Phi}{dt} = \Delta_2 - \Delta_1 + \frac{\mu_2}{3}a_2 - \frac{\mu_1}{3}a_1. \tag{3.2}$$

where the interaction-integrals are given by

$$F_n = 2\kappa_n a_m w_m \int_{-\infty}^{\infty} \text{sech}^2(\psi)\text{sech}^2(\psi')\tanh(\psi')d\psi.$$

$$\text{where} \quad \psi' = \frac{w_m}{w_n}\psi - w_m\Delta\Phi \quad \text{and} \quad n = 1, 2; \; m = 2, 1. \tag{3.3}$$

The reduced system 3.2 conserves the analogue of the "momentum" P (1.4), namely $(\kappa_2[\lambda_1^2/\mu_1^2]w_1^3 + \kappa_1[\lambda_2^2/\mu_2^2]w_2^3)$, which enables us to reduce the system (3.2) to a planar dynamical system. This resulting planar sytem has been analysed in detail by Gottwald and Grimshaw

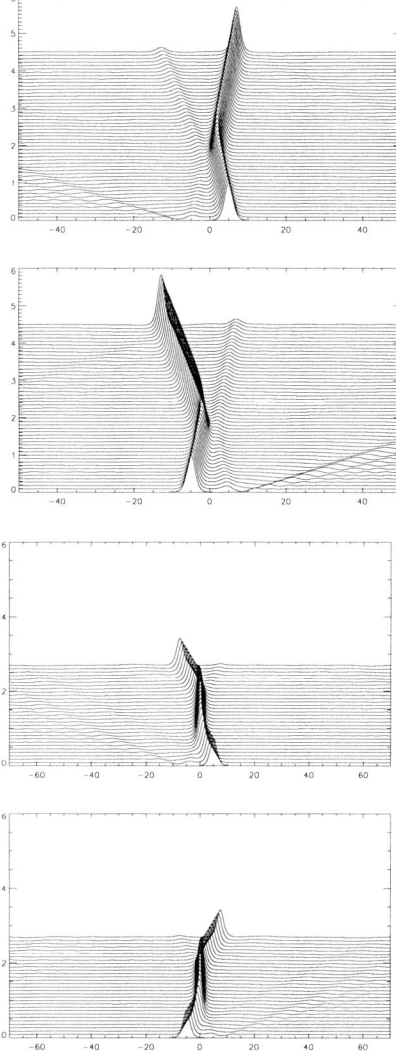

Figure 1 Solitary wave interactions for a saddle point with $\Delta_1 = -2.25, \Delta_2 = 2.25, \mu_1 = \lambda_1 = -1, \mu_2 = \lambda_2 = 1, \kappa_1 = -0.2, \kappa_2 = 0.3$. The top two panels show a repulsive regime with initial amplitudes $a_1 = a_2 = 6.0$ for the upper and lower layer respectively. The lower two panes show a quasi-locked state with initial amplitudes $a_1 = a_2 = 4.44$. The horizontal axis is the space coordinate x and the vertical axis is the time coordinate t.

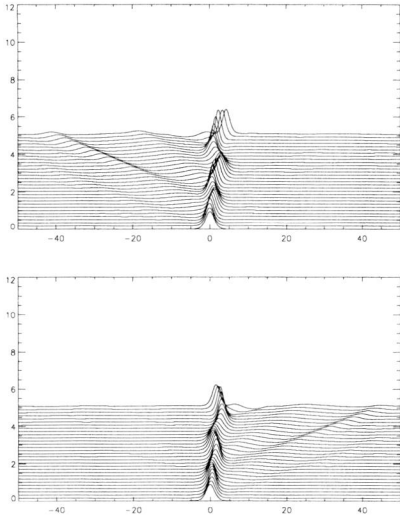

Figure 2 A trapped regime for a stable centre with $\Delta_1 = -1.7, \Delta_2 = 1.8, \mu_1 = \lambda_1 = -1, \mu_2 = \lambda_2 = 1, \kappa_1 = 0.3, \kappa_2 = -0.2$, where the amplitudes of the the solitary waves oscillate about $a_1 = 5.7$ and $a_2 = 5.4$. The axes are as in Figure 1, and the upper and lower panels correspond to the upper and lower layers respectively

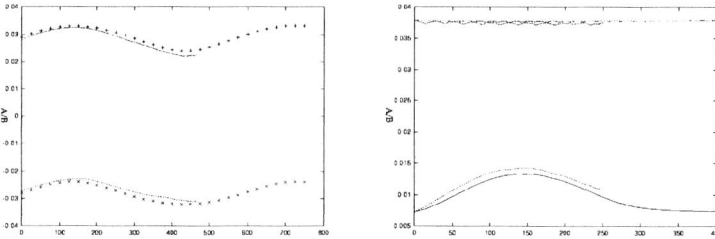

Figure 3 Comparison of the dynamics given by the coupled KdV equations (1.2) and the two-layer quasi-geostrophic system (2.1). The left panel shows a centre, where we plot the amplitude of the solitary waves against the time coordinate; the upper(lower) curves are that for A(B), and the curves which extend for all time are those for the coupled KdV system. The right panel shows a quasi-locked state for a a saddle point configuration.

(1999). Here, we just note that it possesses a critical point $\Delta\Phi = 0$ with $w_1 = w_2 = w_0$ say, which is in fact also an exact solution of the full coupled KdV system (1.2), thus descibing an exact phase-locked solitary wave. A routine stability analysis then shows that this solution is a centre and so stable if $\kappa_1\mu_1\mu_2 < 0$, but is a saddle-point and so unstable if $\kappa_1\mu_1\mu_2 > 0$. Typical scenarios describing these two configurations are shown in Figures 1 and 2, these results being obtained by direct numerical simulation of the coupled KdV system (1.2).

Although the coupled KdV system (1.2) has been systematically derived from the full two-layer quasi-geostrophic system (2.1), there remains a need to verify directly that the scenarios identified in the reduced system can also be found in the full system. To this end we report here some preliminary numerical simulations of the full two-layer quasi-geostrophic sytem. Using the results shown in Figures 1,2 as a guide we simulated a stable solitary wave interaction in (2.1) which displays the "centre" scenario, and also an unstable interaction in (2.1) which displays the "saddle-point" interaction. The results are shown in Figure 3.

References

[1] Baines, P.G. and Mitsudera, H. (1994): On the mechanism of shear flow instabilities, *J. Fluid Mech.* **276**, 327-342.

[2] Craik, A.D.D. (1985): Wave interaction in fluid flows, C.U.P., 332 pp.

[3] Gottwald, G. and Grimshaw, R. (1999): The formation of coherent structures in the context of blocking, *J.Atmos Sci.,* **56**, 3663-3678.

[4] Grimshaw, R., (2000): Models for long-wave instability due to a resonance between two waves, in "Trends in Applications of Mathematics to Mechanics", ed. G. Iooss, O. Gues and A. Nouri, *Monographs and Surveys in Pure and Applied Mathematics, 106, Chapman & Hall/CRC,* 183-192.

[5] Helfrich, K.R. and Pedlosky, J. (1993): Time-dependent islated anomolies in zonal stable, inviscid shear flows I. Derivation of the amplitude evolution equations, *J Fluid Mech.* **251**, 377-409.

[6] Hickernell, F., (1983a): The evolution of large-horizontal scale disturbances in marginally stable, inviscid shear flows I. Solutions of the Boussinesq equation, *Stud Appl Math.,* **69**, 1-21.

[7] Hickernell, F., (1983b): The evolution of large-horizontal scale disturbances in marginally stable, inviscid shear flows II. Solutions of the Boussinesq equation, *Stud Appl Math.,* **69**, 23-39.

[8] Mitsudera, H. (1994): Eady solitary waves: a theory of type B cyclogenesis, *J. Atmos. Sci.,* **57**, 734-745.

[9] Pedlosky, J. (1987): Geophysical Fluid Dynamics, 2nd ed. Springer-Verlag, 710pp.

BAROCLINIC STRUCTURE OF A MODIFIED STOMMEL-ARONS MODEL OF THE ABYSSAL OCEAN CIRCULATION.

P.F. HODNETT and R. McNAMARA
Department of Mathematics and Statistics
University of Limerick, Limerick, Ireland

Abstract

The Stommel and Arons (1960 a,b) model of the circulation in the abyssal ocean is modified by allowing the temperature (homogeneous in Stommel - Arons) to vary with depth and latitude and weakly with longitude. The temperature and vertical velocity are specified at the top of the abyss as functions of latitude and weakly of longitude. The resulting distributions of temperature for the North Atlantic show agreement with the climatological data given in Levitus (1982) when the vertical coefficient of thermal diffusion is $1.25 \times 10^{-4} m^2 s^{-1}$. The vertical, meridional and zonal components of velocity show significant change in structure from the Stommel - Arons expressions.

Introduction

The World Ocean Circulation Experiment (WOCE) is a component of the World Climate Research Programme investigating the role played by the ocean circulation in the earth's climate system. The WOCE observational phase from 1990-1997 has used satellites and *in-situ* physical and chemical measurements to produce a data set of unprecedented scope and precision. There is a consequent need for reliable theory through which to understand the observational data.

For example, until the relatively recent past the understanding of the abyssal ocean circulation resided in the simple model presented by Stommel and Arons (1960a,b). In this model the temperature is assumed to be homogeneous throughout the entire abyss. The model gives uniform upwelling from the abyssal region into the thermocline region of the ocean. In the North Atlantic this upwelling is fed from a southward flowing western boundary current whose source is a downwelling point at the northern extremity of the basin.

More recently Pedlosky (1992) considered the baroclinic structure of the abyssal circulation by allowing the temperature to vary with position in the abyss. However he assumed a particular form for the temperature field where the primary variation is in the vertical modified by a small perturbation which varies with depth, latitude and longitude. The weakness of the model is that (contrary to the model assumption) lateral variations of temperature are not much smaller than vertical variations in the abyss. His model gives a meridional velocity which changes sign with depth in the western ocean agreeing with observations which show Antarctic Bottom water flowing north beneath

North Atlantic Deep water flowing south beneath Antarctic Intermediate water flowing north all beneath the thermocline.

A different approach is adopted by Hodnett and McNamara (2000) in allowing the temperature to vary with depth and latitude and in assuming that the variation with longitude is negligible. The Hodnett and McNamara (2000) model emphasises the substantial changes which occur in the Stommel-Arons model of the abyssal circulation when temperature is allowed to vary with depth and latitude, and yields results consistent with some observations from Worthington and Wright (1970) and with some aspects of the three dimensional numerical model of the North Atlantic circulation given in Bogden et al (1993).

Here we allow for *weak* zonal variation in temperature which was neglected in Hodnett and McNamara (2000) and identify the substantial resulting changes in the vertical and meridional velocity components. The results produced are also compared with some results from Pedlosky (1992).

Equations and Boundary Conditions

The equations which describe steady motion in the abyss are the planetary geostrophic equations which in non-dimensional form are

$$\frac{\partial p}{\partial z} = T, \quad 2u\sin\phi = -\frac{\partial p}{\partial \phi}, \quad 2v\sin\phi = \frac{1}{\cos\phi}\frac{\partial p}{\partial \lambda} \tag{1 a,b,c}$$

$$\frac{\partial w}{\partial z} = v/\tan\phi, \tag{2}$$

$$w\frac{\partial T}{\partial z} + v\frac{\partial T}{\partial \phi} + \frac{u}{\cos\phi}\frac{\partial T}{\partial \lambda} = \beta\frac{\partial^2 T}{\partial z^2}, \tag{3}$$

with thermal diffusion only in the vertical and where (2) is the planetary divergence relation which results from substitution of expressions (1 b,c) for u and v into the continuity equation. In the above equations λ is longitude (positive eastwards), ϕ (latitude, positive northwards), z (vertical distance, positive upwards), p (pressure), T (temperature) and u,v,w are the respective velocity components in the directions λ, ϕ, z increasing. In equation (3) it is assumed that horizontal heat diffusion is negligible in comparison with vertical heat diffusion. It was found, through numerical experimentation, that when horizontal heat diffusion is included at the same order as vertical heat diffusion in the non dimensional equation (3) then its influence on the solution is negligible and on the basis of using the simplest possible model it is set at zero here.

The boundary conditions are
(i) there is an assigned temperature distribution and upwelling velocity at the top of the abyss i.e.
$$T = T_S(\lambda, \phi) \text{ and } w = w_S(\lambda, \phi) \text{ at } z = 0, \tag{4}$$

(ii) temperature is constant and the normal component of velocity is zero at the (flat) ocean floor i.e.
$$T = 0 \text{ and } w = 0 \text{ at } z = -\gamma, \tag{5}$$

(iii) the integrated (from bottom to top of the abyss) mass transport at the eastern boundary is zero i.e.
$$\int_{s=-1}^{s=0} u \, ds = 0 \text{ at } \lambda = \lambda_E. \tag{6}$$

where the vertical coordinate z is replaced by $s = z/\gamma$. Note $\lambda = \lambda_E$ (a line of longitude) is the eastern boundary where $\lambda_E = 5\pi/18$ represents the North Atlantic. Note that while for the Stommel and Arons (1960 a,b) solution u vanishes at the eastern boundary here u satisfies the integrated condition (6).

(iv) the temperature is symmetric at the equator i.e.
$$\frac{\partial T}{\partial \phi} = 0 \text{ at } \phi = 0, \tag{7}$$

Velocity Field

Integration of eqn. (1 a) and substitution of the resulting expression for p into eqns. (1 c) and (2) gives (on applying the boundary conditions for w at $z = 0$ and $z = -\gamma$)

$$v = \frac{\tan \phi \, w_S(\lambda, \phi)}{\gamma} + \frac{\gamma}{\sin 2\phi} \left[\int_{-1}^{s} \frac{\partial T}{\partial \lambda} ds' - \int_{-1}^{0} \int_{-1}^{s'} \frac{\partial T}{\partial \lambda} ds'' ds' \right] \tag{8}$$

$$w = (1+s)w_S(\lambda, \phi) + \frac{\gamma^2}{2\sin^2 \phi} \left[\int_{-1}^{s} \int_{-1}^{s'} \frac{\partial T}{\partial \lambda} ds'' ds' - (1+s)\int_{-1}^{0} \int_{-1}^{s'} \frac{\partial T}{\partial \lambda} ds'' ds' \right] \tag{9}$$

Substitution into eqn. (1 b) and application of boundary condition (6) give

$$-2u\sin\phi = \gamma \left[\int_{-1}^{s} \frac{\partial T}{\partial \phi} ds' - \int_{-1}^{0} \int_{-1}^{s'} \frac{\partial T}{\partial \phi} ds'' ds' \right] + \frac{2}{\gamma} \frac{\partial}{\partial \phi} \left[\sin^2 \phi \int_{\lambda'=\lambda_E}^{\lambda'=\lambda} w_S(\lambda', \phi) d\lambda' \right] \tag{10}$$

When T is constant, above, the Stommel and Arons (1960 a, b) solution results and when $\frac{\partial T}{\partial \lambda} = 0$, with $w_S(\lambda, \phi) = w_S(\phi)$, the Hodnett and McNamara (2000) solution results.

Temperature field

The temperature field is determined through solving the energy equation (3) involving v, w, u as given by eqns. (8), (9), (10) and is a three dimensional problem coupling the temperature, T with the velocity components u, v, w. Stommel and Arons (1960 a,b) assumed the temperature is constant while Hodnett and McNamara (2000) assumed $\frac{\partial T}{\partial \lambda}$ is zero. Here we assume $\frac{\partial T}{\partial \lambda}$ is small but nonzero characterised by a *small* but nonzero parameter ϵ.

Zero Order Problem

This zero-order ($\epsilon = 0$) solution occurs when the assigned temperature distribution, T_0, and upwelling velocity, w_0, at the top of the abyss are independent of λ i.e.

$$T_O = T_S(\phi) \text{ and } w_O = w_S(\phi) \text{ at } s=0. \tag{11}$$

This solution is developed in detail in Hodnett and McNamara (2000) for both the North Atlantic and North Pacific. It is the zero-order element of the solution here and is calculated for the North Atlantic only. When T_O is independent of λ then expressions (8), (9), (10) and eqn. (3) give

$$v_O = w_S(\phi) \tan\phi / \gamma \tag{12}$$
$$w_O = w_S(\phi)(1+s), \tag{13}$$

$$-2u_O \sin\phi = \gamma \left[\int_{-1}^{s} \frac{\partial T_O}{\partial \phi} ds' - \int_{-1}^{0} \int_{-1}^{s'} \frac{\partial T_O}{\partial \phi} ds'' ds' \right] + \frac{2(\lambda - \lambda_E)}{\gamma} \frac{d}{d\phi}\left[w_S(\phi) \sin^2\phi \right], \tag{14}$$

$$w_S(\phi)\left[(1+s)\frac{\partial T_O}{\partial s} + \tan\phi \frac{\partial T_O}{\partial \phi} \right] = \bar{\beta}\frac{\partial^2 T_O}{\partial s^2}, \tag{15}$$

where $\bar{\beta} = \beta/\gamma$ and v_O, w_O are determined independently of the temperature, T_O and eqn. (15) for T_O is linear. When temperature, T_O, has been determined through solving eqn. (15) numerically then u_O is given by expression (14). The solution process involves solving the energy equation (15) numerically for a series of values of $\bar{\beta}$ (starting with $\bar{\beta}$ very small) until a temperature distribution, $T_O(\phi,z)$, acceptably close to the climatological data given in Levitus (1982) is identified. These numerical experiments yield $\bar{\beta} = 0.25$ (gives thermal diffusion coefficient, $k_z' = 1.25 \times 10^{-4} m^2 s^{-1}$ which is an order of magnitude larger than recent observational data suggests). Once T_O is known, u_O can be determined and streamlines can be plotted. Figure 1 shows the streamlines on the ocean floor where the eastward flow between the equator and 30°N is consistent with the eastward spreading of Antarctic bottom water as shown in Worthington and Wright (1970).

Weak Zonal Variation in Temperature

Zonal variation in temperature can be created through zonal variation in assigned temperature distribution and/or upwelling velocity at the top of the abyss. Here we deal only with case where temperature distribution (at $s = 0$) varies with λ but upwelling velocity (at $s=0$) is independent of λ i.e.

$$T = T_s(\phi) + \epsilon(\lambda/\lambda_E)\sin^3 3\phi \text{ at } s = 0, \tag{16}$$

$$w = w_s(\phi) \text{ at } s = 0, \tag{17}$$

where $T_s(\phi)$, $w_s(\phi)$ are as taken in (11), ϵ is a small parameter (to be determined) and $\lambda_E = 5\pi/18$ represents the width of the North Atlantic. The full nonlinear problem is solved iteratively for a series of increasing values of ϵ starting at $\epsilon=0$. The iteration procedure is described in detail in Yuan and Hodnett (2000). It transpires that $\epsilon=0.1$ is the largest value of ϵ for which the iteration procedure converges. When $\epsilon=0.1$, expression (16) gives a temperature distribution (at the top of the abyss) qualitatively similar to the climatological data of Levitus (1982). Here the abyss is taken to lie between 1500m and 4000m. Even the small zonal variation in T resulting from (16) produces dramatic changes in v, w (from their zero order distributions given by expressions (12) and (13). Figure 2 shows the dramatic change in v at the western boundary ($\lambda = 0$) from the depth invariant v_0 (given by (12)) which is also plotted for comparison. Figure 2 also shows that most of the depth variation in v occurs in the upper part of the abyss (except near the equator) which is consistent with some results from Pedlosky (1992).

References

Bogden, P.S., R.E. Davis and R. Salmon 1993. The North Atlantic circulation: combining simplified dynamics with hydrographic data. *J. Mar. Res.*, **51**, 1-52.

Hodnett, P.F. and R. McNamara 2000. A modified Stommel-Arons model of the abyssal ocean circulation. Accepted for publication in Math. Proc. R. Ir. Acad.

Levitus, S., 1982. *Climatological atlas of the world ocean*. NOAA Technical Paper 13, National Oceanic and Atmospheric Administration, Rockville, Maryland.

Pedlosky, J. 1992. The baroclinic structure of the abyssal circulation. *J. Phys. Oceanogr.*, **22**, 652-9.

Stommel, H., and A.B. Arons, 1960a. On the abyssal circulation of the world ocean - 1. Stationary planetary flow patterns on a sphere. *Deep-Sea Res.* **6**, 140-154.

--- and ---, 1960 b. On the abyssal circulation of the world ocean - II. An idealized model of the circulation pattern and amplitude in oceanic basins. *Deep-Sea Res.* **6**, 217-33.

Worthington, L.V. and Wright, W.R., 1970. *North Atlantic Ocean Atlas*, Woods Hole Ocean. Inst., Massachusetts.

Yuan, Y. and P.F. Hodnett, 2000. A simple model of a northern ocean with eastern boundary slope current. Accepted for publication in Math. Proc. R. Ir. Acad.

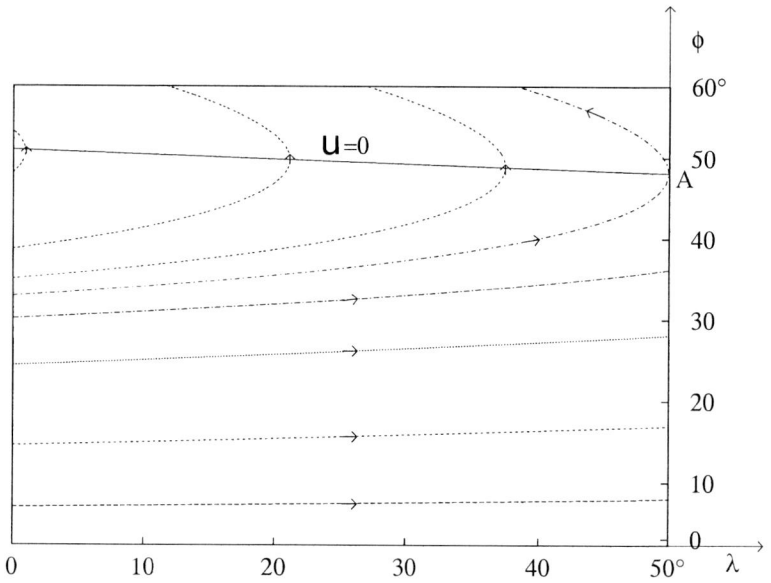

Figure 1. Streamlines on ocean floor.

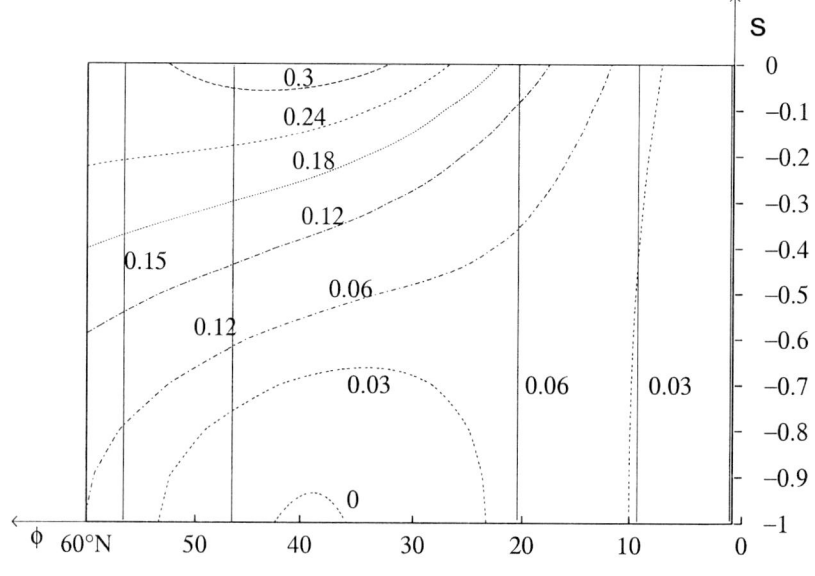

Figure 2. Meridonal velocity v, at the western boundary with v_0 (vertical lines) for comparison.

THE AVAILABLE POTENTIAL ENERGY IN A COMPRESSIBLE OCEAN

Rui Xin Huang
Woods Hole Oceanographic Instititution
Woods Hole, Massachusetts, 02543, U.S.A.

Abstract

Available potential energy (APE) in a compressible ocean is defined as the sum of available gravitational potential energy (AGPE) and available internal energy (AIE). It is shown that AGPE in a compressible ocean is 40 times larger than that calculated from the traditional quasi-geostrophic approximation. In addition, there is also a large amount of AIE. The major sources of AGPE are due to surface cooling and vertical mixing in the interior. Cooling at high latitude also leads to a decrease in AIE and an increase in AGPE, and thus induces a major energy transformation in the compressible ocean. In comparison, the role played by haline forcing is minor.

1 Introduction

Paralleling Lorenz (1955), the available potential energy in the oceans is defined as $\Pi_{QG}^a = -\frac{g}{2} \iiint_V [\rho - \bar{\rho}(z)]^2 \left(\frac{\partial \bar{\rho}_\theta}{\partial z}\right)^{-1} dV$, where ρ is the local density, $\bar{\rho}(z)$ is the global mean density, and $\frac{\partial \bar{\rho}_\theta}{\partial z}$ is the vertical gradient of the global mean potential density. This definition is valid only for the quasi-geostrophic approximation, and it is not valid for the thermohaline circulation. It is found that the application of the traditional definitions of available potential energy and its sources can lead to substantial errors, as shown for the case of an incompressible ocean by Huang (1998). This study is focused on the available potential energy for a compressible ocean, based on the original definition.

2 Finding the Reference State

There have been theoretical studies in meteorology in which the available potential energy is examined using a variational approach, e.g., Dutton and Johnson (1967). Although a variational approach can lead to a well-defined problem mathematically, no practical algorithms have been discussed that can be used to find such a state of minimum potential energy.

In this study we used a computer program to search for the reference state with minimal potential energy. Beginning from the upper surface, the first step is to calculate the potential density of each grid box, using 500 db as the reference pressure. One can sort all grid boxes according to density, with the lightest water parcel sitting on top. To improve the stability of the water column at depth, one can repeat the sorting process by using a deeper reference level at 1500 db. All water parcels lying below 1000 db will be re-sorted. Similarly, re-sorting using reference pressures of 2500 db, 3500 db, and 4500 db will guarantee the stability of the whole water column at any given depth. A simple computer program can be written to sort out the stratification for any given number of reference pressure levels. The accuracy of the solution can be improved by using a small reference pressure interval dp. For $dp < 1$ db, the results are insensitive to the choice of reference pressure levels. Thus, the reference state, which is stably stratified at any given level, can be constructed to any reasonable desired degree of accuracy. The case with bottom topography is slightly more complicated, but it can be handled in an iterative way.

3 The Available Potential Energy

The available potential energy (APE), Π^a, including both the available gravitational potential energy (AGPE) and the available internal energy (AIE), can be defined as

$$\Pi^a = g \int (z - Z) dm + \int (e - e^r) dm + p_a(V - V^r),$$

where z is the vertical position in the physical state, with $z = 0$ defined as the deepest point on the bottom; Z is the vertical position in the reference state; e is the internal energy, superscript r indicates the reference state [internal energy is defined in terms of the Gibbs function $e = G - T\left(\frac{\partial G}{\partial T}\right)_{S,p} - pv$, where G is the Gibbs function and v is the specific volume of seawater (Feistel and Hagen, 1995)]; $dm = \rho dv$ is the mass of each water parcel; p_a is atmospheric pressure at sea level; and V and V^r are the total volume of the ocean in the physical state and in the reference state, respectively. Thus, the total amount of available potential energy consists of three parts, i.e., AGPE, AIE, and the pressure work (PW). For comparison, we calculated the energy balance for three models: the Compressible Model (CM), the Boussinesq Model (BM), and the Quasi-Geostrophic Model (QM).

The reference state should be the state in which the total potential energy of the system is a global minimum. We use an algorithm that can lead to a reference state that is stably stratified and has a global minimum gravitational potential energy. However, we are unable to show that such a state is also a state of global minimum of total potential energy.

In order to overcome the difficulties associated with calculating the total available potential energy, an alternative definition based on enthalpy, defined as $h = pv + e$, has been used to calculate the available potential energy in an oceanic region, such as a warm-core ring, i.e., Bray and Fofonoff (1981). However, it is easy to show that for an ocean with bottom topography, enthalpy is not equivalent to the sum of gravitational potential and internal energy.

4 The Source of AGPE and AIE

The strength of a source due to surface forcing is defined for infinitesimal processes, i.e., $\Pi_s^a = \lim_{\delta t \to 0} \frac{\delta \Pi^a}{\delta t}$. During such infinitesimal processes, the ordering of layers in the reference state does not change. Let us assume that the minimum density difference between adjacent layers in the reference state is $\Delta \rho_{\min} > 0$. Thus, there exists a Δt such that if the time increment $\delta t \leq \Delta t$, the maximum density change in the reference state is smaller than $\Delta \rho_{\min}$; therefore, the ordering of layers in the reference state can be considered unchanged after an infinitesimal time lapse. There are two ways of calculating the strength of the source of AGPE due to surface thermal forcing:

1) Direct sorting: First, calculate the AGPE for the initial state, $\Phi^{a,0}$. Second, heat a surface grid box, say $\Delta T = 0.0001°C$, and calculate the AGPE for the new state, $\Phi^{a,1}$. Then $\Phi^{a,1} - \Phi^{a,0}$ is the source of AGPE due to heating.

2) Infinitesimal analysis: For the case of surface thermal forcing, the potential temperature change is $\dot{\theta}_{i,j,1}$ for the physical state, and $\dot{\theta}_n$ for the reference state. Note that a grid box in the physical space needs a triplet for its identification; the corresponding water parcel in the reference state needs only a single index n for its identification. The change of Gravitational Potential Energy (GPE) in the surface box in the physical state is $\dot{\Phi}_{i,j}^p$ and in the reference state is $\dot{\Phi}_n^{r,1}$. Note that layers above the heated layer in the reference state are pushed upward, and the rate of change in GPE in the reference state is $\dot{\Phi}_n^{r,2}$. The net source of AGPE due to surface thermal forcing is

$$\dot{\Phi}_{i,j}^a = \dot{\Phi}_{i,j}^p - \dot{\Phi}_n^{r,1} - \dot{\Phi}_n^{r,2} = -g\alpha_n^r \dot{\theta}_n \rho_{i,j,1}^p \Delta z_1 (H - Z_n^c) + 0.5g(\alpha\rho)_{i,j,1}^p \Delta z_1^2 \dot{\theta}_{i,j,1}$$

where α_n^r and $\alpha_{i,j,1}^p$ are the thermal expansion coefficients for the corresponding grid box in the reference state and physical state, Δz_1 is the thickness of the grid box in the surface layer, H is the depth of the ocean, Z_n^c is the central level of layer n in the reference state.

At high latitudes the water parcel corresponding to the central level of layer n is far below the free surface, so $\Delta z_1 \ll H - Z_n^c$ and the last term can be neglected, therefore $\dot{\Phi}_{i,j}^a = -g\alpha_n^r \dot{\theta}_n \rho_{i,j,1}^p \Delta z_1 (H - Z_n^c)$.

For the BM (Boussinesq Model), the rate of GPE change in the surface box is due to the GPE change in both the physical and reference states; but the third term is zero due to the non-divergent approximation

$$\dot{\Phi}_{i,j}^{a,BM} = -g(\alpha\rho\dot{\theta})_{i,j,1}^p \Delta z_1 \left[z_{i,j,1}^c - Z_n^c \frac{(\alpha\rho)_n^r}{(\alpha\rho)_{i,j,1}^p} \right].$$

Note that AGPE is proportional to α: in the BM it takes the value on the surface, but in the CM (Compressible Model) it is the value on the sea floor. Since the coefficient of thermal expansion increases greatly with depth, the cooling-induced source of AGPE based on the CM is much larger than that based on the BM.

Similarly, we can calculate the contributions to AGPE and AIE due to surface thermohaline forcing and mixing in the ocean interior. For simplicity, vertical and horizontal mixing with constant coefficients have been used in this study.

5   A Numerical Example

Numerical experiments have been carried out using a 3-dimensional z-coordinate primitive equation model. The model ocean is a $60° \times 60°$ square basin, with a depth of 5.7 km. The horizontal resolution is $4° \times 4°$, and there are 15 layers vertically, with the top layer 30 meters thick. The model is driven by a relaxation boundary condition for temperature, with a reference temperature $25°C$ at the equator linearly decreasing to $0°C$ at the northern boundary at $60°N$.

The haline circulation in the numerical model is driven by the natural boundary condition, with a freshwater flux profile $\omega = \frac{w_0}{\cos\phi}\left(1 - \frac{2\phi}{\phi_0}\right)$, where $w_0 = 0.5$ m y^{-1} is used in the numerical experiment, which gives the maximum evaporation rate 0.5 m y^{-1} at the equator and a maximum precipitation rate of 1 m y^{-1} at $60°N$.

There is no wind stress forcing in the model. The horizontal momentum dissipation and tracer mixing coefficients are 0.5×10^5 m^2 s^{-1} and 1.0×10^3 m^2 s^{-1}, respectively, while the vertical momentum dissipation is 10^{-3} m^2 s^{-1}, and the vertical tracer mixing coefficient is 10^{-4} m^2 s^{-1}. The model is started from an initial state of a homogeneous ocean at rest, and integrated for 3,712 years to reach a quasi-steady state. The temperature calculated from the model is treated as potential temperature, and density is calculated using the original UNESCO equation of state.

Using the sorting program to find the reference state, we have calculated the available potential energy and its sources due to surface thermohaline forcing and internal mixing (Table 1):

Table 1. AGPE and AIE density, in J m^{-3}.

	CM	BM	QM
AGPE	1484.4	1971	33.4
AIE	455.2	1479.3	—

Incidentally, our numerical experiments based on a newly developed oceanic general circulation model in pressure coordinates showed similar results. We have built a model in pressure coordinates that is based on mass conservation and the exact equation of the state and a parallel model that is based on the Boussinesq approximations. These two models were run for a model basin mimicking the North Atlantic, with the same wind stress and thermohaline forcing on the upper surface, and the same mixing parameterization. After 5000 years integration, both models came to a quasi-equilibrium: the meridional overturning rate in the Boussinesq model is 16 Sv, but it is 15.1 Sv in the compressible model. The exact reason for this relatively larger meridional overturning rate in the Boussinesq model is unclear at this time.

The AIE density is 455.2 J m^{-3}, which is 13 times larger than the AGPE density calculated from the quasi-geostrophic approximation. (We did calculate the AIE under the non-divergence approximation; this quantity is, however, meaningless because it cannot be converted into the mechanical energy in the model.)

Although the AGPE density for CM is rather close to that for BM, the source of AGPE is quite different in these two models. As shown in Table 2, the cooling-induced source of AGPE based on CM is about 3 times larger than that based on BM. This large difference is because their definitions are based on the heat expansion coefficient at the sea surface (BM) vs at the sea floor (CM). The source calculated from QM is totally out of order, so it is meaningless.

The most important point is that cooling induces a decrease in AIE, mostly in the deep-water formation site, the NW corner of the basin. Thus, cooling is the major mechanism that is responsible for the energy transfer between the AIE and AGPE, which is one of the major players in the energetics of the thermohaline circulation.

Table 2. Sources of AGPE and AIE due to surface thermal forcing, in mW m^{-2}.

| | AGPE | | | AIE | APE |
	(CM)	(BM)	(QM)	(CM)	(CM)
NW Corner	211	77	1.5	−137	74
SW Corner	0.47	0.36	5.23	−0.01	0.46
Basin Mean	6.23	1.52	0.73	−4.65	1.58

The contribution due to haline forcing is generally much smaller and has a negative sign because the model is in a thermal mode, i.e., water sinks due to cooling at high latitudes, Table 3. The difference between CM and BM is quite small, but their values are much larger than that based on QM.

Table 3. Sources of AGPE and AIE due to surface haline forcing, in mW m^{-2}.

| | AGPE | | | AIE | APE |
	(CM)	(BM)	(QM)	(CM)	(CM)
NW Corner	−12	−12	−0.24	−0.44	−12.4
SW Corner	−0.02	−0.02	−0.18	0.001	−0.02
Basin Mean	−0.61	−0.44	−0.06	−0.008	−0.62

The available potential energy balance for the whole model basin is shown in Figure 1. Note that the balance of AGPE for BM is quite different from that of CM. For example, the energy required for mixing in the physical state is 0.8 mW m^{-2} for CM, but it is 15 mW m^{-2} for BM (about 5 times higher than the level of tidal dissipation observed.) In any case, it is clear that the gravitational potential energy balance in BM is greatly distorted, and it seems impractical to build an energy-conserving numerical model based on the Boussinesq approximations. In addition, the BM may not be able to simulate the time-dependent processes accurately because the model cannot simulate energy transformation processes accurately.

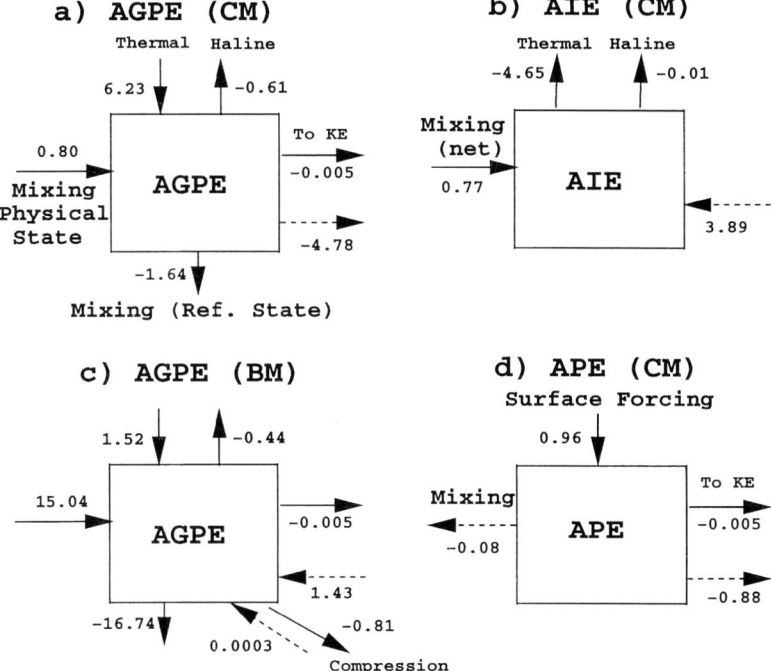

Figure 1: Balance of available potential energy for a model ocean with flat bottom, all fluxes in mW m^{-2}.

Note that both AGPE and AIE have a relatively large residual term, on the order of 4 mW m^{-2}. The nature of such residual terms may be due to the convective adjustment process that was not included in the present calculation; this is left for further study.

Acknowledgments. This study was supported by the U.S. National Science Foundation through grant OCE 96-16950, and a Mellon Independent Study Award from the Woods Hole Oceanographic Institution. This is WHOI Contribution No. 9902.

References

Bray, N. A., and N. P. Fofonoff, 1981: Available potential energy for MODE eddies. *J. Phys. Oceanogr.*, **11**, 30–47.

Dutton, J. A., and D. R. Johnson, 1967: The theory of available potential energy and a variational approach to atmospheric energetics. *Adv. Geophys.*, **12**, 333–436.

Feistel, R., and E. Hagen, 1995: On the Gibbs thermodynamic potential of seawater. *Prog. Oceanogr.*, **36**, 249–327.

Huang, Rui Xin, 1998: Mixing and available potential energy in a Boussinesq ocean. *J. Phys. Oceanogr.*, **28**, 669–678.

Lorenz, E. N., 1955: Available potential energy and the maintenance of the general circulation. *Tellus*, **7**, 157–167.

THE ROLE OF BOTTOM PRESSURE TORQUES IN THE OCEAN CIRCULATION

C.W. HUGHES
Proudman Oceanographic Laboratory
Bidston Observatory, Prenton, CH43 7RA, U.K.
cwh@pol.ac.uk

Abstract. It is shown that, if wind stress is balanced by form stress in a zonal and depth integral of the zonal momentum equation, at each latitude, then wind stress curl is balanced by bottom pressure torque in a zonal integral of the barotropic vorticity equation. It is, therefore, unclear that viscosity is important in western boundary currents.

1. Introduction

If one asks "What balances the zonal integral of zonal wind stress?", the usual answer would be "bottom form stress", *i.e.*, pressure differences from east to west across an ocean basin, or across topographic features in the Southern Ocean. Of course, wind stress is balanced by the Coriolis force within the Ekman layer, as an eastward wind stress forces an equatorward Ekman flux, but mass conservation requires that the same mass of water return poleward at greater depth. The return flow is generally acknowledged to be geostrophic, resulting in the east-west pressure differences which balance the wind stress in a zonal and depth integral, for which Coriolis integrates to zero.

On the other hand, if one asks "What balances wind stress curl in a zonal integral of the barotropic vorticity equation", the most common answer would be "viscosity". Again, the standard picture allows wind stress curl to be balanced by advection of planetary vorticity (βV) in an interior circulation in Sverdrup balance, but mass conservation requires that this meridional mass flux return somewhere so that βV integrates to zero, and some other term in the barotropic vorticity equation must be important in this other region. Studies of ocean basins with vertical sidewalls have led to the conclusion that friction is the missing term, although in some circumstances nonlinearity may also have an influence.

The purpose of this paper is to show that these two standard viewpoints are incompatible. In fact, if the zonal wind stress is balanced by form stress at each latitude, then the zonal integral of wind stress curl must be balanced by bottom pressure torque. This is true even in the vertical sidewall case, but in that case the bottom pressure torque is confined to a delta function at the ocean boundary, forcing friction to play the dominant role in the ocean interior. For an ocean with sloping sidewalls, friction need be important in the zonal integral of the barotropic vorticity balance only if the region occupied by the sloping walls is narrower than a Munk viscous boundary layer.

That is not to say that friction is unimportant in the ocean, it may still be crucial in some form in order to obtain a global balance, but friction is unimportant in the zonal integral of the barotropic vorticity equation.

2. A mathematical identity

The barotropic vorticity equation and the depth and zonally integrated zonal momentum equation may both be derived from the integral over depth of the steady momentum equation. The steady momentum equation may be written in the form

$$\rho f \mathbf{k} \times \mathbf{u} = -\nabla p + \boldsymbol{\tau}_z + \mathbf{a} + \mathbf{b}, \tag{1}$$

where f is the Coriolis parameter, $\mathbf{u}$ is the two-dimensional horizontal velocity, ρ is density, p is pressure, $\boldsymbol{\tau}$ is the viscous stress on a horizontal surface, $\mathbf{a}$ is the divergence of the remaining (lateral) viscous stress, and $\mathbf{b}$ represents terms nonlinear in $\mathbf{u}$. If we integrate this from the ocean floor ($z = -H$) to the surface ($z = \eta$), we may write $(\mathbf{U}, P, \mathbf{A}, \mathbf{B}) = \int_{-H}^{\eta}(\rho \mathbf{u}, p, \mathbf{a}, \mathbf{b})\, dz$ to obtain

$$f\mathbf{k} \times \mathbf{U} = -\nabla P + p_b \nabla H + \boldsymbol{\tau}_0 + \mathbf{A} + \mathbf{B}, \tag{2}$$

where $\boldsymbol{\tau}_0 = \boldsymbol{\tau}_a - \boldsymbol{\tau}_b$, the difference between wind stress and bottom stress, and p_b is ocean bottom pressure. The first two terms on the right hand side of (2) come from the depth integral of the pressure gradient, after taking the gradient operator out of the integral. All pressures are defined as the actual pressure minus an atmospheric pressure which is independent of position. A spatially variable atmospheric pressure would exert a small force on slopes of the ocean surface, which could also be included if it were found to be significant.

An important point is that, as the ocean depth tends to zero ($-H \to \eta$), all of the terms in (2) also tend to zero, being depth integrals of finite quantities. This means that, whereas it may be reasonable to assume that

τ_0 represents the surface wind stress in the ocean interior, where bottom stress should be small, that is not true at the boundary where depth tends to zero and so τ_0 tends to zero, meaning that the bottom stress exactly balances the wind stress.

A zonal integral of (2) from the west (W) to the east (E) of an ocean basin leads to the angular momentum, or zonal momentum balance:

$$f \int_W^E V \, dx = \int_W^E p_b H_x + \tau_0^x + A^x + B^x \, dx, \tag{3}$$

where V is the northward component of $\mathbf{u}$, dx is shorthand for $r \cos \phi \, d\lambda$, where λ is longitude, and a superscript x represents the zonal component of a vector. If the basin is closed to the north or south, or the integral is round a closed contour, then the left hand side must be zero unless there is a source or sink of mass. This leads to a balance between form stress, zonal (wind - bottom) stress, lateral friction, and nonlinear terms. It was noted in the introduction that the first two of these terms are generally considered to balance (with bottom friction negligible):

$$\int_W^E \tau_0^x \, dx \approx -\int_W^E p_b H_x \, dx \tag{4}$$

The barotropic vorticity equation is found by taking the curl of (2). Assuming no sources or sinks of mass ($\nabla \cdot \mathbf{U} = 0$), this gives the barotropic vorticity (BV) equation

$$\beta V = \nabla \times (p_b \nabla H) + \nabla \times \boldsymbol{\tau}_0 + \nabla \times (\mathbf{A} + \mathbf{B}), \tag{5}$$

where $\beta = f_y$. The conventional picture mentioned in the introduction has it that $\nabla \times \boldsymbol{\tau}_0$ and $\nabla \times \mathbf{A}$ are most important in the zonal integral of this equation (for which βV integrates to zero), with $\nabla \times \mathbf{B}$ also having an influence in some cases. While it is generally acknowledged that the bottom pressure torque, $\nabla \times (p_b \nabla H)$ can play a role locally, it is certainly not usual to claim that it can balance the wind stress curl, making $\nabla \times \mathbf{A}$ unnecessary. In fact, since each term in (3) is the line integral of a term in (2), and each term in (5) is the curl of a term in (2), terms in (3) and (5) are related by Stokes' theorem. This means that, if (4) were exactly true, then the zonal integral of (5) would be a balance between bottom pressure torque and $\nabla \times \boldsymbol{\tau}_0$, with $\mathbf{A}$ and $\mathbf{B}$ integrating to zero.

To see this, consider a zonal strip of ocean A, as shown in Fig. 1, bounded by a contour δA. The bounding contour consists of two latitude lines, and two strips of coastline where the ocean depth is zero. If we now perform an area integral of the bottom pressure torque in (5), Stokes' theorem gives

$$\int_A \nabla \times (p_b \nabla H) \, dS = \oint_{\delta A} p_b \nabla H \cdot \mathbf{ds}. \tag{6}$$

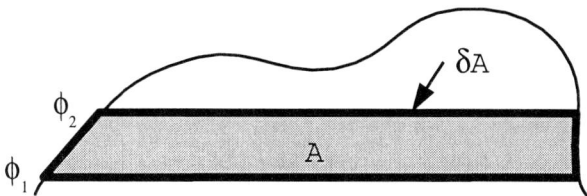

Figure 1. Schematic showing a zonal strip of ocean A bounded by a contour δA which consists of two latitude lines at latitudes ϕ_1 and ϕ_2 and two stretches of coastline at zero depth.

However, p_b is zero at the coastline, and the remaining part of the line integral consists of two zonal integrals:

$$\int_A \nabla \times (p_b \nabla H)\, dS = \int_{\phi_1} p_b H_x\, dx - \int_{\phi_2} p_b H_x\, dx, \qquad (7)$$

where ϕ_1 is the southern bounding latitude of A and ϕ_2 is the northern bounding latitude. Comparing the right hand side of (7) with the form stress in (3), we see that the area integral of bottom pressure torque over a zonal strip is equal to the difference between form stresses evaluated at the bounding latitudes of that strip. Taking the limit as the strip width tends to zero gives

$$\int_W^E \nabla \times (p_b \nabla H)\, dx = -\frac{d}{dy} \int_W^E p_b H_x\, dx, \qquad (8)$$

where W and E represent the positions in longitude of the western and eastern boundaries of the ocean at each latitude. So the zonal integral of bottom pressure torque is simply d/dy of the form stress.

The same argument can be applied to each of the terms in (5), showing that its zonal integral is identically equal to d/dy of the zonal integral of the corresponding term in (3). Clearly, this means that if τ_0 and form stress exactly balance at each latitude, in (3), then $\nabla \times \tau_0$ and bottom pressure torque must exactly balance in any zonal integral of (5). It will be noted that, whereas the line integral of τ_0 along the coast is zero, the line integral of wind stress need not be zero. This is really a question of how the Ekman flux into the coast is balanced, addressed in the vertical sidewall case by Pedlosky (1968). It is only of significance when there is a longshore wind stress.

In practice, although it is generally accepted that wind stress and form stress are the largest terms in (3), they do not balance exactly, as indicated by the approximate equality in (4). In fact it is possible that wind stress and form stress are approximately equal in (3), but that d/dy of these terms could be quite different, since wind stress varies quite slowly with y

whereas form stress, lateral friction, and nonlinear terms may vary rapidly with y. In fact, model diagnostics (Wells and de Cuevas 1995, Stevens and Ivchenko 1997, Gille 1997) show that the largest terms in d/dy of (3) are form stress and the nonlinear term. If we are to answer the question posed in the introduction "What balances wind stress curl in a zonal integral of the barotropic vorticity equation", the answer is that wind stress curl is actually a minor term in that balance, it is only when the zonal integral is over a strip wide enough to smooth out nonlinear effects, typically about 3 degrees of latitude or wider, that wind stress curl becomes a dominant term, and in that case it is clear that the answer, following (4), is that bottom pressure torque must balance wind stress curl.

3. Discussion

The identity described above, between area integrals of terms in the BV equation, and the corresponding terms in the angular momentum equation, highlights the role of bottom pressure torques. This argument also works for the vertical sidewall case, if that is considered as the limiting case for steeply sloping sidewalls. The difference in that case is that the bottom pressure torque is confined to a delta function at the ocean boundary. However, βV cannot occur in such a delta function, the return flow which balances the Sverdup interior circulation must be of finite width, away from the sidewall. Since there is no bottom pressure torque away from the sidewall, this requires a viscous torque to act in the ocean interior. However, if the sidewalls slope over a region wider than the viscous boundary layer, βV must occur over a region where bottom pressure torques are known to exist, and the need for viscous torques evaporates.

This suggests that the primary balance in western boundary currents is between βV and bottom pressure torques, although we must expect the nonlinear term to be significant on small scales.

In conclusion then we see that, on the large scale, if zonal wind stress is balanced by bottom form stress, then wind stress curl is balanced by bottom pressure torque, in a zonal integral. The evidence for the former balance is considerable, coming from scaling analysis (Oort 1985), model studies (Treguier and McWilliams 1990, Wolff et al. 1991, Ponte and Rosen 1994, Stevens and Ivchenko 1997, Gille 1997, Bryan 1997), in-situ studies (Bryden and Heath 1985), and satellite remote sensing measurements (Morrow et al. 1994), while evidence is still wanting even for the contention that Sverdup balance holds in subtropical ocean interiors (Wunsch and Roemmich, 1985). Given this evidence, it seems that we should expect the barotropic vorticity balance of western boundary currents in the real ocean to be dominated by βV and bottom pressure torque, with nonlinear effects important locally.

Again, this is confirmed by model diagnostics from a realistic ocean model, in Pacific western boundary currents (Saunders et al. 1999) and globally (Hughes and de Cuevas 2000).

This, of course, does not mean that viscosity is unimportant to obtaining a global dynamical balance in the ocean, only that it is unimportant in this particular integral balance. It does, however, mean that we should reassess our assumption that the western boundary current is the place where any viscous effects must occur, since it is no longer clear that viscosity is important in western boundary currents.

References

Bryan, F. O., 1997: The axial angular momentum balance of a global ocean general circulation model, *Dyn. Atmos. and Oceans.*, **25**, 191–216.

Bryden, H. L., and R. A. Heath, 1985: Energetic eddies at the northern edge of the Antarctic Circumpolar Current in the Southwest Pacific, *Prog. in Oceanogr.*, **14**, 65–87.

Gille, S. T., 1997: The Southern Ocean Momentum Balance: Evidence for Topographic Effects from Numerical Model Output and Altimeter Data. *J. Phys. Oceanogr.*, **27**, 2219–2232.

Hughes, C. W., and B. A. de Cuevas, 2000: Why western boundary currents in realistic oceans are inviscid: a link between form stress and bottom pressure torques. *J. Phys. Oceanogr., submitted*.

Morrow, R., R. Coleman, J. Church, and D. Chelton, 1994: Surface eddy momentum flux and velocity variances in the Southern Ocean from Geosat altimetry. *J. Phys. Oceanogr.*, **24**, 2050–2071.

Oort, A. H., 1985: Balance conditions in the earth's climate system. *Adv. Geophys. A* **28**, 75–98.

Pedlosky, J., 1968: An overlooked aspect of the wind-driven ocean circulation. *J. Fluid Mech.*, **32**, 809–821.

Ponte, R. M. and R. D. Rosen, 1994: Oceanic Angular Momentum and Torques in a General Circulation Model. *J. Phys. Oceanogr.*, **24**, 1966–1977.

Saunders, P. M., A. C. Coward, and B. A. de Cuevas, 1999: Circulation of the Pacific Ocean seen in a global ocean model: Ocean Circulation and Climate Advanced Modelling project (OCCAM). *J. Geophys. Res.* **104**, 18,281–18,299.

Stevens, D. P., and V. O. Ivchenko, 1997: The zonal momentum balance in an eddy-resolving general-circulation model of the Southern Ocean. *Quart. J. Roy. Meteor. Soc.*, **123**, 929–951.

Treguier, A. M., and J. C. McWilliams, 1990: Topographic influence on wind-driven stratified flow in a β-plane channel: An idealised model for the Antarctic Circumpolar Current, *J. Phys. Oceanogr.*, **20**, 321–343.

Wells, N. C., and B. A. de Cuevas, 1995: Vorticity dynamics of the Southern Ocean from a General Circulation Model, *J. Phys. Oceanogr.*, **25**, 2569–2582.

Wolff, J. -O., E. Maier-Reimer, and D. Olbers, 1991: Wind-driven flow over topography in a zonal β-plane channel: A quasi-geostrophic model of the Antarctic Circumpolar Current, *J. Phys. Oceanogr.*, **21**, 236–264.

Wunsch, C., and D. Roemmich, 1985: Is the North Atlantic in Sverdrup Balance? *J. Phys. Oceanogr.*, **15**, 1876–1880.

QUASI-GEOSTROPHIC POTENTIAL VORTICITY FOR A GENERALISED VERTICAL COORDINATE: FORMULATION AND APPLICATIONS

N. KEELEY
The Met. Office
Joint Centre for Mesoscale Meteorology,
Department of Meteorology, The University of Reading,
P.O. Box 243, Reading, Berkshire, RG6 6BB, U.K.

A.A. WHITE[1]
The Met. Office
Numerical Weather Prediction Division,
London Road, Bracknell, Berkshire, RG12 2SZ, U.K.

1. Introduction

In this contribution we present a formulation of quasi-geostrophic potential vorticity (QGPV) for a generalised vertical coordinate, η. The original motivation for this work was in the possible use of PV as a control variable for variational data assimilation (Var) which would require the inversion of PV on model levels. In addition, we are unaware of any previous formulation of the QG system for a *general* vertical coordinate and present the following as a contribution to the literature on the subject. We find that our formulation is strictly appropriate only for vertical coordinates that are not terrain-following. Ways of overcoming this limitation are suggested.

2. Formulation

We present here an outline of the derivation only. Full details will be published elsewhere (Keeley and White, 2000). Our starting point is the formulation of the hydrostatic primitive equations for a general vertical coordinate formulated by Kasahara (1974).

Before proceeding we define a hydrostatically-balanced reference state that is

---

[1] Corresponding author: email aawhite@meto.gov.uk

a function of our general vertical coordinate η alone, denoted by the subscript 0. The meteorological variables that define this reference state are: specific volume, α; pressure, p and geopotential, Φ:

$$\begin{aligned}\alpha &= \alpha_0(\eta) + \alpha'(x,y,\eta,t) \\ p &= p_0(\eta) + p'(x,y,\eta,t) \\ \Phi &= \Phi_0(\eta) + \Phi'(x,y,\eta,t)\end{aligned} \qquad (1)$$

Hydrostatic balance is expressed as

$$\alpha_0 \frac{dp_0}{d\eta} + \frac{d\Phi_0}{d\eta} = 0 \qquad (2)$$

for the reference state and as

$$\alpha_0 \frac{\partial p'}{\partial \eta} + \alpha' \frac{dp_0}{d\eta} + \frac{\partial \Phi'}{\partial \eta} = 0 \qquad (3)$$

for the deviations α', p', Φ' from the reference state. We may write the horizontal momentum, continuity and thermodynamic equations for frictionless adiabatic flow and a generalised vertical coordinate η as:

$$\frac{D\boldsymbol{v}}{Dt} + (f_0 + \beta y)\boldsymbol{k} \times \boldsymbol{v} + \nabla_\eta(\Phi' + \alpha_0 p') + \alpha' \nabla_\eta p' = 0 \qquad (4)$$

$$\frac{D}{Dt}\left(\frac{\partial p}{\partial \eta}\right) + \frac{\partial p}{\partial \eta}\left(\nabla_\eta \cdot \boldsymbol{v} + \frac{\partial \dot{\eta}}{\partial \eta}\right) = 0 \qquad (5)$$

$$\frac{D\theta}{Dt} = 0 \qquad (6)$$

We are working on a β-plane and the Lagrangian time derivative is:

$$\frac{D}{Dt} = \left[\frac{\partial}{\partial t} + \boldsymbol{v}\cdot\nabla_\eta + \dot{\eta}\frac{\partial}{\partial \eta}\right]$$

We now define a geostrophic wind as:

$$\boldsymbol{v}_g = \frac{1}{f_0}\boldsymbol{k} \times \nabla_\eta(\Phi' + \alpha_0 p') = \boldsymbol{k} \times \nabla_\eta \Psi \qquad (7)$$

We further define a modified ageostrophic flow $\hat{\boldsymbol{v}}_{ag}$ as:

$$\hat{\boldsymbol{v}}_{ag} = \boldsymbol{v}_{ag} - \frac{\alpha'}{f_0}\boldsymbol{k} \times \nabla_\eta p' \qquad (8)$$

and a modified vertical velocity $\hat{w}$ by:

$$\hat{w}\frac{dp_0}{d\eta} = \dot{\eta}\frac{dp_0}{d\eta} + \left(\left.\frac{\partial}{\partial t}\right|_\eta + \boldsymbol{v}_g \cdot \nabla_\eta\right) p' \qquad (9)$$

By using these definitions, equations (2) and (3) and applying QG approximations to equations (4)–(6) we arrive at:

$$\left[\frac{\partial}{\partial t} + \boldsymbol{v}_g.\nabla_\eta\right]\boldsymbol{v}_g + f_0\boldsymbol{k}\times\hat{\boldsymbol{v}}_{ag} + \beta y\boldsymbol{k}\times\boldsymbol{v}_g = 0 \qquad (10)$$

$$\nabla_\eta.\left(\hat{\boldsymbol{v}}_{ag}\frac{dp_0}{d\eta}\right) + \frac{\partial}{\partial\eta}\left(\hat{w}\frac{dp_0}{d\eta}\right) = 0 \qquad (11)$$

$$\left[\frac{\partial}{\partial t} + \boldsymbol{v}_g.\nabla_\eta\right]\frac{\partial\Psi}{\partial\eta} - \frac{\alpha_0}{f_0\theta_0}\frac{d\theta_0}{d\eta}\hat{w}\frac{dp_0}{d\eta} = 0 \qquad (12)$$

These are the QG horizontal momentum, continuity and thermodynamic equations, and we have introduced a reference state potential temperature, $\theta_0(\eta)$. Using the definitions (7)–(9), the hydrostatic balance relations (2) and (3), and equations (10)–(12) we obtain the following conservation equation:

$$\frac{\hat{D}}{Dt}\left\{\nabla_\eta^2\Psi + \beta y - f_0^2\left(\frac{dp_0}{d\eta}\right)^{-1}\frac{\partial}{\partial\eta}\left(\frac{\theta_0}{\alpha_0}\left(\frac{d\theta_0}{d\eta}\right)^{-1}\frac{\partial\Psi}{\partial\eta}\right)\right\} = 0 \qquad (13)$$

The Lagrangian time derivative is now defined as:

$$\frac{\hat{D}}{Dt} = \left[\frac{\partial}{\partial t} + \boldsymbol{v}_g.\nabla_\eta\right] \qquad (14)$$

The conserved quantity:

$$Q = \nabla_\eta^2\Psi + \beta y - f_0^2\left(\frac{dp_0}{d\eta}\right)^{-1}\frac{\partial}{\partial\eta}\left(\frac{\theta_0}{\alpha_0}\left(\frac{d\theta_0}{d\eta}\right)^{-1}\frac{\partial\Psi}{\partial\eta}\right) \qquad (15)$$

is the QGPV for the generalised vertical coordinate, η.

3. Problems

There are, however, a number of problems with this formulation as it stands. The first is general; the choice of reference state is arbitrary, provided that it is a monotonic function of η. This is also the case for the usual pressure coordinate formulation of the QG system, but for our general vertical coordinate system the choice of α_0 in particular will affect the absolute value of the geostrophic wind obtained, which is not the case in pressure coordinates.

The second problem is more serious. Our formulation is not suitable for terrain-following vertical coordinates (as used in most current NWP models) in the presence of orography. There are two reasons for this. The first is a purely practical point: in order to obtain the geostrophic wind from the streamfunction Ψ we are obliged to take the gradient of Φ' and/or p' on a surface of constant η. If η is a terrain-following coordinate there will be large variations in the magnitude of

Φ' and p' from grid-point to grid-point near steep orography, such as Greenland, which finite difference schemes are ill-equipped to cope with. This aspect (which is of course a well-known problem in hydrostatic primitive equation models) can lead to spurious large values of the geostrophic wind over steep slopes.

The second point is a theoretical aspect of the formulation. The reference state for the geopotential (gz) is chosen as $\Phi_0(\eta)$, which means that z is constant on η surfaces in the reference state, i.e. the η surfaces are "flat" in the reference state. If we now consider a case in which there is a mountain, but no flow, we encounter a difficulty. As the reference state has flat η surfaces, such a case can only be described using our formulation of QG by a non-zero Φ'. Consequently, other primed quantities must also be non-zero. The difficulty is that this situation of no flow will be described in our formulation by a geostrophic flow and an equal and opposite part of the modified ageostrophic flow ($\hat{v}_{ag}$, see equation (8)). Not only is this inelegant (expressing zero as $+x - x$) but the formulation has been derived by treating the geostrophic flow and the ageostrophic flow differently, under the assumption that the former is much larger than the latter. Thus, we find that our formulation of QGPV for a generalised vertical coordinate system is not completely general: if orography is present, our formulation is appropriate only for non-terrain following vertical coordinates (e.g. height, pressure, density or potential temperature).

4. Possible Solutions

There are two possible ways forward from this problem. Firstly, one may just ignore it and use the formulation as it stands for a terrain-following vertical coordinate. This seems a rather unsatisfactory way of dealing with the problem, but if one is interested in the inversion of QGPV *increments* (as is the case in most practical applications) at least the practical problem of calculating the gradient of Φ' and/or p' is not so marked.

Secondly, one may take a different approach to the problem and directly transform the pressure (or height) coordinate form of QGPV to a general vertical coordinate. For this formulation, we retain the usual QG definition of streamfunction and the static stability parameter remains a function of pressure, rather than η. We have carried out this transformation and the result will be presented elsewhere (Keeley and White, 2000). The resulting expression is somewhat unwieldy, but should remain amenable to standard finite difference methods.

References

Kasahara, A. (1974) Various vertical coordinate systems used for numerical weather prediction. *Mon. Wea. Rev.*, **102**, 509-522

Keeley, N. and White, A.A. (2000) Remarks on the theory and practice of PV inversion. To be submitted to *Quart. J. Roy. Meteor. Soc.*

ON THE EFFECT OF A SURFACE DENSITY FRONT ON THE INTERIOR STRUCTURE OF THE VENTILATED OCEAN THERMOCLINE

P.LIONELLO
University of Lecce[†]

AND

J.PEDLOSKY
WoodsHole Oceanographic Institution

1. Introduction

This study analyzes the response of the ocean thermocline to a density front at the sea surface. A set of computations has been carried out, solving the thermocline structure that results from a surface density front and analyzing the behavior of the solution as the location of the front at the sea surface varies in the meridional direction. In fact, though theoretical studies often assumes a constant SSD (Sea Surface Density) gradient, a recent observational study (Raffaele and Rudnick, 2000) suggests that much of the surface density variation, even on large scales, is gathered into frontal gradients.

This investigation is based on a model (Lionello and Pedlosky, 2000a and 2000b, here after LP model) which is an extension to a continuous fluid of the dynamics adopted in the "layer" theory of the ventilated thermocline (Luyten et al. 1983). The LP model is, in fact, the limit of a discrete many-layers model for the number of layers tending to infinity, and, consequently, their thickness tending to zero.

The LP model integrates former continuous models of the ventilated thermocline (Killworth, 1987, Huang 1988). Its major progress is that it allows the solution of the direct problem, i.e. the computation of the full 3-dimensional circulation from the SSDD (SSD Distribution), by using the relation between potential vorticity q and Montgomery function π. In fact, q and π are conserved along streamlines, and they are consequently functionally related. In the LP model, an analytical expression for the functional relation between the potential vorticity and the Montgomery function is associated to the prescribed SSDD. The knowledge of their functional dependence reduces the solution of the ventilated thermocline problem to the solution of a second order ordinary differential equation (Pedlosky, 1996).

Since the LP model contains no unventilated regions, its application to the whole basin would require the addition of the two regions called 'shadow zone' and 'unventilated pool' in the LPS ventilated thermocline theory, i.e. the consideration of streamlines departing from the lateral boundaries of the ocean. When the Ekman pumping velocity, $w_E(x,y)$ due to the wind stress is negative, the LP model is appropriate if the total depth of the

[†] e-mail: "lionello@unile.it", address: via per Arnesano, 73100, Lecce, Italy

thermocline vanishes at the northern and eastern boundary and if the basin has no western boundary, i.e. it has no limitation to the west.

2. The model of the continuous ventilated thermocline

The motion of the continuous fluid is assumed frictionless, steady, adiabatic, and geostrophic. The density of the fluid is assumed to increase monotonically with depth and it is used as vertical coordinate. Thus, the fluid is described by the three coordinates x, y, ρ, where the density ρ varies in the range $\rho_{min} < \rho < \rho_{Max}$ (with ρ_{min} and ρ_{Max} the density at the southern and northern boundaries of the gyre, respectively) and x, y are the zonal and meridional coordinates. The thermocline is suspended over a quiescent abyss with density $\rho_A = \rho_{Max} + \Gamma$, and $\Gamma > 0$ gives a finite density step between abyss and thermocline. The function $z(\rho, x, y)$ represents the level of constant density ρ. The absolute value of the level $z(\rho_{Max}, x, y)$ does not corresponds to the total depth of the thermocline $H(x, y)$, because a layer of finite thickness and density ρ_{Max} lays above the thermocline bottom (Lionello and Pedlosky 2000b). The total meridional transport $V_S(x,y)$ follows the Sverdrup relation $\beta V_S(x, y) = f w_E(x, y)$, where β is the meridional gradient of f and $w_E(x, y)$ is the the Ekman pumping, assumed non positive in this study.

The detailed presentation of the equations governing the motion of a continuous geostrophic fluid is available in the literature (Pedlosky, 1996) and it is summarized in former papers (Lionello and Pedlosky 2000b). The pressure within the thermocline can be computed using the hydrostatic relation in density coordinates, and accounting for the density discontinuity between thermocline and abyss, where the fluid is not in motion:

$$p(\rho, x, y) = C - g\rho z(\rho, x, y) + g\Gamma H(x, y) + \int_{\rho_{Max}}^{\rho} gz(\rho', x, y)d\rho' , \qquad (1)$$

where g is the acceleration of gravity, C an arbitrary constant.

The geostrophic balance in density coordinates is given by the relation $\rho f \hat{k} \wedge \mathbf{u} = -\nabla_H \pi$, where f is the Coriolis parameter, $\mathbf{u} = (u, v)$ the fluid velocity, $\nabla_H = (\frac{\partial}{\partial x}, \frac{\partial}{\partial y})$ the gradient on surfaces of constant density, and π the Montgomery function $\pi(\rho, x, y) = p(\rho, x, y) + \rho g z(\rho, x, y)$, that is constant along streamlines. By substituting eq.(1) in the definition of π one obtains,

$$\pi(\rho, x, y) = g\Gamma H(x, y) + \int_{\rho_{Max}}^{\rho} gz(\rho', x, y)d\rho' , \qquad (2)$$

where the constant C has been neglected.

Since the motion is adiabatic, the Ertel potential vorticity $q(\rho, x, y) = -f/\frac{\partial z}{\partial \rho}$ is conserved during the motion after the subduction of the fluid and, therefore, constant along streamlines. Consequently, a function Q exists such that $q(\rho, x, y) = Q(\rho, \pi(\rho, x, y))$.

An important results of former studies is to prove that, in the fully ventilated portion of the thermocline, if the outcrop lines are latitude circles, the product of $q(\rho, x, y)$ and $\pi(\rho, x, y)$ is a function of density only (Killworth 1987, Lionello and Pedlosky 2000b):

$$q(\rho, x, y)\pi(\rho, x, y) = c(\rho) . \qquad (3)$$

This equation gives the functional dependence of the potential vorticity q on the Montgomery function π that can be used for the solution of the thermocline problem. In fact, from the definition of π and of the potential vorticity one has that

$$\frac{\partial^2}{\partial \rho^2}\pi(\rho,x,y) = -\frac{fg}{q(\rho,x,y)} \ . \tag{4}$$

which using eq.(3) becomes an equation for $\pi(\rho,x,y)$ only:

$$\frac{\partial^2 \pi(\rho,x,y)}{\partial \rho^2} = -\frac{fg}{c(\rho)}\pi(\rho,x,y) \ . \tag{5}$$

Eq.(5) is not useful unless the expression for $c(\rho)$ is known. An analytical approximation $c_{an}(\rho)$, that allows a physically meaningful solution of the thermocline problem, has been recently proposed (Lionello and Pedlosky 2000b):

$$c_{an}(\rho) = \frac{g}{f_{Max}}\frac{f_c^2(\rho)}{df_c/d\rho}\left[\Gamma f_{Max} + (\rho_{Max}-\rho)(f_{Max}-f_c(\rho))\right] \ . \tag{6}$$

where f_{Max} is the Coriolis parameter at the Northern boundary of the gyre and $f_c(\rho)$ is the Coriolis parameter along the outcrop line of the fluid with density ρ. The important advantage of this expression is that it allows to construct $c_{an}(\rho)$ from the SSDD $f_c(\rho)$, and therefore to solve the direct problem of the thermocline. It has to be remarked that this expression is an approximation to the (unknown) exact expression, and it is valid if the ratio $a = (\rho_{Max} - \rho_{min})/\Gamma$ is not too large (say $a < 100$, Lionello and Pedlosky 2000a and 2000b))

Eq.(5) should, in general, be solved together with the Sverdrup relation, because the total depth of the thermocline enters the boundary conditions. After some algebra, the Sverdrup relation becomes an equation for the total depth of the thermocline (Lionello and Pedlosky 2000b):

$$G(x,y)\frac{\partial H^2(x,y)}{\partial x} = w_E(x,y) \ , \tag{7}$$

where $G(x,y)$ is a known functional of the isopycnal levels $z(\rho)$ (Lionello and Pedlosky 2000b). Anyway, if the outcrop lines are latitude circles, the Montgomery function can be factorized as (Lionello and Pedlosky 2000a, Lionello and Pedlosky 2000b)

$$\pi(\rho,x,y) = P(\rho,y)H(x,y) \tag{8}$$

and eq.(7) and eq.(5) can be solved separately. Defining the function $\alpha(\rho,y)$ as the fractional depth of the density level ρ, $\alpha(\rho,y) = z(\rho,x,y)/H(x,y)$, eq.(5), reduces to a system of two first order equations,

$$c(\rho)\frac{\partial \alpha}{\partial \rho} = f(y)P(\rho,y) \ , \ \alpha(\rho,y) = \frac{1}{g}\frac{\partial P}{\partial \rho} \tag{9}$$

that can be easily solved by standard Runge-Kutta methods. The solution is local in y, which is a parameter not involved in the differentiation, i.e. the stratification in thermocline is determined separately for each latitude in the gyre, by integrating eqs.(9) beginning from the bottom of the thermocline.

The boundary conditions for eq.(9) are given by

$$P(\rho_{Max},y) = g\Gamma \ , \ \alpha(\rho_{Max},y) = 1 - \frac{f(y)}{f_{Max}} \tag{10}$$

The second condition of eq.(10) means that a layer of constant density ρ_{Max} and finite thickness, which resembles some aspects of realistic solutions of the thermocline with small diapycnal diffusion (Samelson and Vallis 1997), lays at the bottom of the continuous thermocline. This feature derives from the conservation of the shallow water potential vorticity for the most dense moving fluid that has been ventilated at an infinite distance from the eastern boundary of the gyre (Lionello and Pedlosky, 2000b).

3. The propagation of a density front inside the thermocline

In this study, the SSDD front is defined by perturbing a linear SSDD. Its location is specified by the parameter ρ_c which enters in the definition of the Coriolis parameter at the outcrop latitude:

$$f_c(\rho) = f_{Max}\left(\frac{\rho - \rho_{min}}{\rho_{Max} - \rho_{min}} - Ae^{-k_c\xi^2}\right) \text{ with } \begin{cases} \xi = \frac{(\rho_c - \rho)(\rho_c - \rho_{min})}{(\rho_{Max} - \rho_{min})(\rho - \rho_{min})} & \text{if } \rho < \rho_c \\ \xi = \frac{(\rho - \rho_c)(\rho_{Max} - \rho_c)}{(\rho_{Max} - \rho)(\rho_{Max} - \rho_{min})} & \text{if } \rho > \rho_c \end{cases},$$
(11)

where the two constants are defined as $A = 0.05$, $k_c = 300$. This expression corresponds to the sum of a linear SSDD and a localized perturbation, centered at the density ρ_c, that vanishes exponentially at the northern and southern boundary of the gyre. Its expression has been chosen to produce a SSDD with a continuous derivative, linear but for a front, whose position at the sea surface could be moved in the meridional direction across the gyre. The front on the southern side (where the SSDD is steep) of the perturbation is compensated by a plateau (where the SSDD is almost flat), on its northern side. The resulting thermocline structure has been analyzed for different locations of the front, corresponding to $\rho_c{}^i = 0.95 - 0.1 \cdot (i - 1)$ and numbered as *front i*, $i = 1, 5$ in this paper. The value $a = (\rho_{Max} - \rho_{min})/\Gamma = 20$ has been adopted, corresponding to $\rho_{Max} - \rho_{min} = 5Kg/m^3$ and $\Gamma = 0.25Kg/m^3$, values which are roughly representative of the average oceanic conditions. The expression of the Ekman pumping is $w_E(y) = -W_E sin(\pi\frac{y}{L})$, where L is the meridional extension of the gyre. Eqs.(9) have been integrated with the step $\Delta\rho = (\rho_{Max} - \rho_{min})/400$. The Coriolis parameter at the outcrop latitude is shown in the left panel of fig.1. The thick line represents the linear SSDD used a reference.

The right panel of fig.1 shows the function $c_{an}^i(\rho)$ obtained using eq.(11) in the expression of eq.(6). The thick line is the function $c_{an}^L(\rho)$ that corresponds to a linear SSDD. The unit adopted for the scale of $c(\rho)$ is $g\Gamma(\rho_{Max} - \rho_{min})f_{Max}$. The insertion of a perturbation introduces a peak and a trough in correspondence to the front and to the plateau, respectively. The effect of the perturbation is confined to a neighborhood of the density $\rho_c{}^i$, and it vanishes exponentially towards the maximum and minimum density of the gyre.

A meridional section of the thermocline is shown in the left panel of fig.2. The lowermost dashed line shows the level of the bottom of the thermocline $H(x, y)$. The scale of the plot is given by $D = [2\rho_0 f_M^2 LW_E/\Gamma\beta g]^{1/2}$, whose order of magnitude is 10^3 meters for typical ocean conditions. The layer of constant density ρ_{Max} and of finite thickness lays at the bottom of the thermocline, between the dashed line and the lowermost continuous thick line. The thin continuous lines show the isopycnal levels. The upper and lower limits of the front (thick continuous lines in fig.2) correspond to the level of the densities where the SSDD gradient is 1.25 the average SSDD gradient. Analogously, the value 0.75 has been used for the limits of the plateau shown by the dash-dot lines in fig.2.

The right panel of fig.2 shows the thermocline depth in the linear case and the differences due to the effect of the fronts, which are, approximately, one order of magnitude

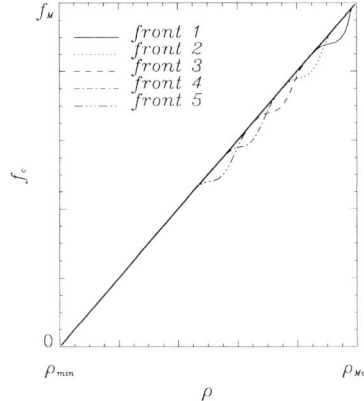 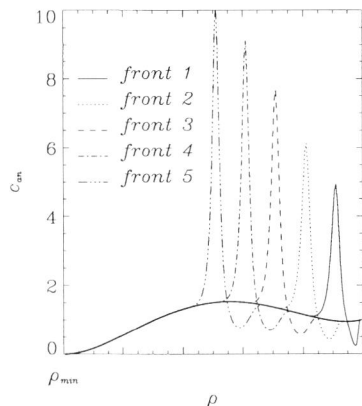

Figure 1. LEFT PANEL: The Coriolis parameter at the outcrop latitude $f_c(\rho)$ for the five SSDDs *front i*, $i=1,...,5$ considered in this study. The continuous thick line shows the linear SSDD. RIGHT PANEL: the five functions $c_{an}(\rho)$, corresponding at the 5 different fronts as function of density ρ. The continuous thick line shows $c_{an}(\rho)$ for the linear SSDD.

smaller. With respect to the linear profile the plateau implies the injection of a more dense fluid and the front of a lighter fluid. The two features compensate each other to bring back the SSDD to the linear distribution. Also their effects on the total depth of the thermocline compensate each other. The injection of more dense fluid deepens the level of the thermocline bottom while the injection of lighter fluid produces a more shallow level, so that no variation is present far from the SSDD perturbation.

The left panel of fig.3 shows the penetration of the 5 fronts, by plotting the level of the isopycnal interfaces that mark their upper and lower limits. The lower dashed line shows the bottom of the thermocline when the reference linear SSDD is used. The maximum penetration depth of the front obviously increases as the outcrop latitude of the front moves to the North.

The right panel of fig.3 shows the thickness of the fronts in the interior of the thermocline. The initial linear rise, the northern side of each curve, corresponds to the outcrop of the front. After their complete subduction, the fronts 3, 4 and 5 shrink, while front 1 and 2 broaden. This is associated with their different location because front 1 and 2 outcrop in a region where the bottom of the thermocline deepens and the whole thickness increases.

4. Conclusions

This study used the LP model for the analysis of the effect of the presence of a SSDD front on the vertical structure of the ocean thermocline. The LP model, though adopting an idealized basin geometry and a simplified dynamics, consistently simulates the penetration of the front inside the ocean. A main result of this study is to show the effectiveness of the LP model for the analysis of the thermocline structure.

The front behavior in the interior of the thermocline depends on the latitude where it has been subducted. A SSD front located near the northern boundary of the gyre

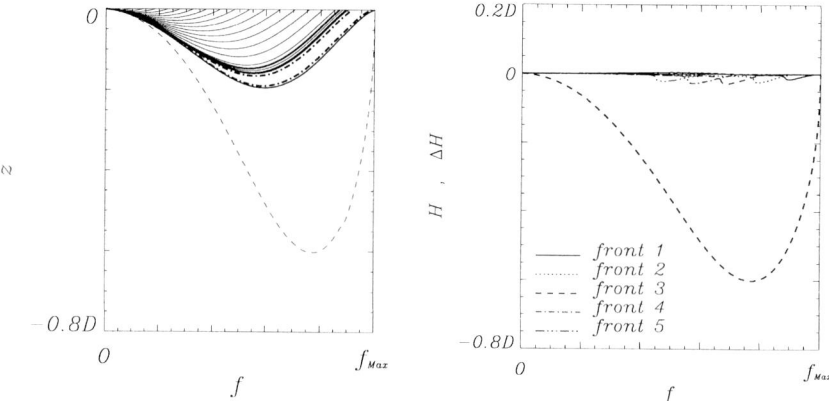

Figure 2. LEFT PANEL: Meridional section of the thermocline for *front 1*. The lowermost dashed line is the bottom of the thermocline, $-H$. The thick line is the isopycnal interface $z(\rho max)$. The remaining lines are $z(\rho_{min} + 0.05i(\rho_{Max} - \rho_{min}))$, , $i = 1, ..., 20$. The limits of the front and of the plateau are marked by the continuous and dash-dot lines respectively. RIGHT PANEL: Effect of the fronts on the total depth of the thermocline as function of f. The lower dashed line represents the bottom of the thermocline when the SSDD is linear. The five curves represents the variation of the level of the bottom of the thermocline due to the effect of the fronts. Negative values indicate a deeper thermocline.

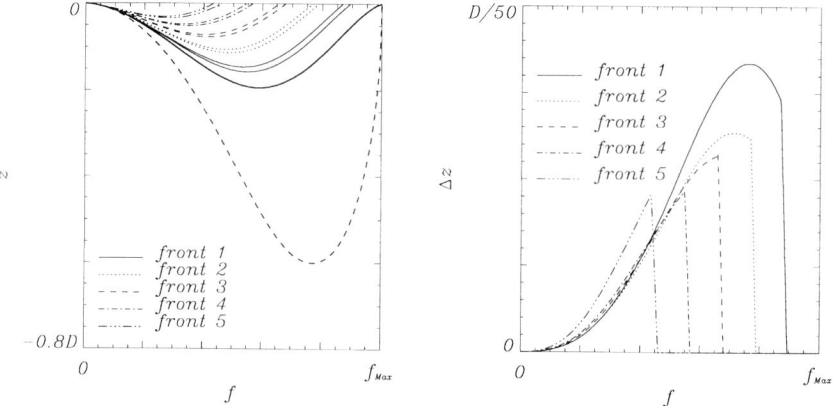

Figure 3. LEFT PANEL: Penetration of the fronts inside the thermocline. The thick line is the isopycnal interface $z(\rho max)$. The lower dashed line represents the bottom of the thermocline when the SSDD is linear. The curves represents the location of the fronts for the five SSDDs analyzed in this study. RIGHT PANEL: Thickness of the five fronts in the interior of the thermocline as function of the Coriolis parameter f. Note that the vertical scale is $D/50$.

broadens while penetrating into the ocean interior, while a SSD front located in the central and southern part of the gyre shrinks. This behavior plainly reflects the behavior of the thermocline bottom. The broadening of the front is associated with the thickening of the thermocline where, in the northern portion of the gyre, its bottom deepens. The shrinking of the front is associated with the reducing thickness of the thermocline, which becomes shallow towards the equator.

These results show that if the fluid motion were analyzed using a passive tracer, the thickness of the layer where the tracer is localized would grow in the northern part of the thermocline and decrease in its southern part. In fact, the thickness of the northernmost analyzed front has an, approximately, 15% increases over 10% of the meridional extension of the gyre. The broadening of the layer where the tracer is localized is not due to cross-diapycnal diffusion, and different growth rates are not related to different diffusion coefficients, but they derive from the dynamical effect of the location of the subduction latitude. In fact, observations (e.g. Ledwell 1993) show that the diffusivity in the thermocline is of the order of 10^{-5} m^2/sec. This means the advective time is much shorter than a diffusion time (the basic premise for the ventilated thermocline) and thus we expect a tracer distribution to be dominated by advection.

This model has moreover identified perturbations associated to the presence of the front in the thermocline structure. The thermocline depth is decreased by the input of lighter fluid corresponding to the front and increased by the input of dense fluid corresponding to the plateau. This changes are not large and they are restricted to a zone downstream the surface location of the SSD perturbation. Away from this region, the thermocline maintains the same bottom depth as for the linear SSDD, in spite of the presence of the front.

References

Huang, R.X., 1988. On boundary value problems of the ideal-fluid thermocline. *J.Phys. Ocean.*, **18**, 619-641.

Killworth P.D. 1987. A Continuously Stratified Nonlinear Ventilated Thermocline, *J. Phys. Oceanogr.*, **17**, 1925-1943

Ledwell,J. Watson, A.J. and Law, C.S. 1993 Evidence for slow mixing across the pycnocline form an open-ocean tracer release experiment. *Nature*, **364**, 701-703.

Lionello P and J. Pedlosky, 2000a. The role of a finite density jump at the bottom of the quasi-continuous ventilated thermocline. *J. Phys. Oceanogr.* **30**, 338-351

Lionello P and J. Pedlosky, 2000b. On the relation between the potential vorticity and the Montgomery function in the ventilated ocean thermocline *J. Phys. Oceanogr.* accepted

Luyten J.R., J.Pedlosky, H.Stommel, 1983. The Ventilated Thermocline. *J. Phys. Oceanogr,*. **13**, 292-309.

Pedlosky J. 1996. *Ocean circulation theory.* Springer-Verlag , Heidelberg, Berlin, Germany 453pp.

Raffaele, F and D.L. Rudnick 2000, Thermohaline variability in the upper ocean, *Journal of Geophys. Res.*, **105**, 16857-16884

Samelson, R.M. and G.K.Vallis, 1997. Large scale circulation with small diapycnal diffusion: the two thermocline limit. *J. Marine Res.* 55, 223-275

Non-hydrostatic Barotropic Instability

Mankin Mak
Department of Atmospheric Sciences
U. of Illinois at Urbana-Champaign
Urbana, IL 61801, USA

1. Introduction

This study is motivated by the need of explaining the formation of a class of common but small atmospheric disturbances known as non-supercell tornado (NST). Their properties have been well documented (e.g. Wakimoto and Wilson ,1989). The local environment that spawns a NST typically has a significant horizontal shear, a boundary outflow, a weak vertical shear and is weakly stratified. It is hypothesized that NST could arise from an instability of a horizontal shear flow with respect to non-hydrostatic columnar disturbances in the presence of implicit buoyancy related to a boundary outflow. The objective of this study is twofold. One is to delineate the nature of non-hydrostatic barotropic instability mechanism per se. The other is to ascertain the extent to which the working hypothesis is valid.

2. Hamiltonian formulation of the model

Non-hydrostatic columnar disturbances in a layer of constant-density fluid over a flat bottom surface are governed by Green-Nagdi (GN) equations (Salmon, 1998) which can be derived by very different methods. Applying the extended form of Hamilton's principle of least action makes the most instructive derivation. The governing equations can be written as

$$\frac{Du}{Dt} = -g\frac{\partial h}{\partial x} - \frac{1}{3h}\frac{\partial}{\partial x}\left(h^2 \frac{D^2 h}{Dt^2}\right) \qquad (1)$$

$$\frac{Dv}{Dt} = -g\frac{\partial h}{\partial y} - \frac{1}{3h}\frac{\partial}{\partial y}\left(h^2 \frac{D^2 h}{Dt^2}\right) \qquad (2)$$

where $\frac{D}{Dt} = \frac{\partial}{\partial t} + u\frac{\partial}{\partial x} + v\frac{\partial}{\partial y}$, (u,v) is the velocity and $h(x,y,t)$ is the local thickness of the layer. The mass continuity equation is

$$\frac{Dh}{Dt} + h\left(\frac{\partial u}{\partial x} + \frac{\partial v}{\partial y}\right) = 0 \qquad (3)$$

The term $-\frac{1}{3h}\nabla\left(h^2 \frac{D^2 h}{Dt^2}\right)$ is the horizontal force on a fluid column associated with the non-hydrostatic pressure.

3. Intrinsic properties of wave motions

For a basic state at rest with a mean depth H_o, wave disturbances in the form of $\sim \exp\{i(kx + \ell y - \sigma t)\}$ exist in this system. Their dispersion relation has three roots,

$$\sigma = \pm \left\{ \frac{gH_o(k^2 + \ell^2)}{1 + \frac{1}{3}H_o^2(k^2 + \ell^2)} \right\}^{1/2} \tag{4a}$$

$$\sigma = 0 \tag{4b}$$

The roots (4a) indicate that the non-hydrostatic process slows down the gravity wave modes, especially for short wavelengths. The root (4b) represents a degenerate vorticity mode in the absence of a basic shear flow.

4. Instability analysis of perturbation in a shear flow

We consider a basic flow that has a velocity scale U_o and a length scale L. The basic thickness of the layer has a scale H_o and two length scales A and B. The domain width is $2Y$. With g being the reduced gravity, we can construct 5 independent non-dimensional parameters such as

$$\eta = \frac{gH_o}{U_o^2}, \quad \gamma = \frac{H_o}{L}, \quad b = \frac{B}{L}, \quad a = \frac{A}{H_o}, \quad \tilde{Y} = \frac{Y}{L}$$

η is the inverse of Froude number. Specifically, we consider the following basic state

$$U(y) = -\tanh(y) \tag{5a}$$

$$H(y) = 1 + a \cdot \tanh\left(\frac{y}{b}\right) \tag{5b}$$

The non-dimensional perturbation equations are

$$\frac{Du'}{Dt} + v'\frac{dU}{dy} = -\eta\frac{\partial h'}{\partial x} - \frac{\gamma^2}{3H}\frac{\partial}{\partial x}\left(H^2 \frac{D}{Dt}\left(\frac{Dh'}{Dt} + v'\frac{dH}{dy}\right)\right) \tag{6a}$$

$$\frac{Dv'}{Dt} = -\eta\frac{\partial h'}{\partial y} - \frac{\gamma^2}{3H}\frac{\partial}{\partial y}\left(H^2 \frac{D}{Dt}\left(\frac{Dh'}{Dt} + v'\frac{dH}{dy}\right)\right) \tag{6b}$$

$$\frac{Dh'}{Dt} + v'\frac{dH}{dy} = -H\left(\frac{\partial u'}{\partial x} + \frac{\partial v'}{\partial y}\right) \tag{6c}$$

The boundary conditions are $v' = 0$ at $y = \pm 8$. We can readily analyze the instability properties of normal mode disturbances in the form of $(u', v', h') = (\hat{u}(y), \hat{v}(y), \hat{h}(y))\exp[i(kx - \sigma t)]$.

5. Impact of non-hydrostatic process on the instability

The results are for the parameter conditions: $\eta = 3$, $\gamma = 4$, $a = 0.25$ and $b = 1$. Figure 1a shows the variation of growth rate of the first six leading eigenvalues (σ_i) with wavelength under hydrostatic condition. There is only one dominant branch of unstable modes for this monotonic shear flow. Figure 1(b) shows the variation of the phase speed (σ_r / k) of all normal modes in the range of ± 1.6 with wavelength. It reveals three distinct sets of normal modes in the system. The phase speeds of the gravity waves lie outside the range of those of most vorticity waves. The dominant unstable modes have slow phase speeds and may be interpreted as arising from wave resonance of vorticity waves. The ensemble of constituent vorticity waves may be thought of in terms of two equivalent counter-propagating waves. The phase-locking between them enables the structure of the normal mode as a whole to remain unchanged in time while the amplitude intensifies exponentially.

The counterpart results under non-hydrostatic condition are presented in Fig. 2. The growth rate of the dominant branch of unstable modes is slightly reduced by the non-hydrostatic effect (~10%). Figure 2a shows that the dominant growth rate curve appears to be made up of two distinct parts, merging together at non-dimensional wavelength D=12. The modes from D= 8 to D=12 may be called *gravity-vorticity hybrid modes* since they arise from resonant interaction between gravity and vorticity waves. The non-hydrostatic effect allows this to occur by sufficiently slowing down the gravity waves. The modes longer than D=12 arise from wave resonance among vorticity waves alone and are referred to as *vorticity modes*.

The structure of a *vorticity mode* (D=18) (perturbation zonal velocity u', meridional velocity v', height h' and vertical velocity at the free surface w') is contrasted with that of a *gravity-vorticity hybrid mode* (D=11). The northern and southern halves of the longer unstable wave (Fig.3, D=18) are essentially similar. The northern and southern halves of the shorter unstable mode (Fig.4, D=11) are qualitatively different. While the southern half of the structure is similar to that of the D=18 mode, the northern half of the structure has a distinctly greater SE-NW tilt in each field. The h' field of the gravity-vorticity hybrid mode is almost three times stronger than that of the vorticity mode (0.55 vs 0.2). The w' field of the hybrid mode is 4.5 times stronger than that of the vorticity mode (0.36 vs 0.08). The overall instability properties of an unstable gravity-vorticity hybrid mode are therefore most reminiscent of those of a NST.

6. Conclusion

The non-hydrostatic process reduces the growth rate of the unstable modes and slows down the gravity waves. The short gravity waves can be sufficiently slowed down that they may resonantly interact with the slow-moving vorticity wave in the shear zone. This leads to the formation of strongly unstable gravity-vorticity hybrid modes. It is justifiable to suggest with caveats that the non-hydrostatic barotropic instability mechanism captures an essential aspect of NST genesis.

7. Acknowledgment

This research is supported by the NSF-USA through grant ATM-9815438.

8. References

Wakimoto, R. and J.W.Wilson, 1989: Non-supercell tornadoes. *Mon. Wea. Rev.,* **117**, 1113-1140.
Salmon, R., 1998: Lecture on Geophysical Fluid Dynamics. Oxford University Press, pp.378.

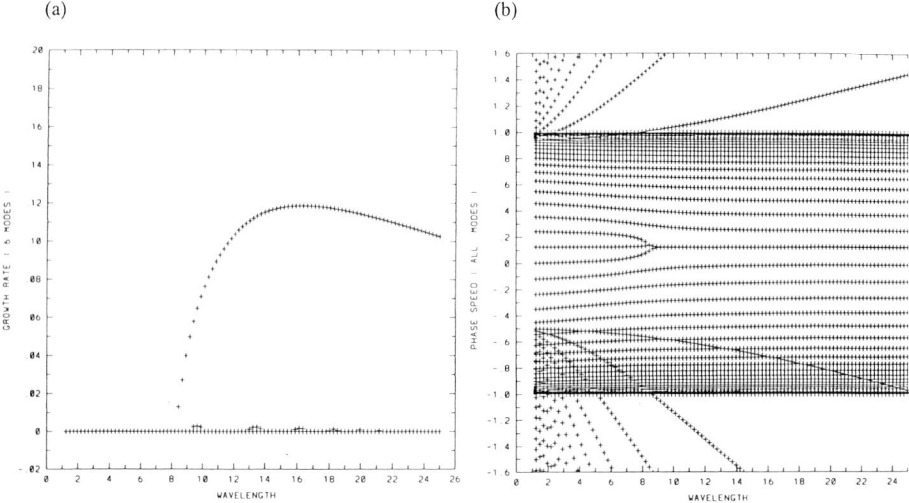

Figure 1. Variation of (a) growth rate of the six leading unstable modes, and (b) phase speed of all normal modes with zonal wavelength under hydrostatic condition for the reference case. Units of (a) and (b) are $L^{-1}U_o = 3x10^{-2}$ s^{-1} and $U_o = 15$ ms^{-1} respectively.

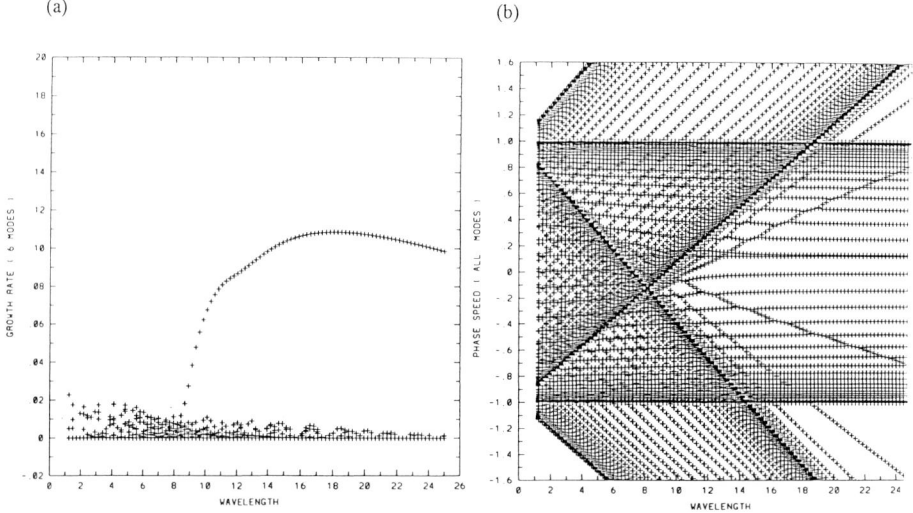

Figure 2. Variation of (a) growth rate of the six leading unstable modes, and (b) phase speed of all normal modes with zonal wavelength under non-hydrostatic condition for the reference case. Units of (a) and (b) are $L^{-1}U_o = 3x10^{-2}$ s^{-1} and $U_o = 15$ ms^{-1} respectively.

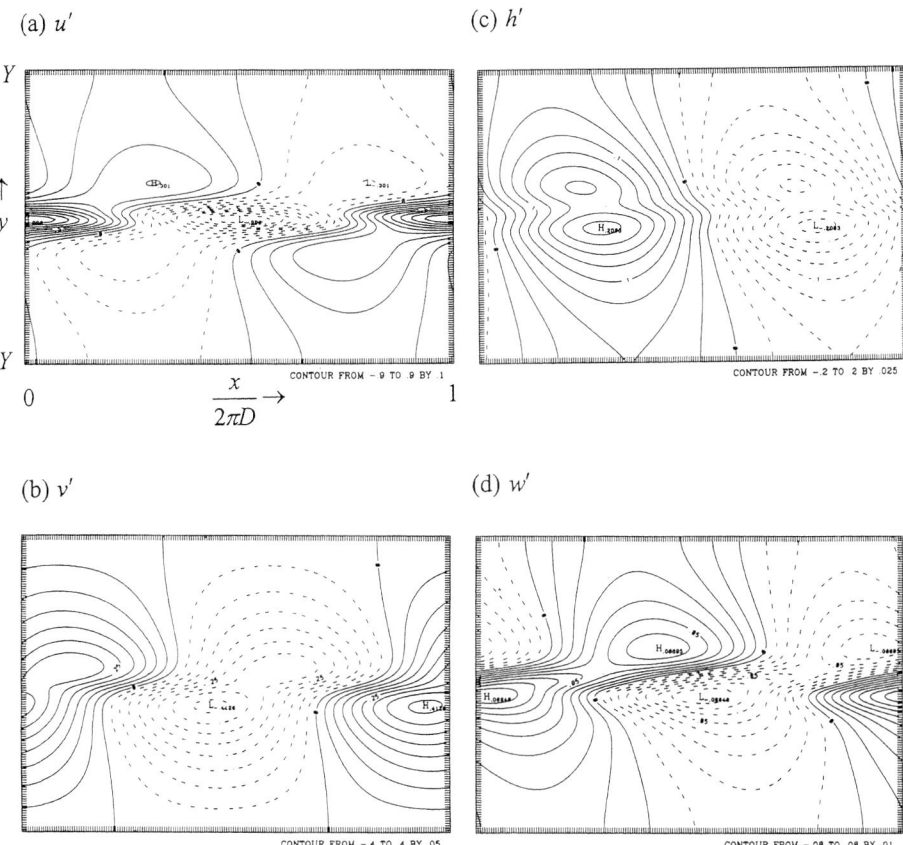

Figure 3. Structure of a sample unstable *vorticity mode* with wavelength D=18 for the reference case : (a) perturbation zonal velocity u', (b) perturbation meridional velocity, (c) perturbation height, and (d) perturbation vertical velocity. Normalized to have $u'_{max} = 1$.

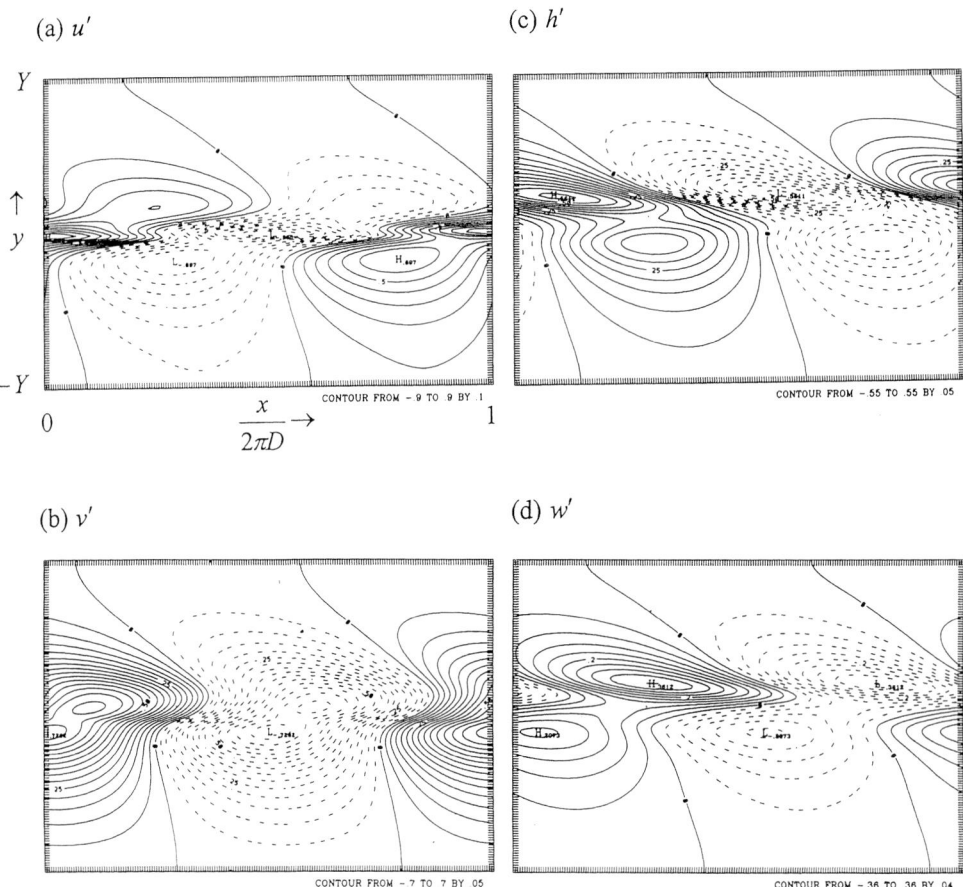

Figure 4. Structure of a sample unstable *gravity-vorticity hybrid mode* with wavelength D=11 for the reference case : (a) perturbation zonal velocity u', (b) perturbation meridional velocity, (c) perturbation height, and (d) perturbation vertical velocity. Normalized to have $u'_{max} = 1$.

THE CSIRO CONFORMAL-CUBIC ATMOSPHERIC GCM

JOHN L. MCGREGOR AND MARTIN R. DIX
CSIRO Atmospheric Research
PB1 Aspendale 3195 Australia

> A grid was devised cubic-conformal
> Which some thought was rather abnormal
> But by techniques semi-Lagrangian
> With conversions Cartesian
> The results turn out better than normal

1. Introduction

Global atmospheric models are usually formulated upon latitude-longitude grids. Near the poles, these grids have disproportionately high resolution, which may severely constrain the time step of integration or require special filtering. Advection problems may also occur near the poles of such grids. Quite recently, a global conformal-cubic grid was devised by Rančić et al. (1996). This grid avoids the disadvantages of latitude-longitude grids, but does require careful selection of numerical techniques to account for the eight vertices of the grid (McGregor, 1996).

At CSIRO, a semi-Lagrangian atmospheric general circulation model (GCM) has been successfully developed on the conformal-cubic grid, complete with physical parameterizations. This paper describes several aspects of the model formulation, and the model performance will be demonstrated for the early part of a simulation being undertaken for the AMIP II atmospheric GCM intercomparison experiment.

2. Formulation

The primitive equations on the grid resemble those on a polar stereographic projection, except that the map factor, m, takes values appropriate to the conformal-cubic grid:

$$\frac{\partial u}{\partial t} + mu\frac{\partial u}{\partial x} + mv\frac{\partial u}{\partial y} + \dot{\sigma}\frac{\partial u}{\partial \sigma} + m\frac{\partial \phi}{\partial x} + mRT\frac{\partial \ln p_s}{\partial x} = fv + N_u \quad (1)$$

      Horizontal   Vertical   Pressure   Coriolis Physics
      advection   advection   gradient   force

$$\frac{\partial v}{\partial t} + mu\frac{\partial v}{\partial x} + mv\frac{\partial v}{\partial y} + \dot{\sigma}\frac{\partial v}{\partial \sigma} + m\frac{\partial \phi}{\partial y} + mRT\frac{\partial \ln p_s}{\partial y} = -fu + N_v \quad (2)$$

      Horizontal   Vertical   Pressure   Coriolis Physics
      advection   advection   gradient   force

$$\frac{\partial T}{\partial t} + mu\frac{\partial T}{\partial x} + mv\frac{\partial T}{\partial y} + \dot{\sigma}\frac{\partial T}{\partial \sigma} - \frac{RT}{c_p\sigma}\frac{\omega}{p_s} = N_T \quad (3)$$

      Horizontal   Vertical   Adiabatic Physics
      advection   advection   term

$$\frac{\partial \phi}{\partial \sigma} = -\frac{RT}{\sigma} \qquad (4)$$

$$\underbrace{\frac{\partial \ln p_s}{\partial t} + mu \frac{\partial \ln p_s}{\partial x} + mv \frac{\partial \ln p_s}{\partial y}}_{\text{Horizontal advection}} + \underbrace{\frac{\partial \dot{\sigma}}{\partial \sigma}}_{\text{Vertical advection}} + \underbrace{m^2 \left[\frac{\partial (u/m)}{\partial x} + \frac{\partial (v/m)}{\partial y} \right]}_{\text{Divergence}} = 0 \qquad (5)$$

where u and v are horizontal velocity components on the grid, T is temperature, ϕ is geopotential height, σ is the normalized-pressure vertical coordinate, p_s is surface pressure, and R is the gas constant. A semi-Lagrangian semi-implicit procedure is adopted, where the horizontal material time derivatives are evaluated along trajectories, whilst the other terms are averaged along the trajectories. The semi-Lagrangian technique (as reviewed by Staniforth and Côté, 1991) allows large advective time steps. The calculation of the departure points follows the procedure advocated by McGregor (1993). To avoid turning problems, which might lead to errors near the vertices, the vector equations for the horizontal wind components (u, v) are solved in terms of the three equivalent equations for the corresponding three-dimensional Cartesian wind components; the updated values are projected back on to the surface of the sphere to give the updated values of (u, v). This Cartesian representation is also used during the calculation of horizontal diffusion. Although weak horizontal diffusion is employed to represent the cascade of energy to smaller scales, the model performs without spurious noise even for zero horizontal diffusion.

As is common for semi-implicit models, the discretized primitive equations are linearized and combined using vertical eigenvectors. This leads to a set of vertically-decoupled Helmholtz equations in terms of a height variable. The grid lends itself to solving each of these Helmholtz equations by efficient three-colour successive over-relaxation; the rationale for the vectorization is similar to that of two-dimensional red-black schemes (e.g., Young, 1971). The three-colouring of the grid is illustrated in Fig. 1 for a very coarse C5 grid (5×5 grid points on each panel). During each iteration all the points of the same colour are updated before proceeding to the other colours.

Figure 1. An example of a conformal-cubic grid. The shading illustrates the three-colour arrangement used for solution of the Helmholtz equations.

Another unusual and attractive feature of the model is the use of reversible staggering for the wind components. All model variables are located and stored at the centre of the grid cells, already illustrated in Fig. 1. However, during the semi-implicit solution procedure it is desirable to transform the wind components so that they are located normal to the mid-points of the cell boundaries, in an Arakawa-C configuration; this is the preferred arrangement for accurate calculation of divergence. A compact procedure has been devised to allow this transformation to be performed reversibly between the mid-points and centres of the cells; the cyclic properties of the conformal-cubic grid are beneficial for this procedure. It can be shown that superior dispersion properties ensue from the reversible-staggering treatment of the winds. A fairly comprehensive set of physical parameterizations has been incorporated into the model, as listed in Table 1.

TABLE I.   Physical parameterizations

GFDL parameterization for long and short wave radiation (Schwarzkopf and Fels, 1991)
Interactive diagnosed cloud distributions
Arakawa/Gordon mass-flux cumulus convection scheme (McGregor et al., 1993)
Tiedtke shallow convection
Evaporation of rainfall
Stability-dependent boundary layer and vertical diffusion, following Louis (1979)
Deformation-based horizontal diffusion (Smagorinsky et al., 1966)
Vegetation/canopy scheme with six layers for soil temperatures and soil moistures.

3. Results

In order to demonstrate the model's capabilitites, we have commenced simulations for the AMIP II intercomparison (for more information see http://www-pcmdi.llnl.gov/amip/), where sea surface temperatures and sea-ice cover are prescribed from 1979 to 1996. A C48 grid is being used, as shown in Fig. 2, having an average resolution of about 200 km; the model has 18 vertical levels.

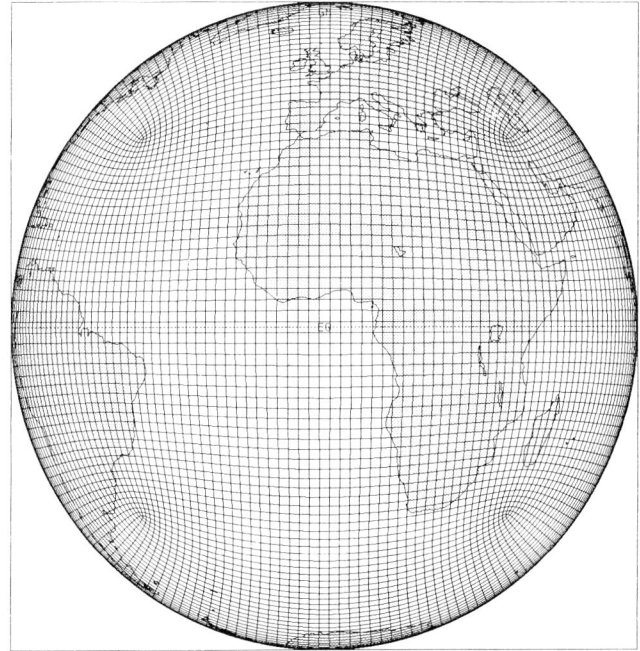

Figure 2.   The C48 conformal-cubic grid.

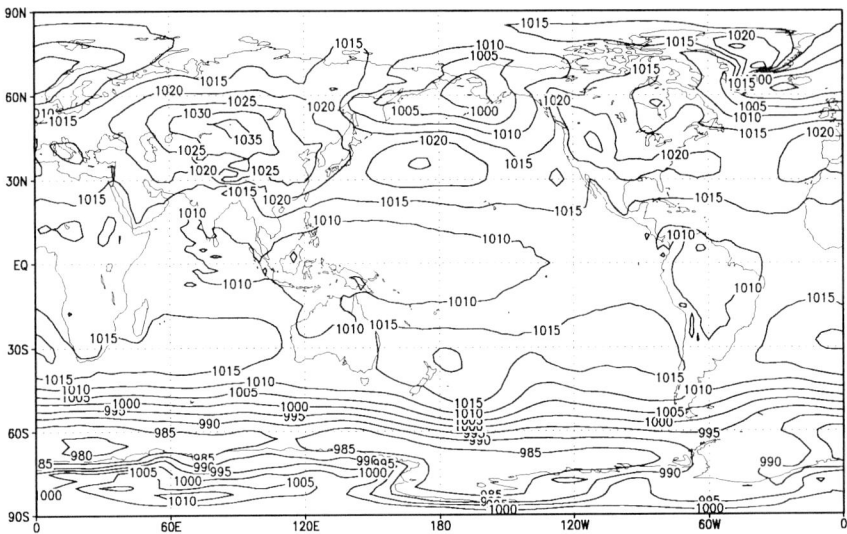

Figure 3a. Mean sea level pressures for November 1979 from the NCEP reanalysis.

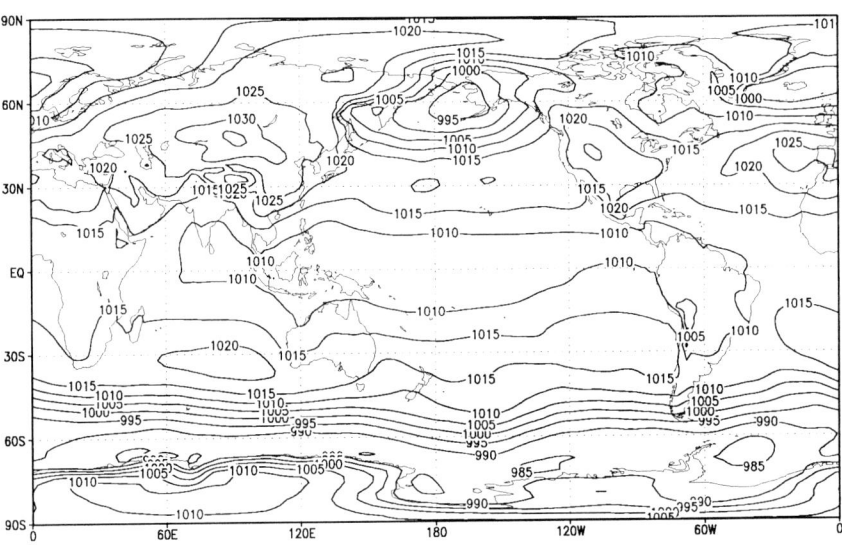

Figure 3b. Mean sea level pressures for November 1979 from the C48 model simulation.

Several months of simulation have been performed so far, starting November 1 1979. Figure 3a shows the average mean sea level pressures for that month from the NCEP reanalysis, whilst Fig. 3b shows the simulated values. There is good agreement for important features. For example, there are strong pressure gradients near Antarctica, there is a well-defined Aleutian low, and an acceptable Icelandic low. Figure 4a show the average observed precipitation for the same month (Xie and Arkin, 1996), whilst Fig. 4b shows the modelled field. Again there is good agreement of the tropical rainfall patterns and the precipitation over each of the continents, although the rainfall over western Australia is somewhat excessive.

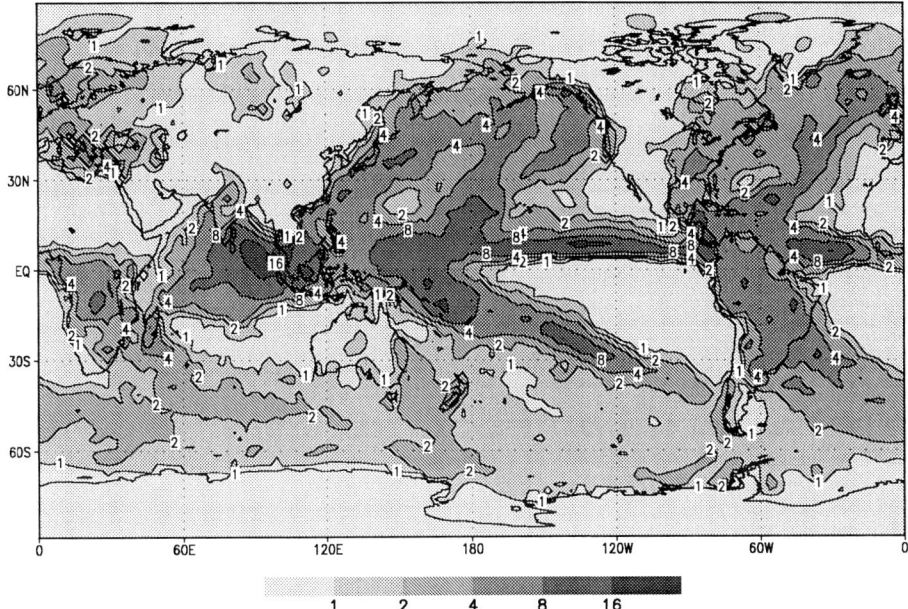

Figure 4a. Rainfall for November 1979 from the Xie and Arkin analysis (mm/day).

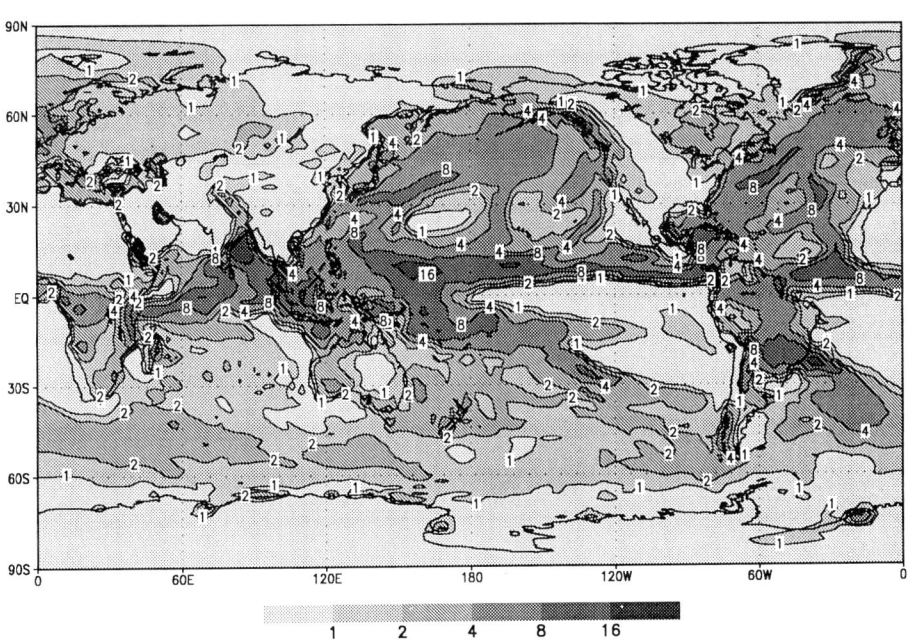

Figure 4b. Rainfall for November 1979 from the C48 model simulation (mm/day).

4. Conclusions

A primitive equations model has been formulated on the conformal-cubic grid, with a fairly complete set of physical parameterizations. Simulations are being undertaken for the AMIP 2 intercomparison. Results so far are very encouraging.

The following advantages are anticipated compared to spectral GCMs (these advantages are probably also available to other grid-point models):

- there is freedom from non-zero ocean orography and other effects caused by the Gibb's phenomenon
- specialized treatment is possible for various horizontal gradient terms near orography
- deformation-based horizontal diffusion schemes can be used.

The following advantages are also anticipated compared to typical grid-point GCMs:

- the grid has relatively uniform resolution, avoiding the requirement for filtering near the poles
- the grid is isotropic
- reversible staggering of the winds is available, providing good dispersion properties
- the grid permits the use of a simple efficient Helmholtz solver
- the panel structure of grid has advantages for code parallelization.

References

Louis, J-F. (1979) A parametric model of vertical eddy fluxes in the atmosphere. *Bound.-Layer Meteor.* **17**, 187–202.
McGregor, J. L. (1993) Economical determination of departure points for semi-Lagrangian models. *Mon. Wea. Rev.* **121**, 221–230.
McGregor, J.L., H.B. Gordon, I.G. Watterson, M.R. Dix and L.D. Rotstayn (1993) The CSIRO 9-level atmospheric general circulation model. *CSIRO Div. Atmospheric Research Tech. Paper No. 26*, 89 pp.
McGregor, J.L. (1996) Semi-Lagrangian advection on conformal-cubic grids. *Mon. Wea. Rev.* **124**, 1311–1322.
Rančić, M., R.J. Purser and F. Mesinger (1996) A global shallow-water model using an expanded spherical cube: Gnomonic versus conformal coordinate. *Quart. J. Roy. Meteor. Soc.* **122**, 959–982.
Schwarzkopf, M.D. and S.B. Fels (1991) The simplified exchange method revisited: an accurate, rapid method for computation of infrared cooling rates and fluxes. *J. Geophys. Res.* **96**, 9075–9096.
Smagorinsky, J., S. Manabe and J.L. Holloway (1965) Numerical results from a nine-level general circulation model of the atmosphere. *Mon. Wea. Rev.* **93**, 727–768.
Staniforth, A. and J. Côté (1991) Semi-Lagrangian integration schemes for atmospheric models – A review. *Mon. Wea. Rev.* **119**, 2206–2223.
Xie, P.P. and P.A. Arkin (1996) Analyses of global monthly precipitation using gauge observations, satellite estimates, and numerical model predictions. *J. Climate* **9**, 840–858.
Young, D.M. (1971) *Iterative solution of large linear systems.* New York: Academic Press, 570 pp.

ON THE ORIGIN OF HELICAL FLOW STRUCTURES

An Introduction to Relativisitc Quantum Fluid Dynamics

N. NAWRI
Department of Meteorology
University of Maryland
College Park, MD 20742
USA
nnawri@essic.umd.edu

1. Introduction

Whether concerned with the very small, as in elementary particle physics, or with the very large, as in cosmology, relativistic quantum theories are at the forefront of modern physics. Since the unifying principles of physics are symmetry considerations, their validity can not depend on specific numerical values, i.e., they can not rely on arbitrary concepts of small and large or fast and slow. If one theory can be proved to be mathematically analogous to another, their solutions must be the same independent of physical scales. In that spirit, special relativity in the widest sense can be defined as *compliance with causality under the assumption that the interaction between events in spacetime is mediated at finite speed.* Similarly, quantum mechanics on a basic level, simply means a transition away from deterministic force laws, or *dynamics* in the classical sense, to wave mechanics in which *kinematical* properties such as energy and momentum play the most important role.

The focus of this paper is on the investigation of symmetry principles that govern the formation of strongly helical flows by establishing analogies with relativistic quantum field theories and cosmology. In a geophysical context the most prominent example of a helical flow structure is the tornado. Tornadolike vortices are known to occur not only in very different geographical regions but also under widely different synoptic situations. The correct description of cloud processes as well as most initial conditions that are needed for a full dynamical investigation of tornadogenesis are unknown. It is hoped, therefore, that a kinematical theory will allow a unified treatment of intense atmospheric vortices and their improved forecasts.

2. Flow Correlation

Considering a small fluid element and neglecting its deformation as it moves with the flow, as for any solid body, two three-dimensional vectors are necessary to completely specify its kinematical state. These are the velocity, v, of the centre of mass and the spin vector describing rotation about it. By making the transition from point to continuums mechanics, the rotational properties of infinitesimally small fluid elements, represented by the vorticity vector, $\boldsymbol{\omega} \equiv \boldsymbol{\nabla} \times \boldsymbol{v}$, become encoded in the spatial dependence of the velocity field.

The kinematical description of a flow can be generalized by introducing scalar quantities such as kinetic energy $K \equiv \boldsymbol{v} \cdot \boldsymbol{v}$, rotational energy or enstrophy $E \equiv \boldsymbol{\omega} \cdot \boldsymbol{\omega}$, and helicity density $h \equiv \boldsymbol{v} \cdot \boldsymbol{\omega}$. Although information about the absolute orientation of velocity vectors is lost, the three scalar variables still contain all information about the relative orientation of velocity and vorticity as well as their magnitudes. With α being the angle between velocity and vorticity, the *alignment* between the two vectors is given by $\cos\alpha = \frac{h}{\sqrt{KE}}$, from which it follows that $h^2 \leq KE$, where the equality holds if and only if α is equal to zero or pi. Defining *chirality*, χ, to be the ratio of helicity to kinetic energy,

$$\chi^2 \leq EK^{-1} \qquad (1)$$

must also be satisfied. The spatial average of alignment can be defined as the correlation coefficient of velocity and vorticity which is a measure of the linear relationship between the two vector fields in a given domain $\mathcal{D}$. For a correlation coefficient of ± 1 velocity and vorticity are perfectly aligned, i.e., there exists a function $f(\boldsymbol{x}, t)$ such that $\boldsymbol{\omega} = f\boldsymbol{v}$ in $\mathcal{D}$. This ideal situation is known as Beltrami flow in which flow parcels can be thought of as particles moving along their spin axis. Scalar multiplying this equation with $\boldsymbol{v}$, f is found to be identical with chirality. A helical flow is defined as one having nonvanishing values of chirality over finite domains, and chirality, by (1), is maximized by Beltrami flows for given values of K and E.

3. Mirror Symmetry

In fluid kinematics it is necessary to distinguish two types of vectors: *true vectors* such as velocity that can be determined without referring to frame dependent conventions by simply connecting two different points in space-time, and *pseudo-vectors* such as vorticity which point in opposite directions if determined by a right- and left-handed observer. While *true scalars* such as kinetic energy and enstrophy are invariant upon frame inversion, a scalar quantity that changes sign under a mirror transformation, such as helicity, is defined as a *pseudo-scalar*.

For the motion of fluids *symmetry of the physical laws under mirror transformations*, or *parity invariance*, can be proved under fairly general conditions. Following Truesdell (1954; pp. 27, 120-121) a continuously differentiable velocity field can locally be represented in terms of three Monge's potentials θ, ϕ, and ψ by $\boldsymbol{v} = \theta \boldsymbol{\nabla} \phi + \boldsymbol{\nabla} \psi$, for which

$$h = \boldsymbol{\nabla} \cdot \psi \boldsymbol{\omega}. \tag{2}$$

Using Gauß' law of integration it can then be shown that, *independent of the dynamics or the material properties of the fluid*, the total (or average) helicity in the whole domain occupied by the fluid must be zero at all times if vorticity is parallel to all finite boundaries and decreases in magnitude sufficiently fast along all unbounded directions. With the no-slip boundary condition for atmosphere and ocean at the Earth's solid surface together with the fact that the atmosphere only has a finite extend these conditions are satisfied. The net result of all forces acting on a geophysical fluid is such that, *in a statistical sense*, neither right- nor left-handed flow structures are generated preferentially, i.e., they conserve parity. True and pseudo-vectors must therefore appear in the dynamical equations in such a way that the effects of a parity transformation cancel out.

However, in an open sub-region of the domain, as the example of a tornado shows, average helicity can be nonzero. In that region helicity can only be changed by transport across the boundaries. Since parity conservation requires that average helicity in the whole domain must be zero *at all times*, locally helicity can only be generated if at the same time a flow structure is generated with an equal amount of helicity but of the opposite sign. Relativity in turn does not allow for instantaneous interactions over finite distances since the speed of propagation of information in the fluid must be finite. The global constraint of helicity conservation together with the requirement of a causal connection between events therefore leads to a local symmetry law: *helical flow structures must locally be generated or destroyed in pairs with opposite signs of helicity*. An example of this pair production of helicity can be observed at the early stages of storm formation when a counter-rotating updraft pair is formed in a vertical shear environment and is split subsequently into two updraft-downdraft couplets with opposite helicity.

This is the same situation known from quantum electrodynamics (QED) where, again due to charge conservation and symmetry requirements, electrons and their positively charged antiparticles, the positrons, are produced or annihilated in pairs by absorbing or emitting a photon. Interpreting, therefore, helicity as electric charge density, (2) is recognized as Gauß' law of electrodynamics with an "electric field" $\boldsymbol{E} \equiv \psi \boldsymbol{\omega}$. Divergence of field lines of $\boldsymbol{E}$ around regions of positive helicity must exactly be compensated for by convergence around regions of negative helicity. Now it is possible to construct a space that is locally helicity-free ("vacuum state") in which the manifold, $\mathcal{M}$, of the nondivergent field lines of $\boldsymbol{E}$ has the topology of a *wormhole* (Wheeler, 1968). The appearance of positive and negative charges in the regular three-dimensional subspace, $\mathcal{S}^3$, and their causal connection is then just a result of this topology. In the following section an expression for $\mathcal{M}$ is derived such that its projection onto $\mathcal{S}^3$ reproduces the divergence of field lines of $\boldsymbol{E}$ consistent with the observed helicity there.

4. Spacetime Curvature

To simplify the analysis, consider a two-dimensional distribution of $\mathcal{S}^3$ with the parameterization $\boldsymbol{x} = \boldsymbol{x}(\tau,\sigma)$ of each surface and corresponding tangent vectors $\boldsymbol{x}_\tau \equiv \boldsymbol{a}$ and $\boldsymbol{x}_\sigma \equiv \boldsymbol{b}$. For $\boldsymbol{a}$ and $\boldsymbol{b}$ to be a coordinate basis, their Lie derivative, $\mathcal{L}_{\boldsymbol{a}}(\boldsymbol{b}) \equiv (\boldsymbol{a}\cdot\nabla)\boldsymbol{b} - (\boldsymbol{b}\cdot\nabla)\boldsymbol{a}$, must vanish (Schutz, 1980; pp. 47-49, 74-78, 81-82). Assume that $\boldsymbol{a}$ and $\boldsymbol{b}$ span the isosurface of a scalar function Φ such that $\boldsymbol{a}\times\boldsymbol{b} = \nabla\Phi$. Then $\boldsymbol{a}$ and $\boldsymbol{b}$ must be nondivergent, i.e., it must locally be possible to write them as $\boldsymbol{a} = \nabla\Theta\times\nabla\Lambda$ and $\boldsymbol{b} = \nabla\Psi\times\nabla\Omega$, where $\Psi = \Psi_0(\Theta,\Lambda)e^{-\omega(\tau-\tau_0)}$, $\Omega = \Omega_0(\Theta,\Lambda)e^{\omega(\tau-\tau_0)}$, $\Phi = \omega\Psi_0\Omega_0$, and ω is a constant in units of τ^{-1}. Let $\Theta = KE$, $\Lambda = h^2$ and $\Psi_0 = \Omega_0 = \sqrt{(KE)}$, then $\boldsymbol{a} = \nabla KE \times \nabla h^2$ and $\boldsymbol{b} = \nabla KE \times \nabla\omega\tau$ which allows the reparameterization $(\tau,\sigma) \to (\omega\tau, h^2)$. An analogy with cosmology can be established if further τ is associated with complex time it (in units of h^{-2}), h^{-2} with radius r, and $(KE)^{-1}$ with $2m$, where m is the mass of a spherically symmetric body. Then the metric

$$(ds)^2 = -\left(1 - \frac{h^2}{KE}\right)(dt)^2 + \left(1 - \frac{h^2}{KE}\right)^{-1}(dh^{-2})^2 \qquad (3)$$

on the coordinate basis of t and h^{-2} has the same apparent singularity for a Beltrami flow as the *Schwarzschild metric* has at the event horizon of a black hole (Hawking, 1984). All nonvanishing components of the curvature tensor are found to be proportional to plus or minus h^6 (r^{-3}). The particles associated with high-helicity flow regions can therefore be thought of as arising from the curvature of the spacetime defined by (3). On the original surface $\Phi = const.$, however, they are realized as the divergence of $\theta\omega$. Since the metric (3) is determined by h^2, positive and negative regions of helicity can only be distinguished by the direction of these field lines.

5. Helitons

While a topological analogy between small-scale helical flows and elementary particles was established in the previous sections it remains to be investigated in how far helical flow structures, once they are created really behave as charged particles in $\mathcal{S}^3$. To see that, assume that the *unobservable perturbations*, v', from the observable spatial mean flow, $v_0 \equiv \langle v \rangle_\mathcal{D}$, in a domain $\mathcal{D}$ can be represented by a superposition of right- and left-handed Beltrami flow components that preserve their identity over a sufficiently long period of time. With the loose definition of a soliton as a spatially localized, coherent object these flow structures can be identified as helical solitons or, simply, *helitons*. It will also be assumed that these "particles" are *quantized excitations* of $\mathcal{M}$ in the sense that helicity increases by a fixed ratio with respect to kinetic energy, i.e., the absolute value of chirality for every flow component is equal to the same constant χ_0. Then $v' = v_+ + v_-$, where the $v_\pm$ may either be right- or left-handed Beltrami flows depending on the handedness of the chosen frame of reference. They satisfy the respective eigenvector equations $\nabla \times v_\pm = \pm \chi_0 v_\pm$.

By definition, v_0 is the net translational velocity of an observer, $\mathcal{O}$, attached to the "centre of mass" of the small-scale flow structures. In the rest frame of that observer the total velocity $v_R = v - v_0 = v_+ + v_-$. Expressing the positively helical flow component in terms of spatial variables, x, with respect to which $\mathcal{D}$ has periodic boundary conditions and the negatively helical flow component in terms of spatial variables $x' \equiv x - \phi$ results in a frame of reference co-moving with the positive heliton, η^+, whereas the negative heliton, η^-, is allowed to drift across the open boundaries of $\mathcal{D}$ depending on the relative phase shift ϕ between the two helitons. It can be shown that, in the rest frame of $\mathcal{O}$, $G \equiv \langle K \rangle_\mathcal{D} + \chi_0^{-1} \langle h \rangle_\mathcal{D}$ does not depend on ϕ, i.e., it does not depend on the existence and position of the negative heliton. Assuming that η^+ represents a quasi-stationary coherent flow structure, G is a constant of motion for $\mathcal{O}$. This conservation law implies that, in the rest frame of $\mathcal{O}$, without forcing or dissipation, helicity increases while kinetic energy decreases as the oppositely charged helitons drift apart. Conversely, kinetic energy is gained if two oppositely charged helitons annihilate reminding of the situation known from QED, where again helicity is interpreted as charge and kinetic energy as the energy of the light quantum plus the kinetic energy of all particles involved.

In Hamiltonian dynamics, *constants of motion are generators of a group of symmetry transformations*. If helitons can be identified as particles, it should be possible to associate a certain symmetry of the Lagrangian describing their interactions with the quantity G that is conserved during these processes. In a phase space where $\langle K \rangle_\mathcal{D}$ corresponds to the angular momentum about a fixed axis and $\langle h \rangle_\mathcal{D}$ to the linear momentum in that direction, it can be shown that the potential has a helical symmetry with pitch $2\pi \chi_0^{-1}$ (Greiner and Müller, 1994; p. 7). Analysis of the equations of motion then shows that helitons, over short distances, indeed behave like charged particles with the unusual property that like charges attract each other whereas the interaction between unlike charges is repulsive. Since oppositely charged helitons are antiparticles, this mechanism prevents self-destruction and may contribute to the formation of large-scale helical flows.

6. Outlook

It was shown that analogies between helical fluid flow and relativistic elementary particle physics and cosmology can easily be established under very general conditions. To be able to derive criteria for the formation of helical flow structures it now needs to be investigated, in how far helitons are generated by mean flow induced curvature of $\mathcal{M}$, and how they in turn affect the observable flow structure. It is known from quantum cosmology that pair production of elementary particles occurs around black holes (Hawking, 1975) and in regions of strong gravitational curvature in general. With the established analogies a further investigation into this subject is therefore promising. Another crucial aspect that could not be discussed here is *uncertainty*. The observable mean flow will in general not show the symmetry required by the physical laws. Spontaneous symmetry breaking due to undetected particle creation must be expected. The unresolved fluctuations of helicity have the same effect on the topology of $\mathcal{M}$ as quantum fluctuations have on the topology of superspace (Wheeler, 1968). Whether due simply to insufficient measurements or intrinsic to Nature, these unobservable fluctuations are responsible for a certain degree of indeterminism. However, due to page limitations an investigation of these aspects must be the subject of future publications.

Acknowledgment. The research reported in this paper was supported by NASA Training Grant NGT5-30196 to the University of Maryland.

References

Greiner, W., and B. Müller (1994) *Quantum Mechanics - Symmetries*. Springer-Verlag, Berlin, Germany.
Hawking, S. W. (1975) Particle Creation by Black Holes, *Com. Math. Phys.*, **43**, pp. 199–220.
Hawking, S. W. (1984) Quantum Cosmology, in *Relativité, groupes et topologie II*, eds. B. S. DeWitt and R. Stora, North-Holland Physics Publishing, Amsterdam, The Netherlands, pp. 335–379.
Schutz, B. F. (1980) *Geometrical Methods of Mathematical Physics*. Cambridge University Press, Cambridge, UK.
Truesdell, C. A. (1954) *The Kinematics of Vorticity*. Indiana University Press, Bloomington, Indiana, USA.
Wheeler, J. A. (1968) *Superspace and the Nature of Quantum Geometrodynamics*. The Benjamin/Cummings Publishing Company, Menlo Park, California, USA.

BAROCLINIC INSTABILITY OF BOTTOM-DWELLING CURRENTS

MATEUSZ K. RESZKA AND GORDON E. SWATERS
Applied Mathematics Institute
Department of Mathematical Sciences
University of Alberta
Edmonton, Alberta, Canada

1. Introduction

Density-driven benthic flows are important in the dynamics of marginal seas, river estuaries and other coastal regions (LeBlond *et al.*, 1991; Price & O'Neil Baringer, 1994). They often occur along sloping continental shelves, flowing with shallower water on their right (in the northern hemisphere). Mesoscale gravity currents, which are to be discussed in this study, arise from a geostrophic balance between down-slope acceleration due to gravity and the Coriolis force, while their dynamics is characterized by lengthscales on the order of the Rossby deformation radius. There is mounting evidence that such flows are subject to instability, which may drastically alter the mean flow and culminate in a series of isolated plumes or eddies (Armi & D'Asaro, 1980; Houghton *et al.*, 1982).

While previous modelling studies often employed the streamtube approximation (Smith, 1975) or the full primitive equations (Jiang & Garwood Jr., 1995), here we present preliminary numerical results using a two layer, frontal geostrophic model (Poulin & Swaters, 1999). The model allows for continuous stratification in the upper layer and relies on the release of gravitational potential energy due to the gradual downslope slumping of the dense fluid. We investigate the nonlinear spatio-temporal evolution of a baroclinically unstable gravity current on a linearly sloping bottom. The model is also applied to fluctuating deep currents in the Strait of Georgia (located between Vancouver Island and mainland British Columbia, Canada), as described in Stacey *et al.* (1988). The reader is referred to Reszka & Swaters, (2000) for a more thorough examination of the above results.

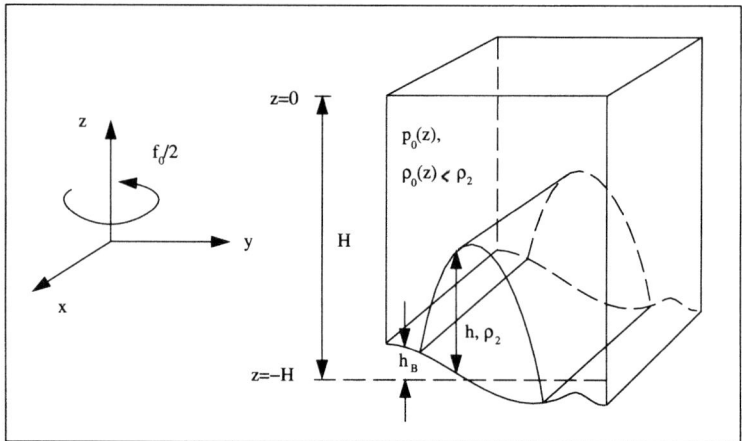

Figure 1. Geometry of the frontal geostrophic model. The upper layer is continuously stratified and the interface is allowed to intersect the topography.

2. Governing Equations

The present model assumes a continuously stratified layer of finite depth overlying a dense, homogeneous layer and sloping (or otherwise varying) bottom topography. The interface between the two layers can intersect the topography (see fig. 1), allowing for isolated patches of dense fluid. The governing equations are derived in an asymptotic expansion of the shallow water equations for the lower layer and the Boussinesq equations for the upper layer. We refer the reader to Poulin & Swaters (1999) for details. The leading order non-dimensional fields on an f-plane are determined by

$$(\Delta\varphi + (N^{-2}\varphi_z)_z)_t + \mu J(\varphi, \Delta\varphi + (N^{-2}\varphi_z)_z) = 0, \tag{1}$$

$$\varphi_{zt} + \mu J(\varphi, \varphi_z) = 0, \quad z = 0, \tag{2}$$

$$\varphi_{zt} + \mu J(\varphi, \varphi_z) + N^2 J(\varphi + h, h_B) = 0, \quad z = -1, \tag{3}$$

$$h_t + J(\mu\varphi + h_B, h) = 0, \quad z = -1, \tag{4}$$

where $\varphi(x, y, z, t)$ is the upper layer geostrophic pressure, $h(x, y, t)$ is the lower layer thickness, $h_B(x, y)$ is the height of the bottom topography, $J(A, B) = A_x B_y - B_x A_y$ and subscripts refer to derivatives unless otherwise specified. Here N is a scaled Brunt-Väisälä frequency while μ is referred to as the interaction parameter and measures the destabilizing effect of baroclinicity relative to the stabilizing influence of topography. In order to gain insight into the basic instability mechanism we assume a simple model configuration, i.e. constant stratification in the upper layer ($N = const.$) and an x-invariant bottom topography ($h_B = h_B(y)$).

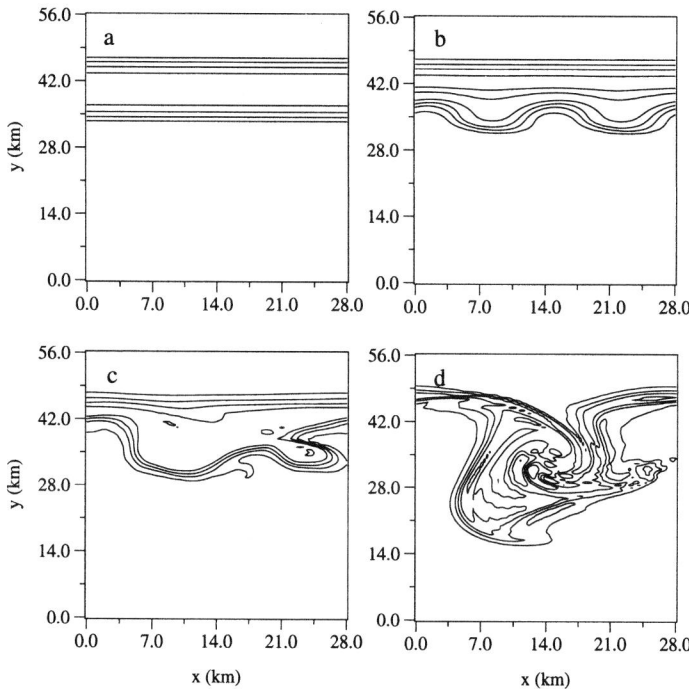

Figure 2. Evolution of lower layer height, h, at a) 0, b) 7, c) 10 and d) 12 days for linearly sloping topography. Contour range is 0–60 m with a contour interval of 15 m.

3. Numerical Solutions

Initially an x-invariant, parabolic profile was specified for the lower layer height. Its half-width was equal to the Rossby radius associated with the upper layer, while its maximum thickness was approximately 10% of the total fluid depth. The upper layer initially consisted of a small-amplitude, random wavefield, which served to excite the instability. The system was integrated forward in time in an x-periodic channel with the no-normal flow boundary condition at the walls. The length of the channel was chosen such that it allowed two or three wavelengths of the most unstable mode. The domain was typically discretized into $120 \times 120 \times 16$ grid nodes, yielding a dimensional resolution of roughly 230 m in the horizontal and 24 m in the vertical.

A simulation with a linearly sloping bottom was performed first. While this is a crude approximation of a continental shelf, it isolates the process of instability we wish to study from additional topographic effects. The evolution of the lower layer thickness for this simulation is given in fig. 2 at day a) 0, b) 7, c) 10 and d) 12. At 7 days the instability is manifested

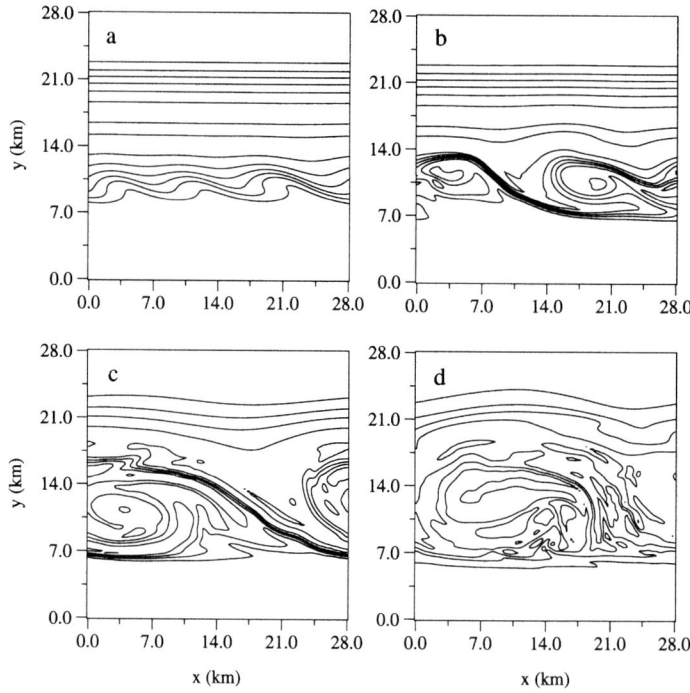

Figure 3. Evolution of lower layer height, h, at a) 8, b) 10, c) 12 and d) 14 days for Strait of Georgia topography. Contour range is 0–60 m with a contour interval of 10 m.

as a growing, periodic deformation of the incropping on the downslope side of the current (the fluid column deepens in the negative y-direction). We estimate e-folding times of about 11 hours and a phase speed of 12 cm/s. The primary source of energy for the growth of perturbations is the potential energy released as the dense fluid slowly descends down the slope (Reszka & Swaters, 2000). The two waves initially seen in the channel then decay and merge into one at 10 days. This longer mode grows until the wave breaks backwards, relative to the direction of the current, and begins to roll up on itself. This process eventually destroys the mean flow (day 12) and results in a coherent patch of dense fluid which rotates anticyclonically.

We note that this simulation was repeated in a channel with twice the length, yielding similar results both qualitatively and quantitatively. The spiral-like cold domes we obtained are similar to the ones observed by Swaters (1998) for a two layer, frontal geostrophic model with homogeneous upper layer (Swaters, 1991). However, the dominant lengthscales associated with the continuously stratified model are invariably smaller (Poulin & Swaters, 1999; Reszka & Swaters, 2000).

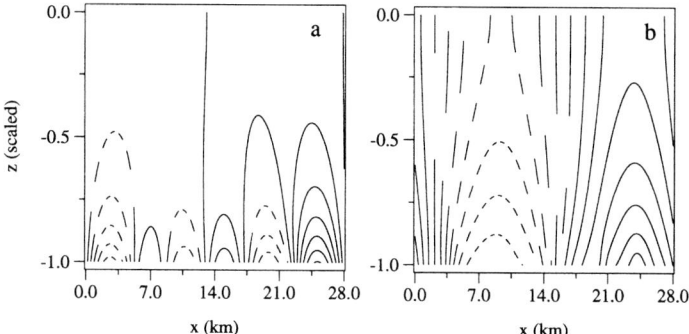

Figure 4. Vertical cross-sections of the nondimensional upper layer pressure at y = 10 km, corresponding to fig. 3a and 3d. Dashed contours correspond to negative values.

The second simulation we present employs bottom topography appropriate for the Strait of Georgia. The topography is comprised of a y-dependent sinusoidal function which forms a trough. Initially, the current rests entirely on one slope of this trough. Low frequency variability within the Strait of Georgia has been attributed to fluctuations and instability of deep currents which propagate along the coast (see LeBlond et al., 1991; Stacey et al., 1988). Fig. 3 depicts four contour plots of the lower layer thickness for this simulation at day a) 8, b) 10, c) 12 and d) 14. In fig. 3a we again see the wavelike deformation of the downslope incropping. It is noteworthy that the fastest-growing wavenumber is about three times the one predicted for the Swaters (1991) model (see Karsten et al., 1995). The e-folding time for this phase of instability is quite short, approximately 7 h. The perturbation moves with the current (i.e. in the negative x-direction) at a speed of roughly 7 cm/s. As the deformations grow, they subsequently encounter the opposite face of the trough ($y < 7$ km) which induces a velocity in the positive x-direction. This, in effect, accelerates the merging of the modes into one anticyclonic gyre which encompasses most of the channel domain (fig. 3c). By 14 days the flow reaches a quasi-steady state (fig. 3d), exhibiting many small-scale filamentous structures.

4. Discussion

In both simulations a train of alternating cyclonic and anticyclonic eddies emerges in the upper layer streamfunction as the instability progresses. These eddies quickly attain a tapered, bottom-intensified shape. Vertical cross-sections of the upper layer pressure at $y = 10$ km are given in fig. 4. Figures 4a and 4b correspond to day 8 and 14, respectively, of the Strait of

Georgia simulation (see fig. 3a and 3d). This vertical structure is consistent with the analysis of Stacey *et al.* (1988) who found evidence of several bottom-intensified vortices in the strait. Surface signatures of deep eddies have also been reported in association with the Denmark Strait Overflow (Bruce, 1995). If such vortices are the result of baroclinic instability, then the present model may provide an appropriate mathematical framework for the dynamics of deep water masses leaving the Denmark Strait.

Acknowledgments: Preparation of this manuscript was supported in part by a PGS A grant awarded to M. K. R. by the Natural Sciences and Engineering Research Council of Canada, a Graduate Fellowship awarded to M. K. R. by the University of Alberta, as well as a research grant awarded to G. E. S. by the Natural Sciences and Engineering Research Council of Canada. A travel grant was also awarded to M. K. R. by the Institute for Geophysical Research.
Author's e-mail: mreszka@acubens.math.ualberta.ca,
WWW address: fluids.math.ualberta.ca/~mreszka/

References

Armi, L. & D'Asaro, E. (1980) Flow structures of the benthic ocean. *J. Geophys. Res.* **85**, 469–484.
Bruce, J. G. (1995) Eddies southwest of the Denmark Strait. *Deep-Sea Research* **42**, 13–29.
Houghton, R. W., Schlitz, R., Beardsley, R. C., Butman, B. & Chamberlin, J. L. (1982) The Middle Atlantic Bight Cold Pool: Evolution of the Temperature Structure During Summer 1979. *J. Phys. Oceanogr.* **12**, 1019–1029.
Jiang, L. & Garwood Jr., R. W. (1995) A numerical study of three-dimensional dense bottom plumes on a Southern Ocean continental slope. *J. Geophys. Res.* **100**, 18,471–18,488.
Karsten, R. H., Swaters, G. E. & Thomson, R. E. (1995) Stability Characteristics of Deep-Water Replacement in the Strait of Georgia. *J. Phys. Oceanogr.* **25**, 2391–2403.
LeBlond, P. H., Ma, H., Doherty, F. & Pond, S. (1991) Deep and Intermediate Water Replacement in the Strait of Georgia. *Atmos.-Ocean* **29**, 288–312.
Poulin, F. J. & Swaters, G. E. (1999) Sub-inertial dynamics of density-driven flows in a continuously stratified fluid on a sloping bottom. I. Model derivation and stability characteristics. *Proc. R. Soc. Lond. A* **455**, 2281–2304.
Price, J. F. & O'Neil Baringer, M. (1994) Outflows and deep water production by marginal seas. *Prog. Oceanog.* **33**, 161–200.
Reszka, M. K. & Swaters, G. E. (2000) Baroclinic instability of benthic currents in a continuously stratified ocean. Submitted to *Can. App. Math. Quart.*
Smith, P. C. (1975) A streamtube model for bottom boundary currents in the ocean. *Deep-Sea Res.* **22**, 853–873.
Stacey, J. W., Pond, S. & LeBlond, P. H. (1988) An objective analysis of the low-frequency currents in the Strait of Georgia. *Atmos.-Ocean* **26**, 1–15.
Swaters, G. E. (1991) On the baroclinic instability of cold-core coupled density fronts on a sloping continental shelf. *J. Fluid Mech.* **224**, 361–382.
Swaters, G. E. (1998) Numerical simulations of the baroclinic dynamics of density-driven coupled fronts and eddies on a sloping bottom. *J. Geophys. Res.* **103**, 2945–2961.

GENERAL PROPERTIES OF BAROCLINIC MODONS OVER TOPOGRAPHY

G. M. REZNIK[1] and G.G. SUTYRIN[2]

[1] *P. P. Shirshov Institute of Oceanology, RAS*
23 Krasikova Street, Moscow, 117256 Russia

[2] *Graduate School of Oceanography, University*
of Rhode Island, Narragansett, RI 02882, USA

Abstract

The properties of solitary topographic Rossby waves (modons) in a uniformly rotating two-layer ocean over a constant slope are analyzed. The modon is described by exact, form preserving, uniformly translating, horizontally localized, nonlinear solution to the inviscid quasigeostrophic equations. Baroclinic modons over topography are found to translate steadily along contours of constant depth in both directions: either with negative speed (within the range of the phase velocities of linear topographic waves) or with positive speed (outside the range of the phase velocities of linear topographic waves). The lack of resonant wave radiation in the first case is due to the orthogonality of the flow field in the modon exterior to the linear topographic wave field propagating with the modon translation speed that is impossible for barotropic modons. Another important property of a baroclinic topographic modon is that its integral angular momentum must be zero only in the bottom layer; the total angular momentum can be non-zero unlike for the beta-plane modons over flat bottom. This feature allows for modon solutions superimposed by intense monopolar vortices in the surface layer to exist.

1. Introduction

Interest in the solitary Rossby waves (so-called modons) is motivated by the important role of the long-lived large-scale coherent vortical structures in the dynamics of the oceans and atmospheres of the Earth and other planets. The problem of the resistance of such long-lived structures to Rossby wave dispersion stimulated theoretical studies of form preserving, horizontally localized, nonlinear solutions to the equations of rotating fluids. A stationary barotropic solution was described by Stern (1975) who introduced the term "modon", and steadily translating barotropic solutions were investigated by

Larichev & Reznik (1976). On the flat beta-plane, the basic modon solution represents a dipolar vortex pair with a characteristic north-south antisymmetry. These dipoles propagate eastward at any speed, or westward at speeds greater than the long-wave speed. It is the nonlinear vortex pair interaction that allows the modon translation speed to be outside the range of linear Rossby waves to avoid resonant radiation. Because of large-scale eddies are typically near geostrophic balance, the quasigeostrophic equations were primarily used in the most of the studies.

Up to now the effects of topography and stratification on the modons were studied separately. E.g., Flierl et al (1980) analyzed a wide variety of modons in a two-layer ocean with flat bottom and discovered that radially symmetric perturbations (so-called riders) of a special form but arbitrary amplitude can be superimposed on the basic solutions. Here we analyze baroclinic modons over a sloping bottom focusing on the *cooperative* effect of the topography and stratification on the modon dynamics. These factors along with the beta-effect play key roles in the dynamics of meso- and large-scale geophysical vortices. For example, they are especially important in the continental slope regions characterized by rather steep slope where topography can be more significant than the beta-effect

2. Model Formulation

We consider quasigeostrophic perturbations steadily propagating with the speed U in the two-layer fluid rotating with a constant angular velocity $f_0/2$ over a sloping bottom. In moving coordinates modons obey the equations

$$J(\Psi_1 + Uy, q_1) = 0, \quad J(\Psi_2 + Uy, q_2 + f_0 sy/h_2) = 0 \tag{1}$$

Here Ψ_1, Ψ_2 are the streamfunctions, subscripts 1, 2 correspond to the upper and lower layers, respectively, s the constant bottom slope, J is the Jacobian; the fluid potential vorticities q_1, q_2 are given by the expressions

$$q_1 = \nabla^2 \Psi_1 + F_1(\Psi_2 - \Psi_1), \quad q_2 = \nabla^2 \Psi_2 + F_2(\Psi_1 - \Psi_2) \tag{2}$$

where $F_i = f_0^2 / g' h_i$, $i = 1,2$ and h_1, h_2 are the constant mean depths of the layers, g' the reduced gravity. The x-axis points along the slope and the y-axis across the slope in the direction of decreasing depth. Thus, without loss of generality the bottom slope s is assumed to be positive.

Integrating (1) along the open streamlines $\Psi_i + Uy = const$, $i = 1,2$ which become straight lines $y = const$ far from the modon center, we obtain

$$\nabla^2 \Psi_1 + F_1(\Psi_2 - \Psi_1) = 0, \quad \nabla^2 \Psi_2 + F_2(\Psi_1 - \Psi_2) - \lambda \Psi_2 = 0 \tag{3}$$

where $\lambda = f_0 s / h_2 U$. One can readily show that the linear system (3) does not possess non-singular localized solutions, therefore *all* streamlines cannot be open. In other words, the streamline pattern must contain at least one domain with the *closed* streamlines and the vorticity distribution within such domain streamlines differs from (3). The domains with closed and open streamlines will be referred to as interior and exterior domains, respectively; the streamline Γ as the separating streamline.

To consider an angular momentum for the localized solution to (1) decaying faster than r^{-2} for $r \to \infty$, we multiply (1) by x and integrate the resulting equations throughout the plane to obtain

$$\int \Psi_2 dxdy = 0, \quad \int \Psi_1 dxdy = \frac{1}{U} \int xJ(\Psi_1, \Psi_2) dxdy \qquad (4)$$

Eqs. (4) are the necessary conditions for the existence of the localized vortices in the two-layer ocean with a sloping bottom. Thus, only the *lower* layer vortex has a zero net angular momentum, the total angular momentum of the localized vortex can be non-zero and the upper layer vortex can be a monopole. We emphasize that these conditions are substantially different from the analogous conditions on the beta-plane where the net angular momentum of the localized vortex must be zero (Flierl at al 1983).

3. Exterior Domain

In the exterior domain the motion is described by the system (3) where the solution can be written in terms of the normal modes

$$\tilde{\Psi}_1 = \Psi_1 + \alpha_1 \Psi_2, \quad \tilde{\Psi}_2 = \Psi_1 + \alpha_2 \Psi_2 \qquad (5)$$

"decoupling" the equations (3) so that

$$\nabla^2 \tilde{\Psi}_1 - p_1^2 \tilde{\Psi}_1 = 0, \quad \nabla^2 \tilde{\Psi}_2 - p_2^2 \tilde{\Psi}_2 = 0 \qquad (6)$$

The mode parameters α_j and p_j are related as follows

$$-p^2 = F_2 \alpha - F_1 = \frac{F_1}{\alpha} - \lambda - F_2 \qquad (7)$$

so that $\alpha_1 > 0$, $\alpha_2 < 0$ while

$$p_1^2 < 0, \ p_2^2 > 0 \text{ for } U < 0 \quad \text{and} \quad p_1^2 > 0, \ p_2^2 > 0 \text{ for } U > 0 \qquad (8a,b)$$

Localized solution for the modes (6) may exist only for $p_j^2 > 0$, thus from (8a) we see that the first normal mode must be zero in the exterior of the modon moving with the negative speed U

$$\tilde{\Psi}_1 = \Psi_1 + \alpha_1 \Psi_2 = 0, \quad \tilde{\Psi}_2 = \Psi_1 + \alpha_2 \Psi_2 \neq 0 \tag{9a}$$

To understand the physical meaning of the condition (9a) we consider the harmonic wave propagating with the same velocity $U < 0$. Obviously, the equations (3)-(8) are valid also for this wave. However, contrary to the modon case, for the harmonic solution in the equation (6b) with $p_1^2 < 0$, the *second* normal mode must be zero

$$\tilde{\Psi}_1 = \Psi_1 + \alpha_1 \Psi_2 \neq 0, \quad \tilde{\Psi}_2 = \Psi_1 + \alpha_2 \Psi_2 = 0 \tag{9b}$$

Comparing (9a) to (9b) we see that the external streamfunction field of the modon moving with the "resonance" negative translation speed $U < 0$ is *orthogonal*[1] to the streamfunction of the harmonic wave propagating with the same velocity. Namely due to this orthogonality the modon does not excite any harmonic waves and remains localized. We emphasize that the conditions (9a) are applied only in the exterior domain; in the modon interior they are not satisfied (see below). One can say that the nonlinearity prohibits the resonant radiation due to the resonant component confined to the modon core. The analogous property is valid for the Rossby modon with baroclinic exterior in two-layer ocean of constant depth (Flierl et al. 1980).

4. Interior Domain.

Following Larichev & Reznik (1976), we consider the simplest case when in both layers the interior domains are circles of the same radius a with linear vorticity distributions

$$q_1 = S_1(\Psi_1 + Uy) + Q_1, \quad q_2 = S_2(\Psi_2 + Uy) - f_0 sy/h_2 + Q_2, \tag{10}$$

where the coefficients S_i, Q_i are determined from the matching conditions at the separating streamline $r = a$.

We now introduce the normal modes

---

[1] The scalar production of the fields $\mathbf{P} = \begin{pmatrix} P_1 \\ P_2 \end{pmatrix}$, $\mathbf{Q} = \begin{pmatrix} Q_1 \\ Q_2 \end{pmatrix}$ is defined here as

$\mathbf{P} \cdot \mathbf{Q} = h_1 \int P_1 Q_1 dx dy + h_2 \int P_2 Q_2 dx dy$

$$T_j = \Psi_1 + \gamma_j \Psi_2, \quad j=1,2 \tag{11}$$

that satisfy the equations

$$\nabla^2 T_j + k_j^2 T_j = -[(1+\gamma_j)k_j^2 U + \gamma_j]y + Q_1 + \gamma_j Q_2; \tag{12}$$

the parameters k_j, γ_j and S_j are related as follows

$$k^2 = F_2 \gamma - S_1 - F_1 = \frac{F_1}{\gamma} - S_2 - F_2 \tag{13}$$

We require the continuity of the functions Ψ_i and their derivatives at the separating streamline:

$$\Psi_i\big|_{r=a-0} = \Psi_i\big|_{r=a+0}; \quad \frac{\partial \Psi_i}{\partial r}\bigg|_{r=a-0} = \frac{\partial \Psi_i}{\partial r}\bigg|_{r=a+0}, \quad i=1,2 \tag{14}$$

In addition to (14) one has to take into account that the circle $r = a$ is a streamline $\Psi_i + Uy = C_i = const$ i.e.,

$$\Psi_i\big|_{r=a+0} = -Ua\sin\theta + C_i, \quad i=1,2 \tag{15}$$

It can be shown that the solution is described by only the axisymmetric component and the dipole proportional to $\sin\theta$. The dipole component $\Psi_i^{(d)}$ and the axisymmetric component $\Psi_i^{(r)}$ will be referred to as the dipole modon and the rider, respectively. The dipole modon exists (if any) independent of the rider; namely the dipole component $\Psi_i^{(d)}$ is the "engine" forcing the solitary wave to move along the *x*-axis. One can readily check also that the conditions (14), (15) provide the continuity of the dipole modon up to the second derivatives. The exact solutions for the modons satisfying the equations (10) and the matching conditions (14), (15) are described by Reznik & Sutyrin (2000).

5. Conclusions.

We have studied the cooperative effect of the stratification and the bottom topography on the modon dynamics using the model of uniformly rotating two-layer ocean with a sloping bottom. Here we considered the general properties of the baroclinic modons; the exact solutions for the modons with piecewise linear dependence of the potential vorticity on the streamfunction are given and analyzed by Reznik and Sutyrin (2000).

Two results are of the most importance. The first result is that the baroclinic modons over a sloping bottom are able to move with the speed within the range of phase velocities of the linear topographic waves. Lack of the resonant radiation in this case is conditioned by the orthogonality of the streamfunction field in the modon exterior to the corresponding field of the linear topographic wave propagating with the modon translation speed. At the same time the streamfunction in the modon core (interior domain) is *not* orthogonal to the linear wave streamfunction i.e., the modon core "contains" the resonant mode. One can say that the nonlinearity prohibits the radiation due to this mode. The solution with similar properties exists also in a two-layer ocean of constant depth on the β-plane. This is the Rossby modon with purely baroclinic exterior found in Flierl et al. (1980).

The second important result making the baroclinic modons over a sloping bottom significantly different from their β-plane analogs (see Flierl et al 1983) is that the total angular momentum of the vortex over a sloping bottom can be non-zero. More exactly, the integral angular momentum in the bottom layer should be zero; the upper layer angular momentum can be non-zero. This feature provides an existence of the modons carrying the axisymmetric riders representing intense monopolar vortices confined to the upper layer. As the rider amplitude is arbitrary the dipole component can be "masked" and the composed modon looks like a monopole. This property is especially interesting since the majority of the observed long-lived vortices are monopoles.

Acknowledgements. This study was supported by Russian Foundation for Basic Research Grant #99-05-64841 and the USA National Science Foundation Grant #ATM-9905209

6. References

Flierl, G.R., Stern, M.E., and Whitehead, J.A. (1983) The physical significance of modons: Laboratory experiments and general integral constraints. *Dyn. Atmos. Oceans*, 7, 233--263.

Flierl, G.R., Larichev, V.D., McWilliams, J.C., and Reznik, G.M. (1980) The dynamics of baroclinic and barotropic solitary eddies. *Dyn. Atmos. Oceans*, 5, 1--41.

Larichev, V.D., and Reznik, G.M. (1976) On the two-dimensional solitary Rossby waves. *Doklady Akad. Nauk SSSR,* **231**, 1077-1079.

Reznik, G.M., and Sutyrin, G.G. (2000) Baroclinic topographic modons. *J. Fluid Mech.*, in press.

Stern, M.E. (1975) Minimal properties of planetary eddies. *J. Mar. Res.*, **33**, 1--13.

ACOUSTIC FILTRATION IN PRESSURE-COORDINATE MODELS
Basic Concepts and Applications in Nonhydrostatic Modeling

REIN RÕÕM
Tartu Observatory, Tõravere 61602, Tartumaa, Estonia
room@aai.ee

AARNE MÄNNIK
Tartu University, Tähe tn. 4, 51010, Tartu, Estonia
aarne@aai.ee

1. Introduction

Pressure (isobaric or p-) coordinates in non-hydrostatic (NH) dynamics with pressure in the role of the vertical coordinate were first introduced by Miller and Pearce (1974), and generalized to sigma-coordinate framework by Miller and White (1984). First numerical applications of the NH model in pressure related coordinates where developed by Xue and Thorpe (1991) for two-dimensional case, and by Miranda and James (1992) for three-dimensional motion.

The main attractions of the pressure-coordinate approach in NH dynamics are: (1) Formal simplicity and aesthetic elegance of equations;
(2) Generality: equations are equally applicable in the scale region ranging from 100 m up to planetary motion (Rõõm and Männik, 1999);
(3) Reduction of development cost (up to 10 times) at the transition from hydrostatic (HS) to NH dynamics in the climate modeling and weather forecast packages due to maintenance of data assimilation, initialization, physical parametrization, and post-processing software.

The most fundamental property of the Miller-Pearce model is its pseudo-anelastic nature: motion is internally non-divergent and volume elements are preserved in the p-coordinate metrics. As a consequence, the internal acoustic mode is filtered. Still, their model is only partially anelastic, as it maintains the external acoustic (Lamb) mode. In the present paper, we describe a modification of the Miller-Pearce approach with additional elimination of the external acoustic mode, which turns the pseudo-anelastic model to a completely anelastic p-coordinate model (AEM). Elimination of the external mode increases numerical stability of the model and provides rather significant enhancement of the time–step. At present, the AEM ideology is applied developing the model NHAD (Rõõm, Miranda, and Thorpe, 2000) and the NH kernel for limited-area, numerical weather prediction model HIRLAM (Männik and Rõõm, 2000; Rõõm, 2000).

2. Equations

We introduce p-coordinates: $x^1 = x$, $x^2 = y$, $x^3 = p$, in the domain $0 < x^1 < L_x$, $0 < x^2 < L_y$, $0 < x^3 < p_s(x^1, x^2, t)$, where t is time and p_s is actual surface pressure, and corresponding velocity components: $v^1 = \dot{x}$, $v^2 = \dot{y}$, $v^3 = \dot{p}$. The pseudo-anelastic equations then are

$$\frac{\partial v^\alpha}{\partial t} = F^\alpha - G^{\alpha\beta} \frac{\partial \phi}{\partial x^\beta}, \quad \alpha = 1, 2, 3, \tag{1a}$$

$$\frac{\partial v^\alpha}{\partial x^\alpha} = 0, \quad \frac{\partial T}{\partial t} = F_T, \tag{1b, c}$$

$$\frac{\partial p_s}{\partial t} = -\nabla \cdot \int_0^{p_s} \mathbf{v} dp. \tag{1d}$$

Here ϕ is NH geopotential fluctuation, G is the metric tensor

$$G^{\alpha\beta} = \begin{pmatrix} 1 & 0 & 0 \\ 0 & 1 & 0 \\ 0 & 0 & p^2/H^2 \end{pmatrix}$$

∇ is the horizontal gradient in p-coordinates, $H = RT/g$ is the scale height, T is the temperature, and F^α, F_T represent components of "hydrostatic" forcing, which in the adiabatic, frictionless case are superpositions of advection-convection, HS pressure-forcing, Coriolis, and energy conversion terms:

$$F^1 = \hat{a}(v^1) - \frac{\partial \varphi}{\partial x} + fv^2, \quad F^2 = \hat{a}(v^2) - \frac{\partial \varphi}{\partial y} - fv^1,$$

$$F^3 = \hat{a}(v^3) + \frac{c_v}{c_p} \frac{(v^3)^2}{x^3}, \quad F_T = \hat{a}(T) + \frac{R}{c_p} \frac{v^3}{x^3}.$$

Here $\hat{a}(u) = -v^\alpha \frac{\partial u}{\partial x^\alpha}$, f is the Coriolis parameter, R, c_v and c_p are thermodynamic constants for dry air, and $\varphi = gh + \int_p^{p_s} \frac{RT}{p'} dp'$ is the hydrostatic geopotential.

3. Acoustic filtration

The purpose of acoustic filtration is to get rid of acoustic waves with maintenance of slow dynamics. In equations (1), internal acoustic mode is already filtered due to the condition (1b). To maintain this condition in time, ϕ must satisfy equation (follows from (1b) after applying time derivation and elimination of velocity tendencies with the help of (1a))

$$\frac{\partial}{\partial x^\alpha}\left(G^{\alpha\beta}\frac{\partial \phi}{\partial x^\beta}\right) = \frac{\partial F^\alpha}{\partial x^\alpha}. \tag{2}$$

This fundamental diagnostic equation for NH geopotential is central for non-hydrostatic model. When equation (2) is applied, the third (vertical) equation in (1a) can be abandoned in favor of diagnostical computation of vertical velocity v^3 from (1b).

To get rid of external mode waves, surface pressure adjustment (initially treated in the flat-surface-case by Miller and White, 1984) must be assumed. This means, first, that dynamics is linearized in respect to surface pressure fluctuation: motion is treated in the fixed domain $0 < p < \bar{p}_s(x^1, x^2)$, where $\bar{p}_s(x^1, x^2)$ is the mean (background) surface pressure distribution, hydrostatic geopotential is presented as $\varphi = gh + \int_p^{\bar{p}_s} \frac{RT}{p'} dp'$, and equation (1d) is linearized:

$$\frac{\partial p'_s}{\partial t} = -\nabla \cdot \int_0^{\bar{p}_s} \mathbf{v} dp,$$

where p'_s is the surface pressure fluctuation in respect with the background state. Secondly, the last equation is replaced (applying time derivative) with the wave equation for p'_s

$$\frac{\partial^2 p'_s}{\partial t^2} = -\nabla \cdot \int_0^{\bar{p}_s} \frac{\partial \mathbf{v}}{\partial t} dp = \nabla \cdot \int_0^{\bar{p}_s} (\mathbf{F} - \nabla \phi) dp.$$

Finally, fluctuative surface pressure adjustment $\partial^2 p'_s / \partial t^2 \Rightarrow 0$ is assumed in the last equation, which results in the two-dimensional diagnostical equation for ϕ

$$\nabla \cdot \int_0^{\bar{p}_s} \nabla \phi dp = \nabla \cdot \int_0^{\bar{p}_s} \mathbf{F} dp. \tag{3a}$$

The physics of this equation is simple: (3a) provides detailed mass balance in vertical air columns for $\forall\, t$, if such balance exists at the initial moment.

The detailed mass balance, in turn, is responsible for complete elimination of external acoustic waves.

4. Boundary conditions for NH geopotential

From the mathematical point of view, relationship (3) is the vertical "boundary" condition for ϕ in the adjusted case, which replaces the common surface condition $\phi|_{p_s} = 0$ of unadjusted dynamics. The surface pressure fluctuation in the adjusted model can be diagnosed from the lower boundary value of ϕ with the help of relationship (for details see Rõõm, 2000)

$$p'_s = \bar{p}_s \cdot \left(\frac{\phi}{RT}\right)_{\bar{p}_s} .$$

The behaviour of ϕ at the top is specified by the integrability condition (*ibid*)

$$\phi \in L_1 : \qquad \int_0^{p_s} |\phi| dp < \infty \quad \text{if} \quad \int_0^{p_s} |\partial F^\alpha / \partial x^\alpha| dp < \infty . \qquad (3b)$$

This condition eliminates exponentially growing (with height) homogeneous solutions of equation (2), the presence of which would otherwise mean an existence of spurious external sources at the infinite high and might cause spurious top-reflections of buoyancy waves.

5. Applications

The AEM equations (1) - (3) are applied for the development of the NH, sigma-coordinate model NHAD (Rõõm, Miranda and Thorpe, 2000), which represents a modification of the NH unadjusted model NH3D (Miranda and James, 1992), and for the development of the NH kernel of the limited-area, hybrid-coordinate, numerical weather prediction model HIRLAM (Rõõm, 2000; Männik and Rõõm, 2000). Both these models make use of explicit-Eulerian time-stepping scheme. An example of orographic wave modeling with the NH HIRLAM is presented in Fig. 1.

The main advantage of the AEM in comparison with the common unadjusted scheme is the increased stability and time–step enlargement due to external mode filtration. At spatial scales $\Delta x < 2$ - 3 km, time-step reaches in explicit-Eulerian mode the Courant-Friedrichs-Lewy limit, and there is up to 60-time enlargement of time-step in comparison with the unadjusted case (from 0.5 s to 30 s at $\Delta x = 1$ km). This stability increase makes the explicit-Eulerian AEM comparable in efficiency with the unadjusted split-explicit or semi-implicit schemes (Table 1). Another advantage of the AEM is the decreased

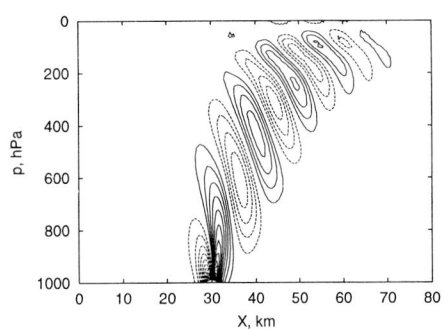 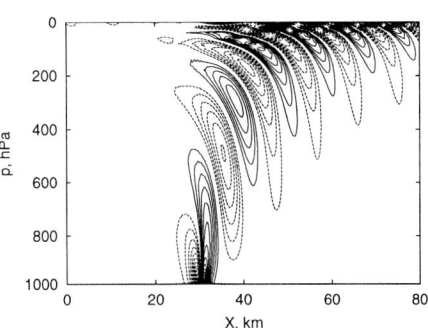

Fig. 1
Nonhydrostatic HIRLAM: Vertical velocity waves at steady flow over
one-dimensional orography.
$a = 1$ km, $h_0 = 200$ m, $U = 15$ m/s, $N = 0.01$ 1/s.
grid $300 \times 50 \times 31$; $t = 2.2$ h; isoline interval 0.05 m/s.

TABLE 1. Estimated CPU time in seconds by Cray-C98 for one-hour simulation with split-explicit Lokal-Modell, semi-implicit MRI-NHM (Saito et al, 1998), and explicit NH HIRLAM

Δx (km)	Δt	Lokal-Modell	MRI-NHM	NH HIRLAM
10	60	36.1	43.3	33
2	60	81.1	42.9	38
2	30	106.8	69.5	69

reflection of buoyancy waves from the top due to condition (3b). Avoiding top reflection entirely is still not possible: spectral smoothing (explicit or semi-implicit), with 3-5 time enhancement of the smoothing rate at the top (5-8 top layers), is recommended for v^1, v^2 and T (Männik and Rõõm, 2000).

References

Männik, A. and Rõõm, R. (2000) Nonhydrostatic adiabatic kernel for HIRLAM. Part II. Anelastic, hybrid-coordinate, explicit-Eulerian model, 48 p., http://apollo.aai.ee/HIRLAM/nhkern2.ps .

Miller, M. J. and Pearce, R. P. (1974) A three-dimensional primitive equation model of cumulonimbus convection, *Q. J. R. Meteorol. Soc.*, **100**, 133–154.

Miller, M. J. and White, A. A. (1984) On the nonhydrostatic equations in pressure and sigma coordinates, *Q. J. R. Meteorol. Soc.*, **110**, 515–533.

Miranda, P. M. A. and James, I. N. (1992) Non-linear three-dimensional effects on gravity-wave drag: splitting flow and breaking waves, *Q. J. R. Meteorol. Soc.*, **118**, 1057–1081.

Rõõm, R. (2000) Nonhydrostatic adiabatic kernel for HIRLAM. Part I. Fundamentals of nonhydrostatic dynamics in pressure-related coordinates, 22 p., http://apollo.aai.ee/HIRLAM/nhkern1.ps .

Rõõm, R., Miranda, P. M. A. and A. J. Thorpe, (2000) Filtered non-hydrostatic models in pressure-related coordinates, *Q. J. R. Meteorol. Soc.*, submitted.

Rõõm R. and Männik A. (1999): Responses of different nonhydrostatic, pressure-coordinate models to orographic forcing, *JAS*, **56**, 2553 - 2570.

Saito, K., Schaettler, U. and Steppeler, J. (1998) 3-D mountain waves by the Lokal-Modell of DWD and the MRI Mesoscale Nonhydrostatic Model, *Papers in Metorol. and Geophys.*, **49**, 7 - 19.

Xue, M. and Thorpe, A. J. (1991): A mesoscale numerical model using the nonhydrostatic pressure-based sigma-coordinate equations: Model experiments with dry mountain flows, *Mon. Weather Rev.*, **119**, 1168–1165.

OSCILLATORY REGIMES OF FORCED ZONAL FLOW

Laboratory and Numerical Simulation

I.A. SAZONOV
Department of Applied Mathematics, University College Cork, Cork, Ireland

AND

S.D. DANILOV, YU.L. CHERNOUS'KO AND V.G. KOCHINA
Obukhov Insitite of Atmosphere Physics RAS
109017 Pyzhevsky 3, Moscow, Russia

Low frequency oscillations of the atmosphere with periods of 30–60 days have been discovered in Tropics and later in the Northern Hemisphere (Madden & Julian, 1971; Anderson & Rosen, 1983). This phenomenon has been addressed by many investigators in attempts to clarify the physics behind it. A related and even broader issue is the low-frequency variability (LFV) of the Earth's atmosphere which is frequently associated with the existence of multiple, systematically recurring, persistent states observed in the real atmosphere (Mo & Ghil, 1988; Yang & Reinhold, 1991; etc.).

After the work by Charney & DeVore (1979) a topographic drag is thought to play an important role in the formation of multiple flow equilibria and the LFV. Topography-assisted coupling between the zonal flow and Rossby waves could be simulated in the framework of other low-order weakly non-linear models. Such simulations predict multiple equilibrium states or even oscillatory regimes (Pedlosky, 1981; Jin & Ghil, 1990; Nathan & Barcilon, 1994; etc.). However the resonance character of coupling dictated by the weakly non-linear approach is at variance with the observed robustness of the LFV. This suggests the use of strongly nonlinear models. Numerical computations (Legras & Ghil, 1985; Yang *et al.*, 1997; etc.) as well as laboratory experiments (Pfeffer *et al.*, 1993; etc.) confirm that the coupling between waves and flow can be observed in a wider parameter range than prescribed by weakly non-linear theory.

The present work is aimed to simulate the topography-assisted LFV in laboratory experiments and numerically. We study oscillatory regimes of vortex dynamics and seek for a low-order model that reproduces this phenomenon and for factors favourable for the robustness of the LFV.

The laboratory set-up consists of a rotating cylindrical vessel (rotation period $T = 1$–3 s) with a sloping bottom and a sink-source system serving to provide a zonal forcing. The flow is confined to an annular channel formed by external ($r_2 = 15$ cm) and internal ($r_1 = 3$ cm) walls inserted into the vessel. The bottom slope is $\beta = 0.19$. The fluid layer depth $H_{\max}$ is maximal in the vicinity of the external wall. It is varied in the range 4–10 cm. More details one can find in Sazonov & Chernous'ko (1998).

The zonal flow is created by injecting water through multiple holes drilled in the external wall and pumping it out through the holes in the bottom near the internal wall. This allows the basic zonal flow to be controlled: its width is determined by the width of the region where the sink holes are located, and its speed by the sink-source pumping rate (as explained in Danilov & Sazonov (1999)). We varied the pumping volume rate in the range $Q = 0$–50 cm^{3}s^{-1}. Forcing was modulated zonally in some experiments (such modulation is natural in the Earth's atmosphere). This was achieved by eliminating sink-source holes within some sectors.

The topography in the form of two diametrically opposing ridges is located across the annular channel. In the zonal direction, each ridge has the form of a half-period of a sine function and occupies about 1/7 of the total azimuth angle. The ridge height varies along the radius so that its top is horizontal.

The flow is visualized by tracer particles floating at the free fluid surface. The 2D velocity field $\mathbf{u}(r, \phi, t)$ is recovered by applying a variant of the correlation image velocimetry (CIV) method. We compute the total angular momentum defined as

$$M(t) = \int_{r_1}^{r_2} \int_0^{2\pi} (\mathbf{r} \times \mathbf{u}(r, \phi, t))\, H(r, \phi)\, d\phi\, dr \qquad (1)$$

where $H(r, \phi)$ is the depth, and the vorticity field.

Without rotation, the flow in the vessel is purely radial and is directed from the periphery to the center. When rotation is applied but the topography is absent, a strong zonal current develops due to the action of the Coriolis force. We checked that this zonal flow is stable over the entire range of the governing parameters of the set-up. In the presence of topography, the flow dynamics change drastically. All of the regimes observed in the experiments can conventionally be divided into four groups.

1. Steady motion without vortices: the zonal flow is perturbed by the ridges but one does not see the closed streamlines which would be

indicative of vortex formation. This regime is observed for sufficiently weak pumping rate. The angular momentum is constant in this regime.
2. Motion with steady or slightly oscillating vortices. For moderate values of pumping rate and rotation frequency, one can observe cyclonic vortices above the ridges that usually oscillate slightly in zonal direction. The angular momentum also exhibits small-amplitude oscillations around some relatively high mean value.
3. Oscillatory motion with recurringly generated and decaying vortices. This regime is observed over a finite (and fairly wide) range of pumping rate and rotation period. The angular momentum oscillates with high amplitude, drops down to zero and even to a small negative value at some instants of time (see Fig. 1). The crests in Fig. 1 are associated with the augmented zonal circulation and the troughs with blocking. This regime resembles LFV in some respects. It is considered below in details
4. Motion which is chaotic in time. The regime is observed for high values of pumping rate. It needs further investigation.

We describe regime 3 in more detail. The sequence of pictures in Fig. 2 shows a typical sequence of flow patterns observed during one quasi-period of oscillation. Corresponding instants of time are indicated by circles in the plot in Fig. 1a. The velocity field in Fig. 2 is represented by arrows, and the vorticity field (cyclonic vorticity only) is shown by shading. Curves correspond to a boundary separating areas with cyclonic and anticyclonic vorticity. The vessel rotates clockwise. The ridges pass in figure's north-east and south-west directions.

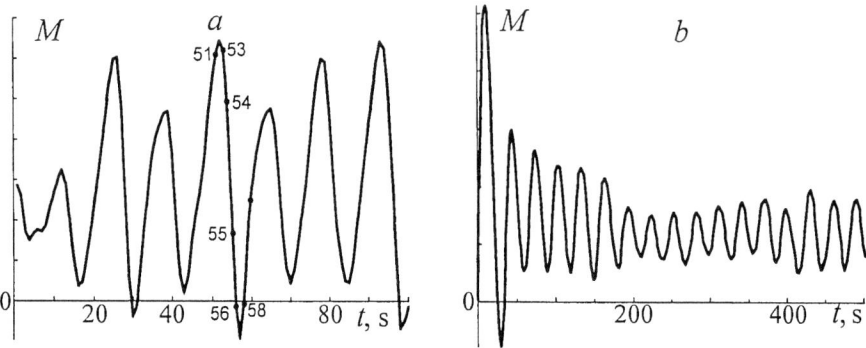

Figure 1. Oscillations of the angular momentum $M(t)$ (a) as measured in the laboratory experiment for a particular set of parameters in regime 3; (b) as found in numerical modelling.

The motion comprises a fascinating series of events. At the beginning of a quasi-period, Fig. 2a, one sees a nearly zonal flow with remnants of vor-

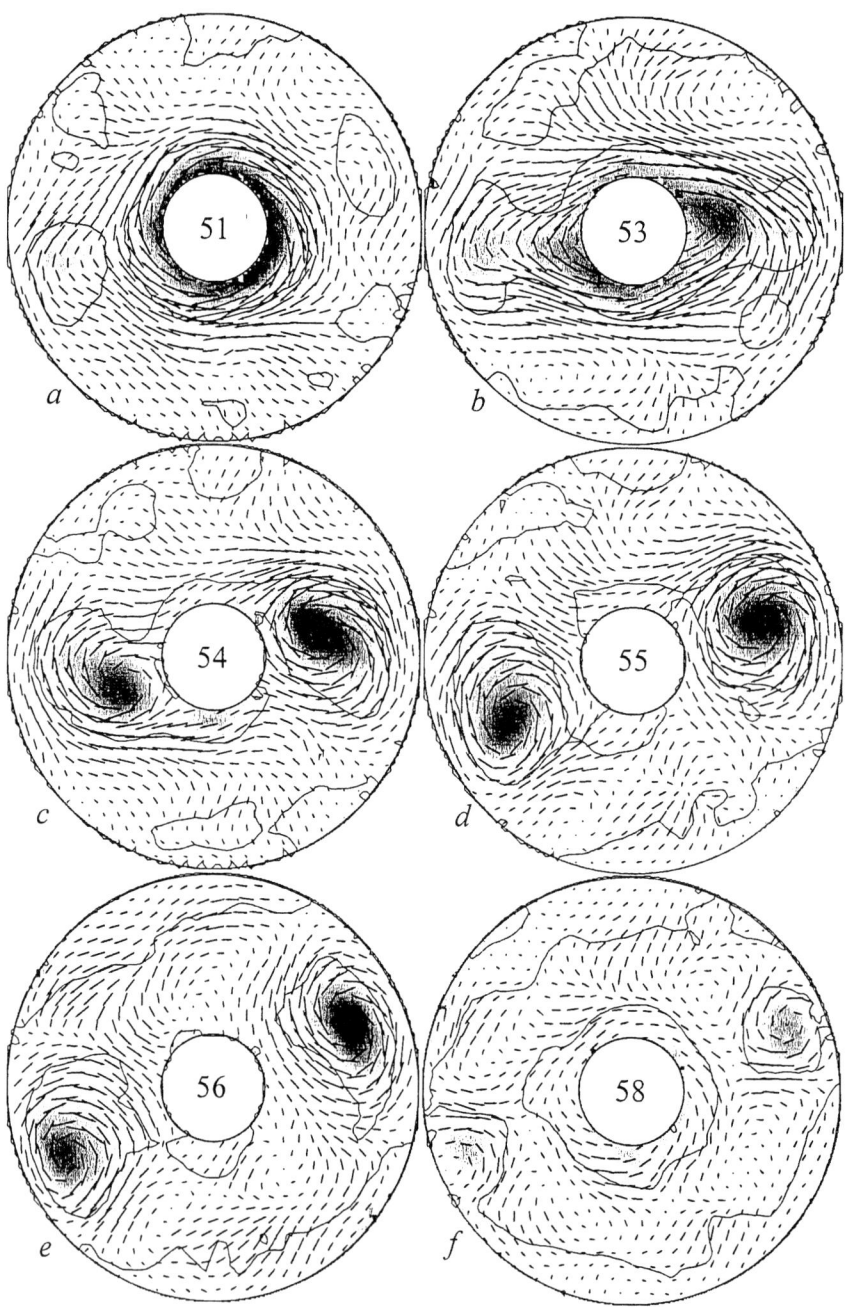

Figure 2. Vortex dynamics during a cycle.

tices from the preceding evolution. The angular momentum is approaching its local maximum. The initial stage of flow instability and Rossby wave generation is shown in Fig. 2b. During the following development, the growing perturbations become so intense that flow streamlines reconnect and isolated vortices form near the inner wall. The zonal jet deviates strongly to the periphery and decays in amplitude (Fig. 2c), while the angular momentum decreases. The cyclones move toward the outer wall (Fig. 2d) and weaken due to dissipative losses at the bottom and outer wall (Fig. 2e). The angular momentum drops to a small negative value. Finally the vortices become weak enough so that they cannot prevent the generation of the zonal flow due to the Coriolis force (Fig. 2f). The angular momentum begins to grow.

This series of events, observed in laboratory, gives a remarkable example of the LFV allowing one to inspect more closely relative roles played by topography, zonal flow and the Rossby wave.

Numerical simulations are based on a quasigeostrophic model. The flow is governed by the potential vorticity (PV) equation:

$$\frac{\partial \xi}{\partial t} + [\psi, \xi] - \frac{f}{H}[\psi, H] = -\lambda \omega - \frac{fq}{H} + \nu \Delta \omega \qquad (2)$$

$$\psi(r_{1,2}) = \psi'_r(r_{1,2}) = 0 \qquad (3)$$

Here ψ is the quasigeostrophic stream function; $\omega = \Delta \psi$ is the relative vorticity; $\xi = \omega - \psi/L_0^2$ is the quasigeostrophic PV ($L_0 = \sqrt{gH_0}/f$ is the Rossby deformation radius with g the acceleration due to gravity, H_0 the characteristic depth of fluid and $f = 4\pi/T$ the Coriolis parameter); $\lambda = \nu/(\delta_E H_0)$ is the Ekman friction coefficient ($\delta_E = \sqrt{2\nu/f}$ is the thickness of the Ekman boundary layer, ν is the kinematic viscosity); q is the pumping due to distributed sinks and sources, per unit area; $[..., ...]$ denotes the Poisson brackets. Although ordinary viscosity is small compared to the Ekman friction at energy-containing scales, we retain it to prevent enstrophy accumulation at the smallest resolvable scales. Apart from influencing boundary conditions at lateral walls, this term is also very important in the sink zone (Hide, 1968; Danilov & Sazonov, 1999).

We use the pseudo-spectral method for integration of the equation with respect to the azimuth angle (e.g. Canuto et al., 1987).

For the zeroth harmonics, equation (2) is in a sense degenerate and we have to deal with ageostrophic velocity components. We find the radial velocity from the continuity equation and then solve the 2D Navier-Stokes equation for zonally averaged zonal velocity.

To integrate the equation with respect to radius we applied the pseudo-spectral method based on Chebyshev polynomials. Some difficulty arises when inverting the Laplace operator to compute the stream function from the vorticity field. To overcome it, we use the idea similar to the so-called 'Derivative Incorporated Boundary Conditions' described by Canuto et al. (1987) but applied directly to the Laplace operator written in matrix form in terms of the polar coordinates.

The numerical simulations qualitatively reproduce the main features of the laboratory experiments. In particular, all of the four regimes of vortex dynamics are recovered. An example of variation of the angular momentum in regime 3 is shown in Fig. 1b.

The introduction of zonal modulation in the sink-source distribution makes broader the parameter range where regime 3 is observed. We have checked this in both laboratory and numerical experiments.

The work was financially supported by the Russian Foundation for Basic Research (projects nos. 96-05-65321, 96-15-98527) and by INTAS-RFBR 95-0988.

The athours wish to thank Prof.F.Dolzhanskii for useful discussions.

References

Anderson, J.R. and Rosen, R.D. (1983) The latitude-height structure of 40–50 day variations in atmospheric angular momentum. *J. Atmos. Sci.* **40**, 1584–1591.

Canuto, C., Hussaini, M.Y., Qurteroni, A. and Zang, T.A. (1987) *Spectral Methods in Fluid Dynamics.* Springer-Verlag, 557pp.

Charney, J.G. and De Vore, J.G., (1979) Multiple equilibria in the atmosphere and blocking, *J. Atmos. Sci.* **36**, 1205–1216.

Danilov, S.D. and Sazonov, I.A., (1999) On a problem of quasi two-dimensional calculation of sink–source flow in a rotating annulas, *Izvestiya RAN, Fiz. Atmos. Okeana* **35**(3), 312–323.

Hide, R. (1968) On source–sink flows in rotating fluid, *J. Fluid Mech.* **32**(4), 737–764.

Jin, F.-F. and Ghil, M. (1990) Intraseasonal oscillations in the extratropics: Hopf bifurcation and topographic instabilities, *J. Atmos. Sci.* **47**(24), 3007–3022.

Legras, B. and Ghil, M. (1985) Persistent anomalies, blocking, and variations in atmospheric predictability, *J. Atmos. Sci.* **42**, 433–471.

Madden, R.A. and Julian, P.R. (1971) Detection of a 40–50 day oscillation in the zonal wind in the tropical Pacific, *J. Atmos. Sci.* **28**, 702–708.

Mo, K. and Ghil, M. (1988) Cluster analysis of multiple planetary flow regimes, *J. Geophys. Res.* **93D**, 10927–10952.

Nathan, T.R. and Barsilon, A. (1994) Low-frequency oscillations of forced barotropic flow, *J. Atmos. Sci.* **51**(4), 582–588.

Pedlosky, J. (1981) Resonant topographic waves in barotropic and baroclinic flows, *J. Atmos. Sci.* **38**, 2626–2641.

Pfeffer, R.L., Kung, R., Ding, W. and Li, G.-Q. (1993) Barotropic flow over bottom topography—experiments and nonlinear theory, *Dyn. Atmos. Oceans* **19**, 101–114.

Sazonov, I.A. and Chernous'ko, Yu.L. (1998) Vortex regimes in the zonal flow over a topography in a β-channel, *Izvestiya. RAN, Fiz. Atmos. Okeana* **32**(1), 18–24.

Yang, S. and Reinhold, B. (1991) How does the low-frequency variability vary? *Mon. Wea. Rev.* **119**, 119–127.

Yang, S., Reinhold, B. and Källén, E. (1997) Multiple weather regimes and baroclinically forced spherical resonance, *J. Atmos. Sci.* **54**(11), 1397–1409.

IMPROVING CLIMATE SIMULATIONS IN THE TROPICAL OCEANS

A. Schiller
CSIRO Marine Research
GPO Box 1538, Hobart 7001, Tasmania, Australia

1. Abstract

A better understanding and prediction of sea-surface temperature (SST) is one of today's key research areas in climate variability. Results presented here focus on an ocean model with special emphasis on the Indo-Pacific Oceans, including the Indonesian throughflow. Three main improvements contribute to the state-of-the-art performance of the model: a hybrid mixed-layer scheme for the upper boundary layer of the ocean, a parameterization for tidal mixing in the Indonesian archipelago, and an atmospheric boundary layer model to calculate improved heat and freshwater fluxes for ocean-only models. The resulting patterns of mixed-layer depths, water mass transformations due to tidal mixing in the Indonesian Seas and heat flux anomalies are all in good agreement with observations.

2. Introduction

Sea-surface temperature affects all maritime atmospheric processes, and it determines the extent of atmospheric convection and monsoon rain in tropical regions. In particular the SST and the associated thermocline structure of the western Pacific, the Indonesian region and the eastern Indian Ocean have a significant impact on the atmosphere over Australia. The patterns of rainfall over Australia show a strong relationship to SST in these regions. Forecasts of these correlations with coupled ocean-atmosphere models crucially depend on the ability of the ocean submodel to accurately simulate processes that generate and maintain SST anomalies. Essential components for realistic SST simulations are a reasonable mixed-layer model and appropriate surface fluxes. The latter is achieved through application of an atmospheric boundary layer model, where diagnosed surface heat fluxes are based on air temperature (and relative humidity) which respond realistically to changes in SST.

The Indonesian Throughflow (ITF hereafter) allows warm Pacific water to flow into the Indian Ocean. Strong tidal flow in the Indonesian region causes enhanced mixing of deep and surface waters, modifying the water masses and SST on their way to the Indian Ocean. The modified SST influences the extent of atmospheric convection and monsoon rain in the eastern Indian Ocean.

We use the Australian Community Ocean Model Version 2 (ACOM2)[1] with a tropically-enhanced grid resolution. The model is a synthesis of Australian enhancements and additions to the GFDL Modular Ocean Model Version 2 (MOM2) [*Pacanowski*, 1995]. The model is forced with 1982 to 1998 monthly mean boundary conditions using a blended wind stress product based on Florida State University [*Stricherz et al.*, 1992] and NCEP reanalysis data [*Kalnay et al.*, 1996].

3. Atmospheric Boundary Layer Model

Apart from sea-ice at high latitudes, an ocean model needs to know the fluxes of momentum, heat and freshwater (and perhaps boundary conditions for some other tracers). Unfortunately, the accuracy of observations, such as satellite-derived products improve only gradually over time. In the meantime, ocean modellers have to rely on pure parameterizations or on a mixture of parameterizations and observations-based formulation of upper ocean boundary conditions. We follow the latter path by using a simple global atmospheric boundary-layer model (ABLM). The model is based on *Kleeman and Power's* [1995] (KP hereafter). In contrast to the traditional relaxation parameterization [*Haney*, 1971], which is a way of prescribing SST with a fixed air temperature, these models permit a non-local response of air temperature to air-sea heat fluxes and thus allow for a limited degree of predictability in SST, though winds still have to be prescribed.

The use of an atmospheric boundary-layer model, while still reasonably simple to interpret, takes proper account of surface heat fluxes and evaporation during the period investigated. The ABLM consists of a single-layer model atmosphere (boundary-layer as well as a portion of the cloud layer) that is in contact with the surface. Air potential temperature is treated as a prognostic variable. The temperature tendency equation includes the effects of horizontal and vertical advection; horizontal diffusion of transient eddies; turbulent sensible heat exchange with the surface; plus a term representing the radiative cooling of the atmospheric boundary layer. The model atmosphere also predicts land temperatures (with a very short time constant), which allows investigation of atmospheric transport processes from land to ocean. See KP for a detailed discussion of the individual terms.

The main difference between the original ABLM by KP and our modified version is the use of monthly mean input data for relative humidity rather than a fixed value of 0.8. Furthermore, we used the latent heat together with precipitation from the NCEP reanalysis [*Kalnay et al.*, 1996] to calculate freshwater fluxes.

---

[1]http://www.marine.csiro.au/acom

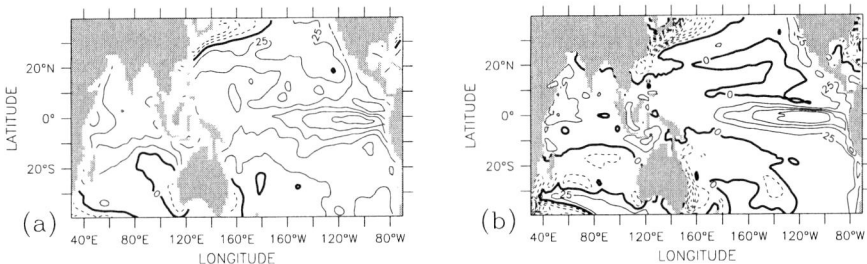

Fig. 1: *Climatological annual mean net surface heat flux from the Southampton Oceanography Centre [Josey et al., 1998] (a) and from the ABLM (b). Contour Interval = 25 W/m².*

The annually averaged heat fluxes from the Southampton Oceanography Centre's climatology [*Josey et al.*, 1998] and the ABLM are depicted in Fig. 1. There is reasonable agreement for all dominant features, like the heat gain of the tropical ocean with its maxima in the eastern Pacific. It is important to note that part of the discrepancies are independent of the ABLM. They are due to the difference between the NCEP net solar shortwave radiation, which is input to the ABLM, and the climatology from the Southampton Oceanography Centre. This mismatch can be as large as other model errors; together these errors locally exceed 50 W/m².

4. Mixed-layer Parameterization

Recent improvements to numerical vertical mixing schemes offer a variety of alternatives with reasonable simulations of observed mixed layer dynamics. We decided to adopt the approach of *Chen et al.* [1994], where vertical mixing and vertical friction are parameterized by a one-dimensional mixing scheme. It was implemented for level models by *Power et al.* [1995]. Strong mixing is assumed to occur within a bulk mixed-layer, as in the *Niiler and Kraus* [1977] model. Below the mixed-layer, internal mixing is parameterized by a gradient Richardson number-dependent mixing based on observations by *Peters et al.* [1988]. Their observations show less mixing at higher Richardson numbers than the more widely used *Pacanowski and Philander* [1981] mixing scheme. The new parameterization particularly improves the model performance at lower latitudes, e.g. lower upwelling velocities along the equator; less prominent SST cold tongue in the eastern Pacific: both improving agreement with observations. The hybrid structure of this mixing scheme allows its application to high latitudes (where mixing is strongly influenced by high wind speeds, and thus the Niiler-Kraus part dominates); and also to the equatorial ocean (where vertical mixing is predominantly determined by large vertical current shears, and thus the gradient Richardson number part dominates).

Fig. 2 shows the mixed-layer depth in the Indian Ocean in May for model and observation [Rao et al., 1989]. Rao et al. defined the MLD to be the depth at which temperature was 1°C lower than that at 10 m depth. The model's mixed-layer depth of Fig. 2b is defined the same way, except that the temperature of the top level was substituted for the temperature at 10 m. Comparison of Fig. 2a and b suggests that the model is generally quite successful in simulating the observed seasonal patterns of mixed-layer depth, with maxima and minima of mixed-layer depth in essentially the correct places and of similar value.

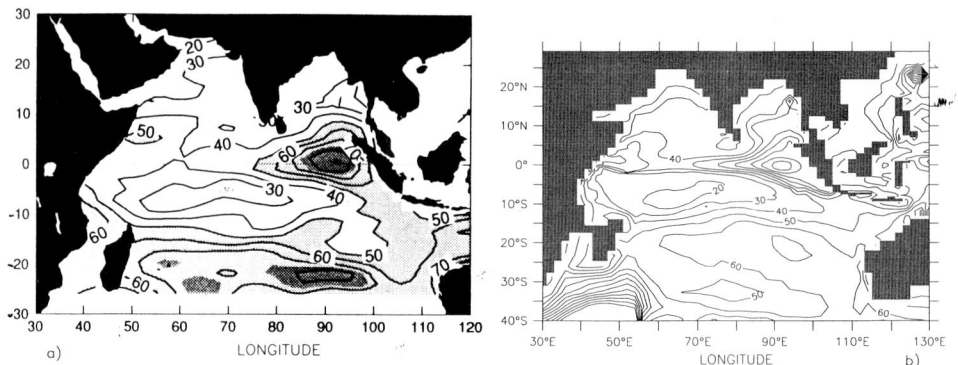

Fig. 2: *Seasonal mixed-layer depths for May: (a) observed [Rao et al., 1989], (b) model [Schiller et al., 1998]. Contour Interval = 10m.*

5. Tidal Mixing in the Indonesian Archipelago

An analysis by *Ffield and Gordon* [1996] suggests that the centre of the tidal-mixing effect on SST is in the Banda Sea. To simulate this observed feature, the vertical mixing coefficients for diffusion and viscosity were increased within the whole water column in the Indonesian area. This was done by defining a spatial Gaussian function similar to the sum of *Ffield and Gordon's* [1996] figures, which describes the spatial shape of the additional vertical mixing. The centre of the additional "tidal mixing" in the model in the Banda Sea has a maximum value of 2×10^{-4} m^2/s. This value gradually decreases as the distance from the Banda Sea increases (Fig. 3). The simulated tidal mixing is independent of time (i.e. no attempt was made to resolve the timescales associated with its physical origin); the only concern is its larger timescale effects on water mass properties. This empirically based parameterization will be subject to improvements as more observations become available.

Incorporating a parameterization of tidal mixing in the Indonesian region leads to more accurate SST and heat flux patterns there. The SST error decreases by more than 0.5°C due to tidal mixing in the Timor Passage, which at the same time reduces the difference between modelled and observed SST [*Reynolds and Smith*,

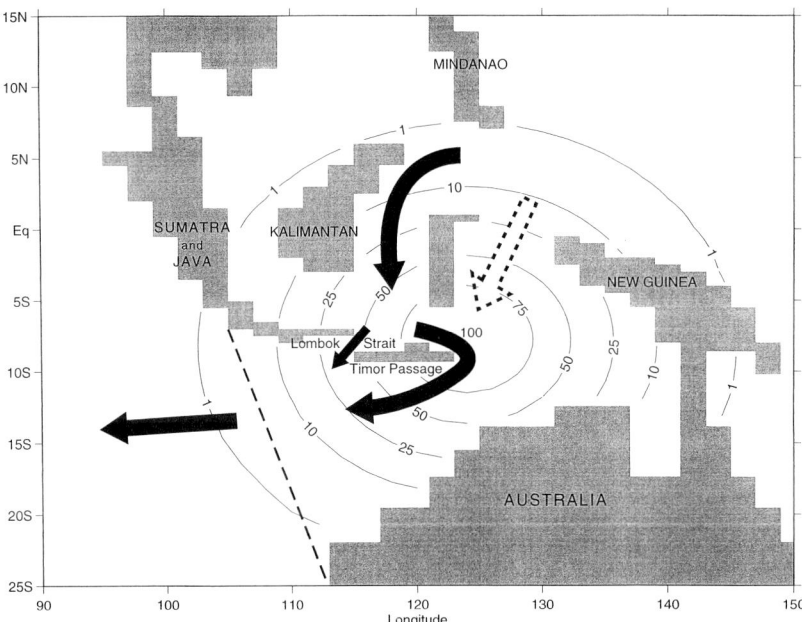

Fig. 3: *Surface topography of the modelled Indonesian archipelago [Schiller et al., 1998]. The two gaps which allow Pacific water masses to enter the Indian Ocean are Lombok Strait and Timor Passage (as indicated by arrows). The isolines denote additional vertical diffusion/viscosity which has been used to simulate tidal mixing. Values are in percent relative to its maximum value of 2×10^{-4} m^2/s in the center of the Banda Sea.*

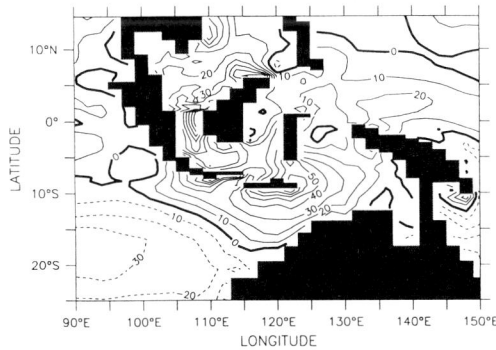

Fig. 4: *Annual mean net surface heat flux of experiment with tidal mixing [Schiller et al., 1998]. Contour Interval = 10 W/m^2.*

1994] to less than 0.3°C in that area. Since SST affects the latent, sensible and longwave radiative heat fluxes, tidal mixing causes a strong increase in ocean heat uptake, which exceeds 60 W/m² south of Timor, leading to an annual average heat flux of over 80 W/m² there (Fig. 4). Heat flux climatologies [e.g., *Josey et al.*, 1998] support this model result.

Effects of tidal mixing are advected along the throughflow into the Indian Ocean, with its strongest signal in the Timor Passage and on the Australian shelf. There, potential temperature differences exceed 1.5°C just below the mixed layer and velocity increases by about 2 cm/s (20%).

6. References

Chen, D., A. J. Busalacchi, and L. M. Rothstein, The roles of vertical mixing, solar radiation, and wind stress in a model simulation of the sea surface temperature seasonal cycle in the tropical Pacific Ocean, *J. Geophys. Res.*, *99*, 20345–20359, 1994.

Ffield, A., and A. L. Gordon, Tidal mixing signatures in the Indonesian Seas, *J. Phys. Oceanogr.*, *26*, 1924–1937, 1996.

Haney, R. L., Surface thermal boundary condition for ocean circulation models, *J. Phys. Oceanogr.*, *1*, 241–248, 1971.

Josey, S. A., E. C. Kent, and P. K. Taylor, *The Southampton Oceanography Centre (SOC) ocean - atmosphere heat, momentum and freshwater flux atlas*, p. 30 & figs, Southampton Oceanography Centre, 1998, Report No. 6.

Kalnay, E., and 21 other authors, The NCEP/NCAR 40-year reanalysis project, *Bull. Am. Met. Soc.*, *77*, 437–471, 1996.

Kleeman, R., and S. B. Power, A simple atmospheric model of surface heat flux for use in ocean modeling studies, *J. Phys. Oceanogr.*, *25*, 92–105, 1995.

Niiler, P. P., and E. B. Kraus, One-dimensional models of the upper ocean, in *Modeling and prediction of the upper layers of the ocean*, edited by E. B. Kraus, pp. 143–172, Pergamon Press, Oxford, 1977.

Pacanowski, R. C., MOM2 documentation user's guide and reference manual, Version 1.0, GFDL Technical Report No. 3, 1995.

Pacanowski, R. C., and S. G. H. Philander, Parameterization of vertical mixing in numerical models of tropical oceans, *J. Phys. Oceanogr.*, *11*, 1443–1451, 1981.

Peters, H., M. C. Gregg, and J. M. Toole, On the parameterization of equatorial turbulence, *J. Geophys. Res.*, *93*, 1199–1218, 1988.

Power, S. B., R. Kleeman, R. A. Colman, and B. J. McAvaney, Modeling the surface heat flux response to long-lived SST anomalies in the North Atlantic, *J. Climate*, *8*, 2161–2180, 1995.

Rao, R. R., R. L. Molinari, and J. F. Fiesta, Evolution of climatological near-surface thermal structure of the tropical Indian Ocean. 1. Description of mean monthly mixed layer depth, and sea surface temperature, surface currents and surface meteorological fields, *J. Geophys. Res.*, *94*, 10801–10815, 1989.

Reynolds, R. W., and T. M. Smith, Improved global sea surface temperature analyses using optimum interpolation, *J. Climate*, *7*, 929–948, 1994.

Schiller, A., J. S. Godfrey, P. C. McIntosh, G. Meyers, and S. E. Wijffels, Seasonal near-surface dynamics and thermodynamics of the Indian Ocean and Indonesian throughflow, in a global ocean general circulation model, *J. Phys. Oceanogr.*, *28*, 2288–2312, 1998.

Stricherz, J., J. O'Brien, and D. Legler, *Atlas of Florida State University tropical Pacific winds for TOGA 1966-1985*, Florida State University, 256 p., 1992.

SYNCHRONEITY OF THE LOW- FREQUENCY PLANETARY WAVE DYNAMICS AND ITS USE TO CREATE A MODEL FOR THE NUMERICAL MONTHLY WEATHER FORECASTING

D.M. SONECHKIN
Hydrometeorological Research Center of Russia
Bolshoy Predtechensky lane 9/13, Moscow 123242, Russia,
E-mail: dsonech@rhmc.mecom.ru

1.Introduction

From the practical point of view the long-range weather forecasting (LRWF) is one of the most important problems in meteorology. Although the predictability theory gives the indication that day-to-day weather variations can be forecasted well for about three-four weeks ahead, in reality the predictability limit of such weather variations is about one week at present. Moreover, because of interactions of short- and long-term weather variations the predictability limit for temporally averaged weather characteristics is almost of the same value. Instabilities of atmospheric processes is the main reason for these limitations, and meteorologists are therefore trying to reduce these instabilities in models in order to represent the long-term weather variations as direct responses of the atmosphere to boundary forcing and thereby enhance the predictability of these long-term variations. Unfortunately, it is not possible to reach this goal without the loss of the qualitative similarity between the modeled and real-world atmospheric dynamics.

A possible way to forecast beyond the present-day predictability limit can be in a forced adjustment of the modeled short- and long-term atmospheric processes. This adjustment can be regarded as analogous to the filtering of high-frequency gravity-inertia waves from the quasi-geostrophic system of equations, which have been used so successfully for short-range weather forecasting. This idea was applied for the first time in the dynamical-stochastical laboratory of the Hydrometeorological Research Center of the USSR (Russia now).

2. The dynamical-stochastical approach to the long-range weather forecasting

The first goal of the research of the laboratory was to understand the consequences for atmospheric low-frequency variability of filtering the instabilities of the high-frequency motions. After consideration of some very simple (toy-) models with strange attractors (Sonechkin, 1984) it was recognized that the reduction of dimensions of these attractors is mainly caused by temporal synchronizations of the modeled planetary wave movements in the extratropical westerlies. Later, the phenomenon of such synchronizations was recognized in the real-world atmospheric dynamics (Vlasova et al., 1989; Sonechkin et al., 1993).

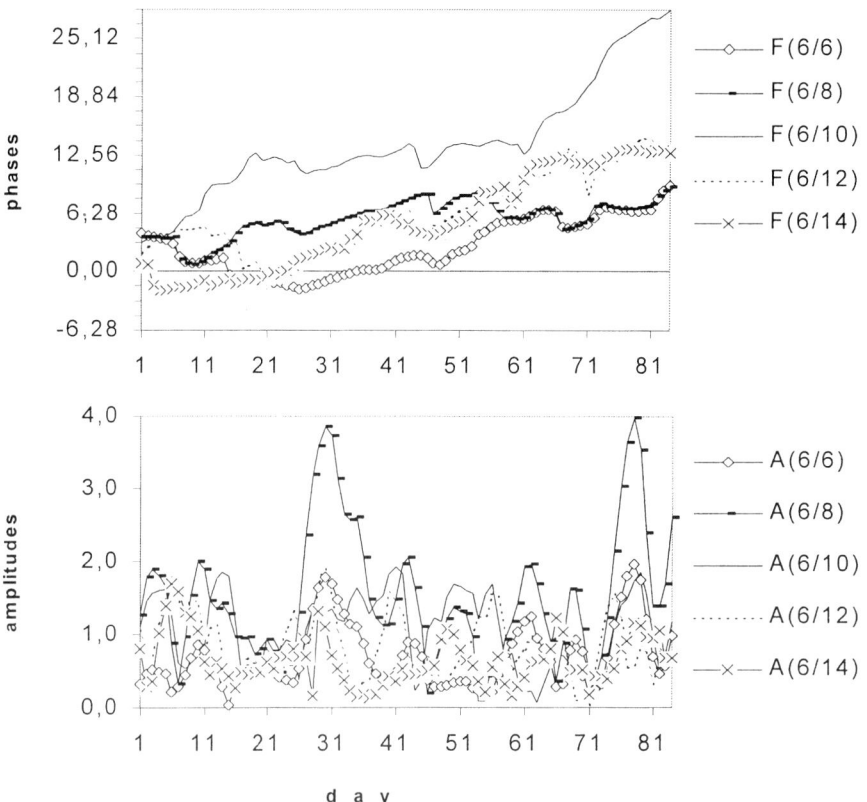

Fig. 1 The time series of amplitudes (A(m/n)) and phases (F(m/n)) of the planetary waves with the same zonal wave number m=6 and the meridional wave numbers n=6, 8, 10, 12, and 14 in the moving five-day- mean 500 hPa geopotential height field of the Northern Hemisphere for the period of 01.12.1995 – 29.02.1996.

It was found that some fingerprints of such synchronizations are seen even in the day-to-day planetary wave movements as some statistical correlations between instantaneous values of the phases of the wave pairs characterized by the same zonal wave number. Similar correlations were found for pairs of the respective wave amplitudes. What is more important, these phase synchronizations look like almost functional relationships between the pairs of phases when the dynamics of the five-day-mean geopotential height field is considered. An illustration of such character of the wave movements in the moving five-day-mean 500 hPa geopotential height field of the Northern Hemisphere is shown in Fig. 1.

The variations of the values of phases (F(m/n)) and amplitudes (A(m/n)) of the waves with the zonal wave number $m=6$ and the meridional wave numbers $n=6, 8, 10, 12$, and *14* over the period 1 December 1995 – 29 February 1996 are represented in this figure. One can see that the variations of all wave phases are rather smooth, and the differences between their simultaneous values are almost constant in time. Some seldom distortions of the smoothness connected with fast jumps of the difference on the $\pm \pi$ value are probably induced by errors in the phase estimations on the base of the objective analysis of the geopotential height field used. These estimations can be especially poor for the time moments when at least one of the amplitudes of a compared wave pair is very small. In contrast, the amplitude values vary much more sharply. Nevertheless, these amplitude variations seem to be more or less synchronous for the different waves the graphs of which are shown in Fig. 1.

Fig. 2 gives another illustration of the synchroneity phenomenon. A temporal sequence of the simultaneous phase values of *F(3/3)* and *F(3/5)* for consequent five-day periods of the 500 hPa geopotential height field of the Northern Hemisphere during the entire 1998 year is shown in this figure. One can see that the simultaneous values of the phases are almost equal ones to others in the majority of five-day periods. Thus, the sequence graph coincides well with the bisector of the figure. The downward deviation of the graph from the bisector on the 2π value seen in the middle third of the sequence may also be treated as a consequence of several large errors in the phase estimation in the time moments of the very small wave amplitudes.

Going back to our toy-model experiments, it must be stressed that some breaks of the synchronization in the modeled dynamics were found. These breaks implied a many-dimensional, very chaotic, i.e. unstable and unpredictable, transient behavior of some fragments of model trajectories up to the time moments when the end-point of such a trajectory did not draw closer the model attractor. Moreover, such a fragment can be relative close to the attractor during a time period. But the end-point of this fragment can leave very far away from the attractor during a next period in order to find the true path actually to draw closer the attractor again during followed times. It implies reiterated temporary predictability losses in the toy-model considered.

One predictability loss of such kind is already well known for GCMs of the primitive equations. It is caused by some nonrealistic gravity wave oscillations during the initial 6-12 hour long period of these equations integration. As well known, these nonrealistic oscillations are induced by inevitable errors of initial data. A forced adjustment of the wind and pressure fields during some initial time steps of the primitive equation integration (an initialization procedure) excludes this first predictability loss. Our conjecture was that a similar initial transient disharmony in the planetary wave motions is just why the present-day predictability limit position on the time axis (about one week) is much more earlier than the predictability theory indicates (about three-four weeks). To overcome this limit we have proposed to initialize the long-range weather forecasting models by means of a forced adjustment of the planetary wave phases with respect to the synchroneity of their low-frequency movement recognized above. Thus, an idea of the filtered hydrodynamical model for the monthly weather forecasting has been formulated.

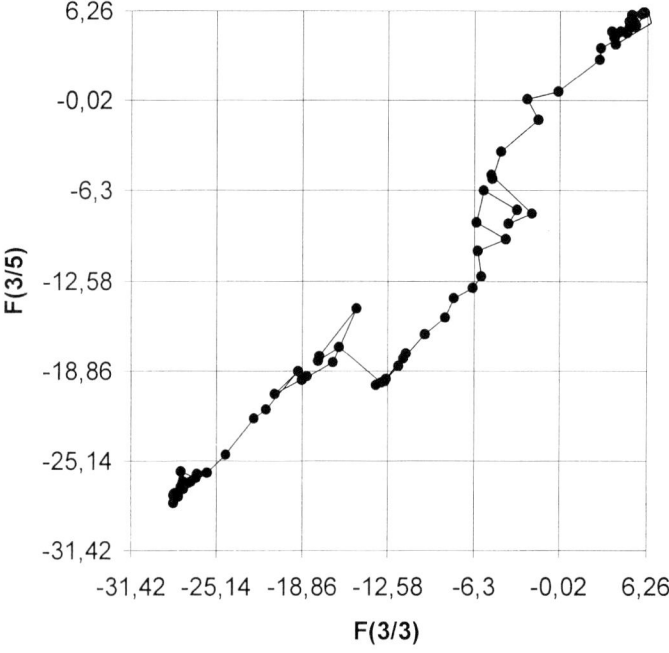

Fig. 2 The sequence of the simultaneous phase values of the wave pair with the zonal wave number m=3 and the meridional numbers n=3 and 5 in the five-day averaged 500 hPa geopotential height field of the Northern Hemisphere during the entire 1998 year. The thin line connects consequent points of the sequence, and the initial five-day-point is located at the upper right edge of the figure.

3.The filtered model

In order to exclude transient nonsynchronized, very unstable and unpredictable planetary wave motions it is possible to put the phase differences of the proper wave pairs to be equal to zero during the entire time period of the forecasting model integration. For the first probe of this idea we chosen a spectral form of the barotropic vorticity equation with zonal flow components Z_n^0, wave amplitudes A_n^m, and wave phases F_n^m of the stream function field, expanded on the spherical harmonic functions, as model variables

(Sonechkin et al., 1995). A typical feature of the form is the presence of the following nonlinear terms in the right sides of the equations for the evolution of zonal components:

$$K_{n1n2n3}^{0m2m3} A_{n2}^{m2} A_{n3}^{m3} \sin(F_{n2}^{m2} - F_{n3}^{m3}), m2 = m3. \qquad (1)$$

Here K_{n1n2n3}^{0m2m3} are the coefficients of the nonlinear interactions between one zonal component and two waves with the same zonal wave number. Similar nonlinear terms exist in the right sides of the equations for the evolution of wave amplitudes:

$$K_{n1n2n3}^{m10m3} Z_{n2}^{0} A_{n3}^{m3} \sin(F_{n1}^{m1} - F_{n3}^{m3}), \quad m1 = m3, \qquad (2a)$$

$$K_{n1n2n3}^{m1m2m3} A_{n2}^{m2} A_{n3}^{m3} \sin(F_{n1}^{m1} + F_{n2}^{m2} - F_{n3}^{m3}), \quad m1 + m2 = m3. \qquad (2b)$$

The equations for the evolution of wave phases include nonlinear terms of the forms:

$$K_{n1n2n3}^{m10m3} Z_{n2}^{0} \frac{A_{n3}^{m3}}{A_{n1}^{m1}} \cos(F_{n1}^{m1} - F_{n3}^{m3}), \quad m1 = m2, \qquad (3a)$$

$$K_{n1n2n3}^{m1m2m3} \frac{A_{n2}^{m2} A_{n3}^{m3}}{A_{n1}^{m1}} \cos(F_{n1}^{m1} + F_{n2}^{m2} - F_{n3}^{m3}), \quad m1 + m2 = m3. \qquad (3b)$$

There are also linear terms in the latter equations which describe the planetary waves displacement by westerlies:

$$K_{n1 n2}^{m10} Z_{n2}^{0}. \qquad (4a)$$

Among these, the main Rossby-Haurwitz term

$$K_{n1\ 1}^{m10} Z_{1}^{0} = -\frac{2 + \sqrt{3} Z_{1}^{0}(n1(n1+1) - 2)}{n1(n1+1)} \qquad (4b)$$

is the main contributor to this displacement. Because our earlier fulfilled investigations (Sonechkin, 1984) have demonstrated that the topographic drag is one of the important drivers of the planetary wave dynamics in the weekly-monthly time scale the quasi-linear terms like

$$K_{n1 n2 n3}^{0m2m3} AH_{n2}^{m2} A_{n3}^{m3} \sin(FH_{n2}^{m2} - F_{n3}^{m3}), \quad m2 = m3 \qquad (5)$$

are added to the equations for the evolution of zonal components. Here AH_n^m, FH_n^m are the amplitudes and phases of the Earth topography. Some respective terms created from (2) and (3) are also added to the equations for the evolution of wave amplitudes and phases. The trigonometric co-factors in (1) – (3) make the values of these terms to be fast oscillating in time. As a result, small errors of initial data, especially of the initial phase values, grow fast during the model integration. In order to prevent this, the terms like (1) and (2) with sines as co-factors are completely rejected from the model equations as well as the nonlinear terms with cosines like (3b). The reason of the latest decision is that during a rather long time period of one week or so, the values of the arguments of the trigonometric co-factors in (3b) vary over entire interval from 0 up to 2π. Therefore, their temporal mean contribution to the evolution of the respective model variables is near zero. Assuming $\cos(F_{n1}^{m1} - F_{n3}^{m3}) = 1$ the nonlinear terms like (3a) are transformed into the following less oscillatory terms

$$K_{n1n2n3}^{m10m3} Z_{n2}^0 \frac{A_{n3}^{m3}}{A_{n1}^{m1}}, \quad m1 = m3 \tag{6}$$

4. Summary

The vorticity equation transformed by such manner was called the filtered model of the atmospheric low-frequency dynamics. This model are regularly integrated with real initial data beginning from 1993. These integrations reveal that the model, despite its spectral resolution is very low (the maximal zonal wave number is equal to 10 only), is capable to depict the dynamics of the spatial patterns of the time averaged 500 hPa geopotential height field for the Northern Hemisphere extratropics. Beginning from 1995 such integrations are carried out each day for 40 days ahead with some success (Sonechkin et al., 1995). The mean H500-field of the 16-20 days of the current month preceding the forecasted one is used as an initial data field for these integrations. The output product of the integration is the five-day-mean H500-fields calculated for eight consequent five-day periods. The forecasted H500-fields of the 3-8 periods, i.e. 1-5, 6-10, 11-15, 16-20, 21-25, and 26-30 days of any forecasted month, are disseminated to several European countries in real time. A grid of 5x10 degrees per a box for the North Atlantic-Europe area is used to represent these forecasted fields. The sequences of the five-day-mean H500-fields obtained as the outputs of the integrations fulfilled each day during the period from 21 up to 30 day of the current month constitute an ensemble of lagged forecasts. This ensemble constitutes a subject of the synoptic interpretation in terms of surface air temperature variations within the forecasted month for some cities of Russia and Central Europe.

References

Sonechkin, D.M. (1984) *Stochasticity in atmospheric general circulation models*, Hydrometeoisdat, Leningrad, - in Russian.
Sonechkin, D.M., Vlasova, I.L, and Zimin, N.E. (1993) Long-period oscillations of zonal flow and wave amplitudes in the 500-hPa geopotential field, *Russian Meteorology and Hydrology* **8**, 21-27.
Sonechkin, D.M., Samrov, V.P., and Zimin, N.E. (1995) The model averaged with respect to planetary wave phases reveals the ability to overcome the weekly predictability limit, *Mon. Wea. Rev.* **123**, 2461-2473.
Vlasova, I.L., Zimin, N.E., and Sonechkin, D.M. (1989) Relaxation oscillations and synchronization of Rossby wave phases, *Soviet Meteorology and Hydrology* **11**, 24-31.

PHYSICAL MECHANISMS OF NONLINEAR EQUILIBRATION OF A BAROCLINICALLY UNSTABLE JET OVER TOPOGRAPHIC SLOPE

G.G. SUTYRIN, I. GINIS, and S.A. FROLOV

Graduate School of Oceanography, University of Rhode Island, Narragansett, 02882 USA

Abstract

Spatio-temporal evolution of meanders on a baroclinically unstable jet over a topographic slope is investigated using pulse asymptotics and numerical simulations. An unperturbed jet is prescribed by a potential vorticity front in the upper layer overlaying intermediate layers with weak potential vorticity gradients and a quiescent bottom layer over a positive (same sense as isopycnal tilt) cross-stream topographic slope.

An initially localized meander evolves into a wave packet growing and propagating downstream. The pulse asymptotics of linear waves allows to characterize the structure of amplifying baroclinic wave packets by spatio-temporal modes which grow exponentially along some rays x/t=const but decay along other rays. In a fully nonlinear numerical solution the instability growth is compensated by the nonlinear terms and the central part of wave packet saturates. The upstream and downstream development of the disturbance near the leading and trailing edges of the wave packet obeys the linear wave theory.

For a weak bottom slope of 0.002 the growth rate is only 10% less than that for a flat bottom. Nevertheless, meanders over a flat bottom are able to pinch off resembling warm- and cold-core rings, while in the presence of a weak bottom slope the maximum amplitudes of meanders and associated deep eddies saturate without eddy shedding.

Two physical mechanisms are important to understand the effects of topographic slope:
- It efficiently controls the nonlinear meander growth via constraining the development of associated deep eddies. The bottom slope modifies the evolution of deep eddies and causes their phase displacement in the direction of the upper layer troughs/crests, thus limiting growth of the meanders.
- Behind the wave packet deep eddies form a nearly zonal circulation which stabilizes the jet. The main equilibration mechanism is homogenization of the lower layer potential vorticity by deep eddies.

1. Introduction.

Many studies in the past decades investigated physical mechanisms of meandering and eddy detachment in baroclinic jets such as the Gulf Stream. These studies have suggested that fluctuations in the Gulf Stream path are caused by baroclinic-barotropic instability processes which are influenced by the beta-effect and bottom topography. Although conventional linear stability theory can predict reasonably well typical wavelengths and phase speeds of growing meanders, nonlinear effects dominate the evolution of large amplitude meanders and ring detachment. Most often the growing instability is studied by constructing an initial flow with along-stream periodic meanders because it allows to use the normal mode method for linear analysis and simple periodic boundary conditions for numerical simulations. This approach is not strictly applicable to oceanographic currents such as the Gulf Stream where the meanders can grow in space downstream of Cape Hatteras. Generally, jet perturbations may arise from an initially localized in space pulse excitation which develops as a wave packet growing and propagating downstream.

A spatio-temporal analysis of pulse asymptotics provides an important conceptual distinction between the so-called absolute and convective instabilities (e.g., Farrel 1982; Pierrehumbert and Swanson, 1995). In the former, the disturbance continues to grow in time at every point which has been reached by the traveling pulse; in particular, it continues to grow at the point of excitation. In the latter, the instability is carried along the stream so that despite its constantly increasing amplitude, at any fixed point the disturbance eventually decreases; in particular, at the point of the excitation the effects of perturbation are eliminated after a finite time. Examples of the nonlinear evolution of such pulse excitation was studied numerically by Swanson and Pierrehumbert (1994) in a two-layer model, Bush et al. (1995) in the nonhydrostatic Boussinesq equations, and Flierl (1999) in a simplified two-layer hybrid model.

In most of above mentioned studies the ocean bottom was assumed to be flat and therefore the results have limited application to the Gulf Stream. The aim of this study is to consider the effects of a topographic slope on nonlinear evolution of a spatially localized pulse in a Gulf Stream-type baroclinic jet.

2. Model Formulation

The numerical model used in this study is the Princeton Ocean Model (POM). The POM is a three-dimensional, primitive equation model with complete thermohaline dynamics. It has an ocean bottom following, sigma vertical coordinate system and a free surface. The latest version of the model is described in detail by Mellor (1998).

For this study the POM is configured as a zonal channel with two open boundaries on the inflow and outflow sides where special boundary conditions are applied to minimize boundary effects. The rectangular channel is 2400 km long and 1200 km wide. The average water depth is set to be 5000 km. The horizontal resolution is 10 km x 10 km that is one third of the mean baroclinic Rossby radius and 1/30 of the dominant meander wavelenth. There are 20 vertical levels with higher resolution in the upper ocean.

A Gulf Stream-type, baroclinic jet in the mean state is initialized as a potential vorticity (PV) front because the stability properties of a jet are strongly linked to its PV distribution which also has an important property of Lagrangian conservation in inviscid flows. A Gulf Stream-type jet is represented by PV gradients in five layers (Logoutov et al., 2000) which are interpolated onto the POM sigma-levels. Although the main PV-front is symmetric in the upper layer, the jet structure in the model is asymmetric due to large difference in the upper layer depth to the south and to the north of the jet, which is in qualitative agreement with the observed Gulf Stream structure.

A set of numerical experiments was conducted with the same initial density field and boundary conditions either for a weak uniform slope (0.002) or for a flat bottom, and either at the f-plane or at the β-plane. A small amplitude, vertically uniform perturbation of the jet stream position, with a Gaussian alongstream profile centered at $x = 0$, was introduced at time $t = 0$, and the subsequent evolution was calculated for 60 days The details of numerical results are desribed by Sutyrin et al. (2001).

3. Physical Mechanisms

An initially localized meander evolves into a wave packet growing and propagating downstream in agreement with the pulse asymptotics. When nonlinear effects come to play, meanders growing over a flat bottom are able to pinch off resembling warm- and cold-core rings, while in the presence of a bottom slope, the maximum amplitudes of meanders and associated deep eddies saturate with no eddy shedding. Through baroclinic instability, the available potential energy is transformed into eddy kinetic energy, while the equilibrated deep eddies eventually release their kinetic energy to the mean zonal flow that stabilizes the jet behind the packet peak. This process is similar to what was obtained for baroclinic life cycles in the atmospheric jets (e.g., Simmons and Hoskins, 1978).

The physical mechanisms of different pulse evolution over the flat and sloping bottom can be elucidated by considering the lower layer PV transformation (Fig. 1). As shown in the right panels in Fig. 1, in the presence of topographic slope the region with PV gradient favorable for development of baroclinic instability between the PV extrema in the lower layer is much narrower than in the case of a flat bottom. Correspondingly, the zonally averaged lower layer PV behind the packet peak becomes nearly homogenized over the sloping bottom while for the flat bottom the PV gradient is only slightly reduced is seen in Fig. 1. The width of a zone favorable for baroclinic instability is controlled by the locations where the initial thermocline slope is equal to the sum of the topographic slope, S, and the ratio

$$S_b = \beta \frac{H-Z}{f}$$

where H is the ocean depth, Z is the thermocline depth in the lower layer, f is the Coriolis parameter, and β is the meridional gradient of f. Baroclinic instability occurs when $S + S_b$ is less than the maximum thermocline slope, $S_{max} \approx 0.0075$ in all cases. Thus, the supercriticality of the flow can be characterized by

$$\gamma = 1 - \frac{S + S_b}{S_{max}}$$

The maximum supercriticality ($\gamma = 1$) is for the flat bottom and $\gamma = 0.73$ for the slope 0.002 at the f-plane. In our experiments at the β-plane, $S_b \approx 0.001$, corresponding to $\gamma = 0.87$ for the flat bottom and $\gamma = 0.6$ for the slope bottom.

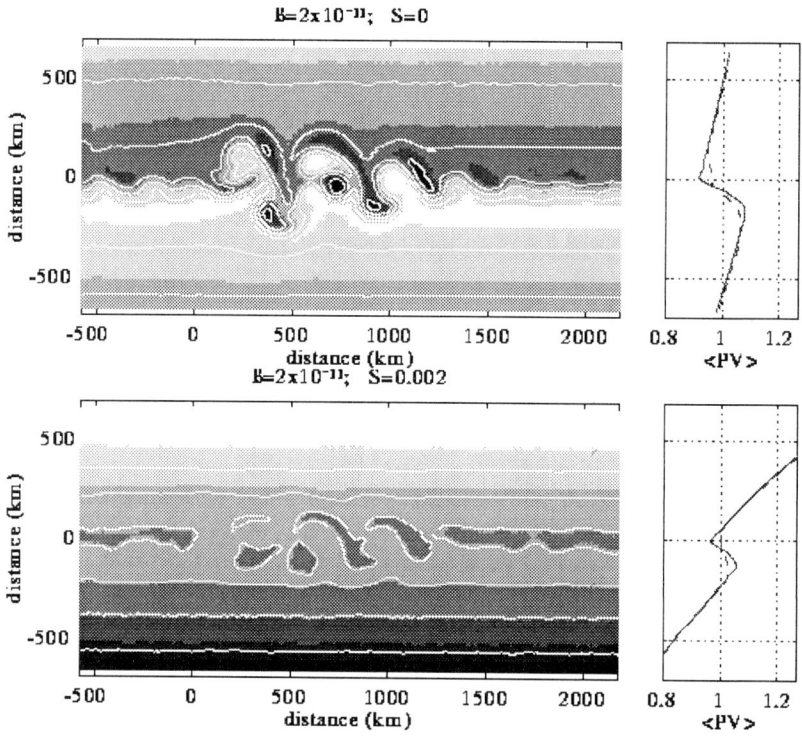

Fig. 1 The lower layer potential vorticity at time = 40 days for the flat bottom (upper panel) and sloping bottom (lower panel) on the β-plane. The right plots show zonally averaged potential vorticity profiles behind the wave packet peak at time = 0 (solid line) and time = 40 days (dashed).

The results can be classified according the supercriticality of the flow. For high $\gamma > 0.87$, meanders growing over a flat bottom are able to pinch off, resembling warm- and cold-core rings, while in the presence of a bottom slope ($\gamma < 0.73$), the maximum amplitudes of meanders and associated deep eddies saturate with no eddy shedding. In the flat bottom case the linear growth rate is only about 10% larger than for the sloping bottom. Nevertheless, at the nonlinear stage the slope efficiently facilitates the saturation of meander growth via constraining the development of deep eddies underneath the jet. The three-dimensional spatio-temporal development of meanders in the sloping bottom case is qualitatively similar to the baroclinic instability amplification considered by Swanson and Pierrehumbert (1994) in a simple two-layer QG model on the beta-plane with a flat bottom. In their case of supercriticality ($\gamma = 0.6$) the beta-effect alone efficiently narrowed the zone favorable for baroclinic instability in the lower layer.

In the sloping bottom cases, meander amplitudes decay behind the packet peak while the deep eddies merge and form a nearly zonal flow. This flow increases the barotropic transport at the warm side and generates a recirculation that is more pronounced at the cold side. The cross-stream asymmetry of the zonal flow in the lower layer beneath the frontal zone found in these cases is qualitatively consistent with the numerical results obtained by Wood (1988) and Boss and Thompson (1999) for periodic perturbations on the beta-plane with a flat bottom. Thus, in the presence of a slope a mature wave packet leaves behind an equilibrated state with asymmetrically modified mean jet structure. The main equilibration mechanism is a homogenization of the lower layer PV by the deep eddies. The width of the homogenized zone is narrower for larger slope and/or on the beta-plane.

4. Comparison with Observations.

Our results demonstrate that the baroclinic instability in all cases is predominantly convective as there is no noticeable upstream propagation of the perturbations. Meanders grow as they propagate downstream and their properties correspond to results of the analysis of observational data (Lea and Cornillon 1996). The dominant wavelength is nearly the same in all our experiments (approximately 300 km). The maximum growth rate at the linear stage is found to be close to the value of 0.14 day^{-1} predicted by a two-layer baroclinic instability model over topography (Orlansky 1969). The smaller value of observed growth rate (0.09 day^{-1}) found by Lee and Cornillon (1996) for meanders below the average amplitude may be due to effects of the deep flow in the Gulf Stream system which is initially absent in our experiments.

The role of a topographic slope in our idealized experiments exhibit several features that are consistent with Gulf Stream observations:
1) Meanders intensify as a topographic slope relaxes, supporting observations that meander amplitudes increase downstream of Cape Hatteras where the Gulf Stream flows into the deep ocean over a weakly sloping bottom that decreases towards the Grand Banks.
2) A topographic slope modifies the evolution of the deep eddies and causes the phase displacement of the deep eddies in the direction of the upper layer troughs/crests, thus

limiting growth of the meanders. This phase-locking of the meanders with the deep eddies underneath agrees qualitatively with the observational data at the SYNOP Central array. 3) The increase of the barotropic transport at the warm side of the jet and the generation of a recirculation on the cold side of the jet is consistent with observations in the Gulf Stream system downstream of Cape Hatteras (Johns et al. 1995).

Acknowledgements: This work was supported by the Office of Naval Research.

5. References

Boss, E., and Thompson, L. (1999) Mean flow evolution of a baroclinically unstable potential vorticity front. *J. Phys. Oceanogr.*, **29**, 273--287.

Bush, A.B.G., McWilliams, J.C. and Peltier, W.R. (1995) The formation of oceanic eddies in symmetric and asymmetric jets. Part I: Early time evolution and bulk eddy transports. *J. Phys. Oceanogr.*, **25**, 1959--1979.

Farrell, B. F. (1982) Pulse asymptotics of the Charney baroclinic instability problem. *J. Atmos. Sci*, **39**, 507--517.

Flierl, G. R. (1999) Thin jet and contour dynamics models of Gulf Stream meandering. *Dyn. Atmos. Oceans*, **29**, 189--215.

Johns, W. E., Shay, T.J., Bane, J.M. and Watts, D.R. (1995) Gulf Stream structure, transport, and recirculation near 68° W. *J. Geophys. Res*, **100**, 817--838.

Simmons, A.J., and Hoskins, B.J. (1978). The life cycles of of some nonlinear baroclinic waves. *J. Atmos. Sci*, **35**, 414—423.

Lea, T., and Cornillon, P. (1996) Propagation of Gulf Sttream meanders between 75° and 45° W. *J. Phys. Oceanogr.*, **26**, 225--241.

Logoutov, O.G., Sutyrin, G.G., and Watts, D.R. (2000) Potential vorticity structure across the Gulf Stream: Observations and a PV-gradient model. *J. Phys Oceanogr.*, **30**.

Mellor, G.L. (1998) Users guide for a three-dimensional, primitive equation, numerical ocean model. *Atmos. and Oceanic Sciences Program*, Princeton Univercity, 39 pp.

Orlansky, I. (1969) The influence of bottom topography on the stability of jets in a baroclinic fluid. *J. Atmos. Sci*, **26,** 1216--1232.

Pierrehaumbert, R.T. and K. L. Swanson, K.L. (1995) Baroclinic Instability. *Ann. Rev. Fluid Mech.*, **27**, 419--467.

Sutyrin, G.G., Ginis, I. and Frolov, S.A. (2001) Equilibration of the Gulf Stream meanders and deep eddies over a sloping bottom. *J. Phys. Oceanogr.*, **31.**

Swanson, K. and R. T. Pierrehumbert, R.T. (1994) Nonlinear wave packet evolution on a baroclinically unstable jet. *J. Atmos. Sci*, **51**, 384--396.

Watts, D.R., Tracey, K.L, Bane, J.M. and Shay, T.J. (1995) Gulf Stream path and thermocline structure near 74° W and 68° W. *J. Geophys. Res*, **100**, 18,291-18,312.

Wood, R. A. (1988) Unstable waves on oceanic fronts: Large amplitude behavior and mean flow generation. *J. Phys. Oceanogr .*, **18**, 775--787.

EVOLUTION OF NEAR-SINGULAR JET MODES

G. E. SWATERS
Applied Mathematics Institute
University of Alberta
Edmonton, AB, T6G 2G1, Canada

1. Introduction

The purpose of this contribution is to very briefly describe some simulations we have done concerning the evolution of modal disturbances to a planar jet in which the phase velocity of the perturbation is initially equal to the maximum jet velocity. For a more complete discussion of this work see Swaters (1999, 2000). As is well known, the perturbation stream function for this configuration is algebraically singular at the jet maximum unlike the logarithmic singularity of a critical layer in a monotonic shear flow.

Perhaps surprisingly, our work clearly shows a "long" time scale oscillation in the underlying modal amplitude even when the numerical simulation is properly initialized with the leading order linear solution and first higher harmonic as determined by weakly-nonlinear asymptotics. A proper theoretical explanation for this behavior remains to be developed.

Our simulations are initialized by perturbing a well-known "jet," the *Bickley* jet (Bickley, 1937), for which there is an explicit solution for an algebraically singular perturbation (Howard and Drazin, 1964).

2. Problem formulation

The nondimensional, inviscid, incompressible two-dimensional Navier-Stokes equations can be written in the form

$$\triangle\psi_t + \psi_x \triangle\psi_y - \psi_y \triangle\psi_x = 0, \qquad (1)$$

where the notation is standard.

The Bickley jet stream function, given by,

$$\psi = \psi_o(y) = -\tanh(y), \quad -\infty < y < \infty, \qquad (2)$$

with corresponding velocity field

$$\mathbf{u} = \mathbf{u}_o(y) = (U_o(y), 0) = (\text{sech}^2(y), 0),$$

is an exact solution to (1).

If we assume a perturbed Bickley jet solution to (1) of the form

$$\psi = \psi_o(y) + \{\varphi(y)\exp[ik(x-ct)] + c.c.\},$$

where k and c are the x-direction wavenumber and complex-valued phase velocity, respectively, where c.c. means complex conjugate and neglect the quadratic perturbation terms, we obtain the Rayleigh stability equation

$$(U_o - c)(\partial_{yy} - k^2)\varphi - U_{o_{yy}}\varphi = 0, \qquad (3)$$

which is solved subject to $|\varphi| \to 0$ as $|y| \to \infty$.

Howard and Drazin (1964) found the singular neutral mode solution for (3) given by

$$\varphi = D\frac{\coth(y)}{\cosh^3(y)} \quad \text{for} \quad (c, k) = (1, 3),$$

where D is a free amplitude constant. Our goal is to examine the finite-amplitude evolution of a near-singular mode for which

$$k = 3 \quad \text{and} \quad c = 1 - \varepsilon, \quad \text{where} \quad 0 < \varepsilon \ll 1.$$

3. Weakly-nonlinear asymptotics

It is convenient to introduce the fast phase and slow space-time variables, given by, respectively

$$\theta = x - (1-\varepsilon)t, \quad (X, T) = \varepsilon^2(x, t), \quad \tau = \varepsilon^3 t.$$

In the outer regions, where $|y| \gtrsim O(1)$, the solution to (1) can be written in the form

$$\psi(\theta, y, X, T, \tau) = -\tanh(y) + \varepsilon^3 \varphi(\theta, y, X, T, \tau). \qquad (4)$$

with the straightforward asymptotic expansion

$$\varphi(\theta, y, X, T, \tau; \varepsilon) \simeq \left[\varphi^{(0)} + \varepsilon\varphi^{(1)} + \varepsilon^2\varphi^{(2)}\right](\theta, y, X, T, \tau) + O(\varepsilon^3).$$

After substantial algebra (see Swaters, 2000), it can be shown that, as $y \to 0$ and to $O(\varepsilon^2)$, the outer solution has the asymptotic form

$$\varphi \simeq A(X, T, \tau)\exp[3i\theta]\left\{\left[\frac{2}{3y} - \frac{7}{9}y + \frac{307}{540}y^3 - \frac{7717}{22680}y^5 + O(y^7)\right]\right.$$

$$+\varepsilon\left[\operatorname{sgn}(y)\left(\frac{128}{15}y^2+\frac{448}{225}y^4+\frac{9616}{4725}y^6\right)\right.$$

$$+\ln|y|\left(\frac{8}{5y}-\frac{28}{15}y+\frac{307}{225}y^3-\frac{7717}{9450}y^5\right)$$

$$\left.+\frac{2}{15y^3}+\frac{29}{15y}-\frac{1843}{900}y-\frac{89627}{22680}y^3-\frac{1200377}{504000}y^5+O\left(y^7\right)\right]$$

$$+\varepsilon^2\left[\frac{2}{35y^5}+\left[\frac{49}{375}-\frac{2i\left(A_T+A_X\right)}{45A}\right]\frac{1}{y^3}+\frac{8\ln|y|}{25y^3}\right.$$

$$\left.+\left[\frac{4132}{875}-\frac{8i\left(A_T+A_X\right)}{15A}\right]\frac{\ln|y|}{y}+\frac{48\ln^2|y|}{25y}\right]+O(1)\bigg\}+c.c.. \quad (5)$$

Examination of (5) suggests secular behavior when $y \simeq O(\sqrt{\varepsilon})$.
In the region where $y \simeq O(\sqrt{\varepsilon})$, the solution to (1) is in the form

$$\psi\left(\theta,\chi,X,T,\tau\right)=-\tanh\left(\sqrt{\varepsilon}\chi\right)+\varepsilon^{\frac{5}{2}}\widetilde{\varphi}\left(\theta,\chi,X,T,\tau\right), \quad (6)$$

where $\chi = y/\sqrt{\varepsilon}$. It follows from (5) that, in the region $y \simeq O(\sqrt{\varepsilon})$, $\widetilde{\varphi}$ has the form

$$\widetilde{\varphi}\simeq\varphi^{(0)}+\varepsilon\,\varphi^{(1,0)}+\varepsilon\ln\varepsilon\,\varphi^{(1,1)}+\varepsilon^2\ln^2\varepsilon\,\varphi^{(2,0)}+\varepsilon^2\ln\varepsilon\,\varphi^{(2,1)}+O\left(\varepsilon^2\right).$$

For our purposes here, all we need are $\varphi^{(0)}$ and $\varphi^{(1,0)}$. Full details can be found in Swaters (2000). The solution for $\varphi^{(0)}$ is

$$\varphi^{(0)}\left(\theta,\chi,X,T,\tau\right)=A\Phi\left(\chi\right)\exp\left(3i\theta\right)+c.c., \quad (7)$$

with

$$\Phi\left(\chi\right)=\chi+\frac{1}{2}\left(1-\chi^2\right)\ln\left(\frac{\chi+1}{\chi-1}\right),$$

where

$$\ln\left(\frac{\chi+1}{\chi-1}\right)=\begin{cases}\ln\left|\frac{\chi+1}{\chi-1}\right| & \text{for } |\chi|>1,\\ \ln\left|\frac{\chi+1}{\chi-1}\right|+\delta\pi i & \text{for } |\chi|<1,\end{cases}$$

where δ ranges from 0 (the nonlinear critical layer) to 1 (the viscous critical layer).

The solution for $\varphi^{(1,0)}$ can be written as

$$\varphi^{(1,0)}=A^2F\left(\chi\right)\exp\left(6i\theta\right)+O\left(\exp\left(3i\theta\right)\right)+c.c., \quad (8)$$

$$F\left(\chi\right)=\frac{\chi}{4\left(1-\chi^2\right)}-\ln\left(\frac{\chi+1}{\chi-1}\right)+\frac{\chi}{4}\ln^2\left(\frac{\chi+1}{\chi-1}\right).$$

4. Numerical simulation

Equation (1) was solved numerically as the system

$$q_t + J(\psi, q) = \frac{1}{R_e}\Delta q, \qquad (9)$$

$$\Delta \psi = q, \qquad (10)$$

where q is the vorticity and R_e is the Reynolds number. We assume a Reynolds number of $R_e = 3.125 \times 10^8$ to effectively smooth out very high wave number features without significantly altering, over the time scales of interest here, the flow evolution.

The numerical procedure we use is a second-order accurate 256×256 finite-difference leap-frog technique (see, e.g., Swaters, 1989, 1999, 2000) in which the Jacobian term is finite differenced using the Arakawa (1966) scheme. To suppress the development of the computational mode a Robert filter (Asselin, 1972) is applied at each time step with a coefficient of 0.005. The stream function was obtained at the end of each time step by inverting (10) using a direct solver.

Our simulations are done in a periodic channel domain, denoted as Ω, given by

$$\Omega = \{(x, y) \mid |x| < x_L, \; |y| < y_L\},$$

in which y_L is chosen so as to have no noticeable effect on the transverse evolution of the perturbation stream function. The stream function satisfies Dirichlet boundary conditions on $y = \pm y_L$ and is smoothly periodic along $x = \pm x_L$. The value of the vorticity on $y = \pm y_L$ was updated using second-order accurate one-sided interior domain differences.

The initial condition is a linear superposition of the leading order near-singular $k = 3$ mode and the leading order $k = 6$ harmonic, as determined by (7) and (8), and can be written as

$$\psi(x, y, t) = -\tanh(y) + \{A\varphi_3(y)\exp(3i[x - (1 - \varepsilon)\tau]) \\ + A^2\varphi_6(y)\exp(6i[x - (1 - \varepsilon)\tau]) + c.c.\}, \qquad (11)$$

with $\tau = 0$ or Δt as needed, and where $\varphi_3(y)$ is the spatially uniformly valid leading order solution for the near-singular $k = 3$ mode given by

$$\varphi_3(y) = \frac{2\varepsilon^3}{3}\left[\operatorname{sech}^3(y)\coth(y) - \frac{1}{y}\right] + \varepsilon^{\frac{3}{2}}\left[\sqrt{\varepsilon}y + \frac{\varepsilon - y^2}{2}\ln\left(\frac{y + \sqrt{\varepsilon}}{y - \sqrt{\varepsilon}}\right)\right],$$

and where $\varphi_6(y)$, which describes the leading order transverse structure of the $k = 6$ harmonic, is given by

$$\varphi_6(y) = \varepsilon^3\left\{\frac{\varepsilon y}{4(\varepsilon - y^2)} - \sqrt{\varepsilon}\ln\left(\frac{y + \sqrt{\varepsilon}}{y - \sqrt{\varepsilon}}\right) + \frac{y}{4}\ln^2\left(\frac{y + \sqrt{\varepsilon}}{y - \sqrt{\varepsilon}}\right)\right\}.$$

We assume $A = \delta = 1.0$ and $\varepsilon = 0.05$. Thus $c = 0.95$ and the critical levels are located at $\pm\sqrt{\varepsilon} \simeq 0.22$. We choose $x_L = y_L = 4\pi/3$. With our grid spacing we had about 13 grid points in between the critical levels, i.e., in the region $|y| < \sqrt{\varepsilon}$, at least initially, for each value of x.

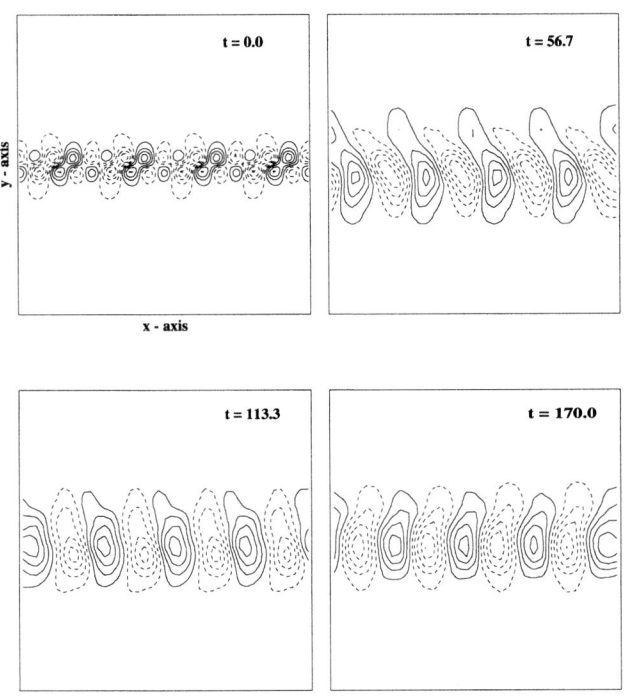

Fig. 1. Contour plots of the perturbation stream function.

In Fig. 1 we show four contour plots of the perturbation stream function for $t = 0.0$, 56.7, 113.3 and 170.0, respectively. The solid and dashed lines correspond to positive and negative stream function values, respectively. The perturbation stream function remains rather stable over the integration although there appears to be some dilation in the individual high and lows.

As a measure of the long time variability in the underlying modal envelope, we computed the area-averaged perturbation kinetic energy normalized by its initial value and then subtracting out the slight linear trend, giving what we call the "residual" $\langle KE \rangle$ (see Swaters, 1999, 2000).

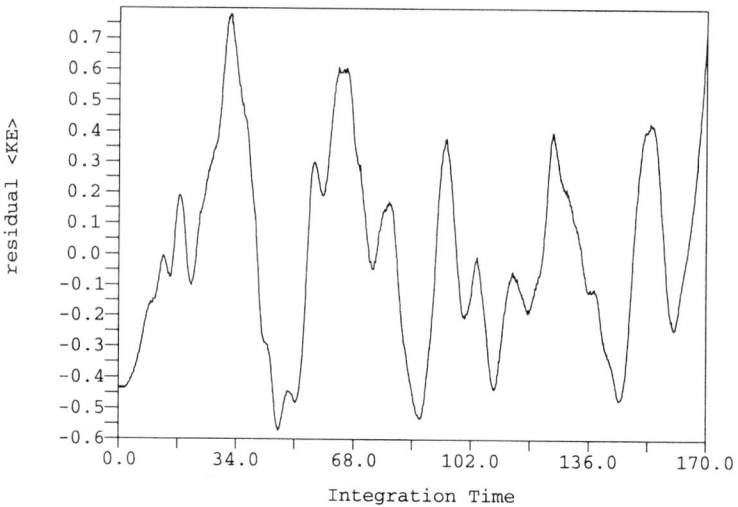

Fig. 2a.

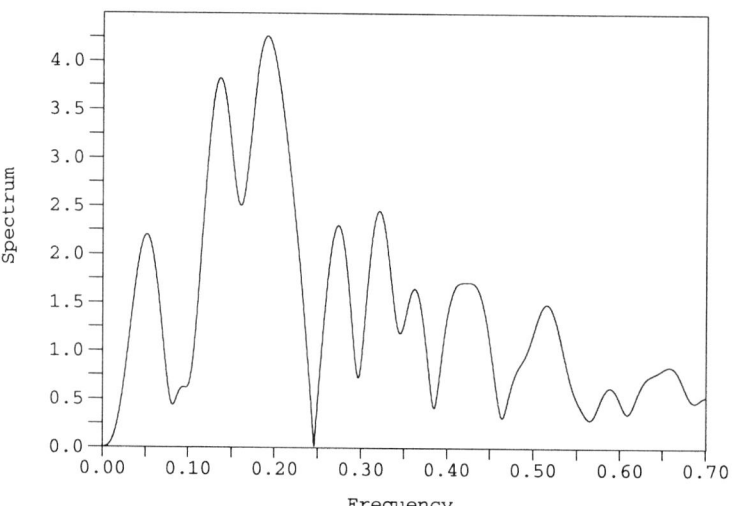

Fig. 2b.

In Fig. 2a we show the residual $\langle KE \rangle$ versus time. One can see that there is a dominant contribution with a period of about 32 time units. This is a longer time scale than the period associated with the underlying fast phase oscillations which is about $2\pi/(kc) \approx 2.2$ time units. In Fig. 2b we show the power spectrum associated with the residual $\langle KE \rangle$. The highest peak is located at a frequency of about 0.19 which corresponds to a period of about 32 time units. This energy peak appears to be somewhat broad with a secondary peak at about frequency 0.13.

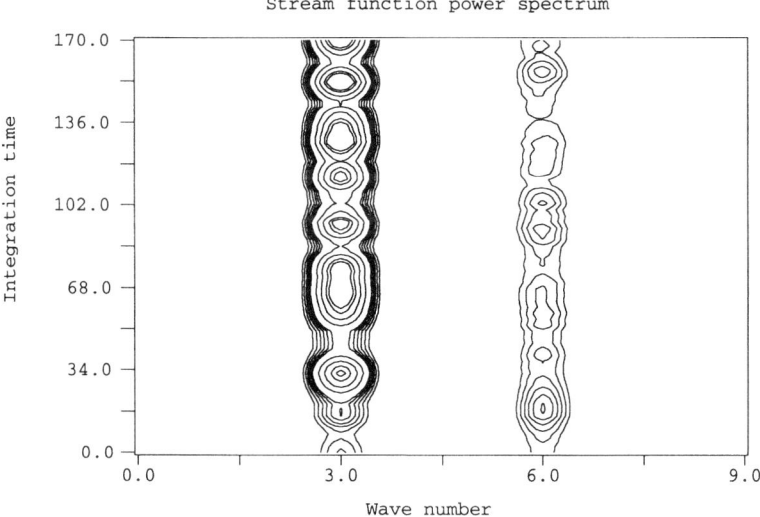

Fig. 3.

To further examine this long time variability we computed a wave number power spectrum for the perturbation stream over time. Because there was little variation as a function of y, it was convenient to average the resulting spectra over y (see Swaters, 1999, 2000) to come up with a power spectrum for the perturbation stream function which is a function of the x-direction wave number and time alone, denoted as $\mathcal{S}(k,t)$.

In Fig. 3 we present a contour plot of $\mathcal{S}(k,t)$. The peak located at $k=3$ corresponds to the contribution from the term proportional to A in (11). The peak located at $k=6$ corresponds to the contribution associated with the term proportional to A^2 in (11). One can see the slow time modulation of the peaks. The time scale of this modulation is consistent with the variability seen in Fig. 2a.

It is interesting, and perhaps somewhat unexpected, that our simulations suggest a relatively stable "slow" neutral oscillation in the perturbation stream function amplitude. A remaining challenge is to derive nonlinear amplitude evolution equations which have oscillatory solutions for these near-singular modes.

References

Arakawa, A. (1966) Computational design for long term numerical integration of the equations of fluid motion: Two-dimensional incompressible flow. Part I. *J. Comput. Phys.* **1**, 119-143.

Asselin, R. A. (1972) Frequency filter for time integrations. *Mon. Wea. Rev.* **100**, 487-490.

Bickley, W. G. (1937) The plane jet. *Phil. Mag.* **23** (7), 727-731.

Howard, L. N. and Drazin, P. G. (1964) On the instability of parallel shear flow of inviscid fluid in a rotating system with variable Coriolis parameter. *J. Math. Phys.* **43**, 83-99.

Swaters, G. E. (1989) A perturbation theory for the solitary drift-vortex solutions of the Hasegawa-Mima equation. *J. Plasma Phys.* **41**, 523-539.

Swaters, G. E. (1999) On the evolution of near-singular modes of the Bickley jet. *Phys. Fluids* **11** (9), 2546-2555.

Swaters, G. E. (2000) Finite-amplitude development of near-singular modes of the Bickley jet. Submitted to *Can. Appl. Math. Quart.*

LINEAR RESONANCE, WKB BREAKDOWN, AND THE COUPLING OF ROSSBY WAVES OVER SLOWLY VARYING TOPOGRAPHY

RÉMI TAILLEUX[1] and JAMES C. McWILLIAMS[2]
(1) LMD, Université Pierre & Marie Curie, Paris 6
Case courrier 99, 4, Place Jussieu, 75252 Paris Cédex, FRANCE
(2) IGPP, University of California Los Angeles
405, Hilgard Avenue, Los Angeles, CA 90095-1567, USA

1. Introduction

WKB theory has been the main analytical tool to investigate the effects of a slowly-varying topography on barotropic and baroclinic Rossby waves in a horizontally uniform stratified ocean for over three decades now, starting with the pioneering work of Rhines (1969). During this period, the main focus was on understanding how the topography modifies the vertical structure of the linear normal modes and the local dispersion relationship (e.g., Straub (1994) and references therein), with little or no attention for their propagation speed and direction. This latter issue was tackled only recently by Killworth and Blundell (1999) (KB99 thereafter), who computed rays and wave amplitudes for baroclinic Rossby waves propagating over realistic topographies (albeit considerably smoothed ones) for the five ocean basins, motivated by the recent analysis of TOPEX/Poseidon altimeter data by Chelton and Schlax (1996). Although KB99's main result is that large scale topography little impacts the phase speeds and wave amplitudes of first-mode baroclinic Rossby waves (compared with standard, linear, normal-mode theory), in Tailleux and McWilliams (2000) (TMC hereafter), we questioned the assumption of no scattering between WKB modes upon which KB99's calculations rely. The support for our claim is illustrated in Fig. 1 which compares the amplitude of annual-period baroclinic Rossby waves excited along an eastern boundary computed by a direct numerical method (left panel) and by means of WKB theory as in KB99 (right panel).

While the WKB solution displays very simple features, in particular a well-defined path for the wave energy reminiscent of the solutions of KB99, the real solution suggests that the wave energy actually splits between different directions when it reaches the bottom of the ridge, so that WKB must breakdown somewhere. The analysis of ray equations shows that this breakdown is neither due to the presence of caustics, nor to the violation of scale separation that WKB requires. This paper is meant to show that the relevant theoretical framework to interpret the results of Fig. 1 is what is called mode conversion theory (Grimshaw and Allen, 1979; Flynn and Littlejohn, 1994). This theory asserts that action splits between two rays at points where two branches of a dispersion relationship come close together, as depicted schematically on the left panel of Fig. 2, the amount of action exchanged depending on the closeness and curvatures of the two branches. At such a point, the wavenumbers and frequency of the wave under study locally satisfy

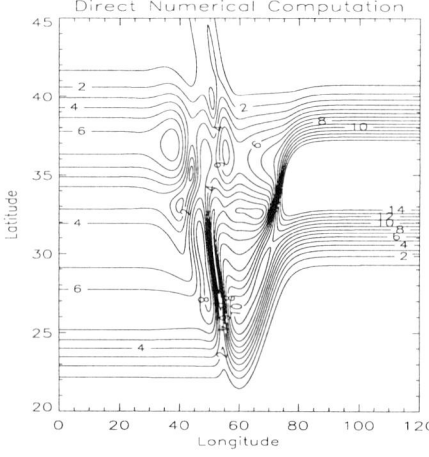 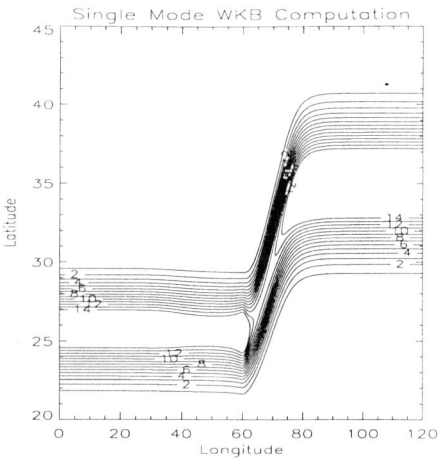

Figure 1: Root-means square of the interface displacement: direct numerical simulation (left panel) versus single mode WKB calculation (right panel).

approximately the other dispersion relationship, so that the phenomenon corresponds to linear resonance. Recently, Vanneste (2000) pointed out relevant references in the present context to explain the empirical results of Hallberg (1997) about localized coupling of surface and bottom-intensified Rossby modes over topography. In this paper, we go farther by showing an explicit example of how mode conversion theory can be used in the case depicted in Fig. 1, by using the theory to compute the successive splitting due to mode conversion of an initial baroclinic ray issued from the eastern boundary. Such a calculation allows us to determine how many different wave packets constitute the solution at any given point. This issue is indeed fundamental to determine how one should interpret the westward propagating features seen in the altimeter data from TOPEX/Poseidon. So far, previous theoretical interpretations have been pursued under the assumption that only one wave packet for which one seeks the characteristics is predominant. The present work suggests that this may often be incorrect.

2. The model

The model uses the two-layer, rigid-lid Planetary Geostrophic Equations that we investigated numerically and theoretically in TMC. In the inviscid and unforced case, these equations can be written under the form

$$S\frac{\partial p}{\partial t} + \left[D_\phi \frac{\partial p}{\partial \phi} + D_\theta \frac{\partial p}{\partial \theta}\right] = 0 \qquad (1)$$

where $p = (p_1, p_2)^T$ are the upper- and lower-layer pressure perturbations, while ϕ and θ denote longitude and latitude respectively. The pressure perturbations are related to

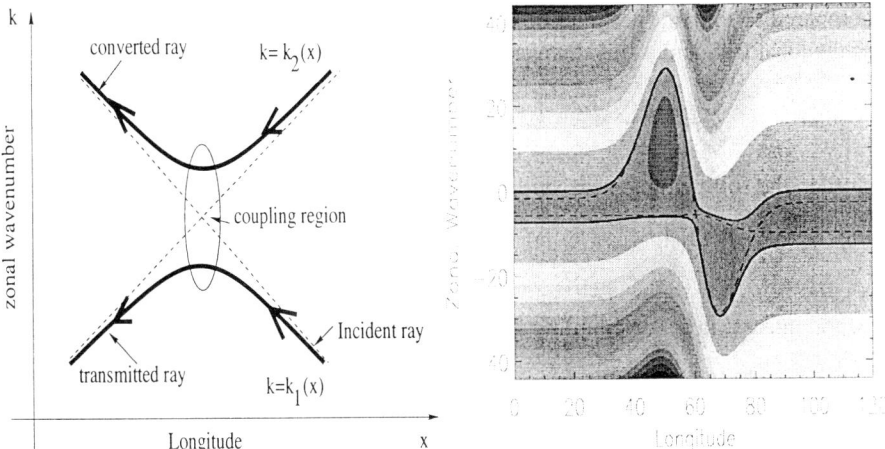

Figure 2: (left panel) Schematic idealized depiction of a standard mode conversion problem. The two thick lines represent two westward propagating rays, while the crossing dotted lines indicate the saddle point of the determinant $\mathcal{D}$ where mode conversion occurs. (right panel) Contour lines of the determinant $\mathcal{D}$ for the problem corresponding to figure 1, showing the ray trajectories and the two saddle points (bottom and top of the ridge's east flank) where mode conversion occurs.

the surface elevation η and the layers' interface displacement η_2 by $p_1 = g\eta$ and $p_2 = g(\eta + \epsilon\eta_2)$, where g is the Earth's gravitational acceleration and $\epsilon = (\rho_2 - \rho_1)/\rho_0$ is a dimensionless measure of the density difference across the two layers. The matrices S, D_ϕ and D_θ are given by

$$S = \begin{pmatrix} 1 & -1 \\ -1 & 1 \end{pmatrix}, \quad D_\phi = \begin{pmatrix} U_1^\phi & 0 \\ 0 & U_2^\phi \end{pmatrix}, \quad D_\theta = \begin{pmatrix} U_1^\theta & 0 \\ 0 & U_2^\theta \end{pmatrix} \quad (2)$$

where the vectors $\mathbf{U}_i = (U_i^\phi, U_i^\theta)$, $i = 1, 2$, are based on the potential vorticity H_i/f of each layer as follows:

$$U_i^\phi = \frac{\gamma}{R_a^2 \cos\theta} \frac{\partial}{\partial \theta}\left(\frac{H_i}{f}\right), \quad U_i^\theta = -\frac{\gamma}{R_a^2 \cos\theta} \frac{\partial}{\partial \phi}\left(\frac{H_i}{f}\right), \quad (3)$$

where $f = 2\Omega\sin\theta$ is the Coriolis parameter, Ω is the Earth's rotation, R_a is the Earth's radius, and $\gamma = g\epsilon$. H_1 and H_2 are the mean thickness of the upper and lower layers respectively. The computational domain is rectangular, is 120° wide, spans 50° in latitude starting from $10°N$, and has a longitudinally varying Gaussian ridge in its center.

3. Results

We seek an approximate solution of (1) by taking p under the form of the WKB ansatz, $p = (p_0 + \epsilon p_1 + \ldots)e^{i\sigma/\epsilon}$, and expanding the result as a series expansion in ϵ assumed to be small compared to unity. The function σ is the classical phase function, from which one defines the local frequency ω and wavenumber $\mathbf{k} = (k_\phi, k_\theta)$ as follows:

$$\omega = -\frac{\partial \sigma}{\partial t}, \qquad k_\phi = \frac{\partial \sigma}{\partial \phi}, \qquad k_\theta = \frac{\partial \sigma}{\partial \theta}. \tag{4}$$

At leading order, the following generalized eigenvalue problem is obtained:

$$\begin{pmatrix} \mathbf{U}_1.\mathbf{k} - \omega & \omega \\ \omega & \mathbf{U}_2.\mathbf{k} - \omega \end{pmatrix} p_0 := M\, p_0 = 0. \tag{5}$$

Of fundamental interest in mode coupling theory is M's determinant, given by

$$\det(M) = \mathbf{U}_1.\mathbf{k}\mathbf{U}_2.\mathbf{k} - \omega(\mathbf{U}_1 + \mathbf{U}_2).\mathbf{k} = \mathcal{D}(\mathbf{k}; \phi, \theta). \tag{6}$$

In order to compute rays for the system, one first needs the dispersion relationship. The latter is obtained by solving $\mathcal{D} = 0$, yielding

$$\omega = \frac{(\mathbf{U}_1.\mathbf{k})(\mathbf{U}_2.\mathbf{k})}{(\mathbf{U}_1 + \mathbf{U}_2).\mathbf{k}} = \Omega(\mathbf{k}; \phi, \theta). \tag{7}$$

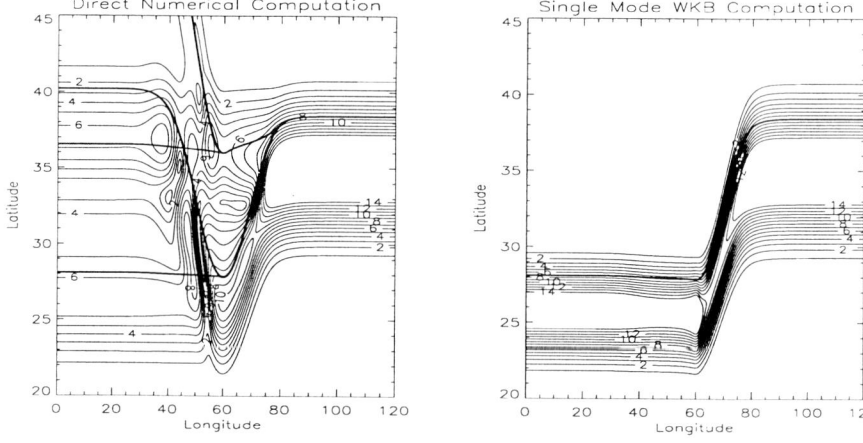

Figure 3: A ray and its subsequent bifurcations computed from one-dimensional linear mode coupling theory, superimposed on the exact numerical computation (left panel) compared with the single-mode WKB prediction (right panel)

According to classical ray theory, e.g. Lighthill (1978), the rays are computed by solving the following Hamiltonian system:

$$\frac{d\phi}{dt} = \frac{\partial \Omega}{\partial k_\phi} = c_{g,\phi}, \qquad \frac{d\theta}{dt} = \frac{\partial \Omega}{\partial k_\theta} = c_{g,\theta}, \tag{8}$$

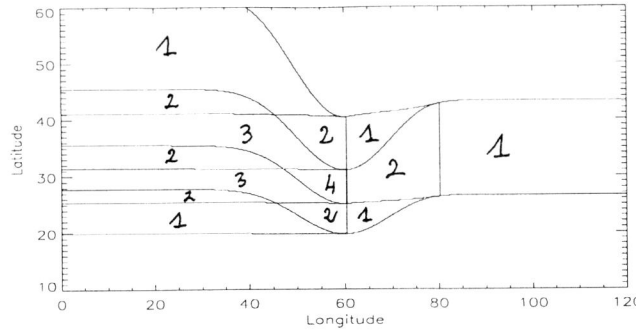

Figure 4: Number of rays constituting the solution at any given point, showing that an interpretation in terms of a single ray is always wrong on the western flank of the ridge.

$$\frac{dk_\phi}{dt} = -\frac{\partial \Omega}{\partial \phi}, \qquad \frac{dk_\theta}{dt} = -\frac{\partial \Omega}{\partial \theta}, \qquad (9)$$

while the wave amplitude is determined by the wave action conservation equation,

$$\mathrm{div}(\mathbf{c}_g A^2) = 0, \qquad (10)$$

where $\mathbf{c}_g = (c_{g,\phi}, c_{g,\theta})$ is the group velocity. The system (8-10) of five ordinary differential equations can be greatly simplified by using ϕ as a pseudo time coordinate, and by using the dispersion relationship (7) (which can be seen as a second-order polynomial in k_ϕ for fixed k_θ and θ) to diagnose k_ϕ. This reduces (8-9) to the following two equations:

$$\frac{d\theta}{d\phi} = \frac{c_{g,\theta}}{c_{g,\phi}}, \qquad \frac{dk_\theta}{d\phi} = -\frac{1}{c_{g,\phi}} \frac{\partial \Omega}{\partial \theta}. \qquad (11)$$

Figure 1 was obtained by integrating (11-10) numerically, with $k_\theta = 0$ (i.e., $\sigma(\phi_E, \theta) = 0$) along the eastern boundary, and by stepping the equations westward. At each step, k_ϕ is diagnosed as the root of $\mathcal{D}$ corresponding to the baroclinic mode, the other root corresponding to the barotropic mode. For a given ray, k_θ and θ can both be regarded as a function of ϕ, and it is then possible to regard M's determinant $\mathcal{D}(k_\phi, k_\theta(\phi), \phi, \theta(\phi))$ along one particular ray as a function of ϕ and k_ϕ, for which Fig. 2 depicts a contour plot. The solid lines highlight the roots of $\mathcal{D}$, with the lower and higher curves corresponding to the baroclinic and barotropic mode respectively. The two thin lines show the saddle points of $\mathcal{D}$, which are the locations where WKB theory breaks down according to linear mode coupling theory. At these points, the action splits between a "converted" and a "transmitted" ray, as depicted schematically on the left panel, and one needs to start a new ray. Figure 3 (left panel) shows one ray and its successive bifurcations, superimposed on the numerical solution reproduced from Fig. 1 (left panel). Mode coupling occurs at all latitudes near the bottom and top of the ridge. The successive mode conversion account qualitatively for the all the features observed in the numerical solution, and contrasts with the single-mode WKB solution shown for comparison on the right-panel. The determination of the mode coupling area allows to determine how many waves of different origins constitute the solution at any given point. Figure 4 shows that the solution can be constituted of either one single or two rays on the eastern flank of the ridge, but that it can be constituted up of four rays on the western flank of the ridge. In the western part of the basin, this maximum number is only three. These results suggest that it may be questionable to interpret the observations of altimetric data of Chelton and Schlax (1996) with theoretical results from single mode WKB theory as most recent studies have done so far.

4. References

1. Chelton D. B., Schlax M.G. 1996: Global observation of oceanic Rossby waves. *Science*, **272**, 234-238

2. Flynn W. G., Littlejohn R.G. 1994: Normal forms for linear mode conversion and Landau-Zener transitions in one dimension. *Annals of Physics*, **234**, 334-403

3. Grimshaw R., Allen J.S. 1979: Linearly coupled, slowly varying oscillators. *Studies in Applied Mathematics*, **61**, 55-71

4. Hallberg R. 1997: Localized coupling between surface and bottom-intensified flow over topography. *J. Phys. Oceanogr.*, **27**, 977-998

5. Killworth P.D., Blundell J.R. 1999: The effect of bottom topography on the speed of long extratropical planetary waves. *J. Phys. Oceanogr.*, **29**, 2689-2710

6. Leblond P.H., Mysak L.A. (1978) *Waves in the ocean.* Elsevier, New-York

7. Lighthill J. (1978) *Waves in fluids.* Cambridge University Press

8. Rhines P.B. 1969: Slow oscillations in an ocean of varying depth. Part I: abrupt topography, *J. Fluid Mech.*, **37**, 161-189

9. Straub D.N. 1994: Dispersion of Rossby waves in the presence of zonally varying topography. *Geophys. Astrophys. Fluid Dyn.*, **75**, 107-130

10. Tailleux R., McWilliams J.C. 2000: Acceleration, creation, and depletion of wind-driven, baroclinic Rossby waves over an ocean ridge. *J. Phys. Oceanogr.*, **30**, 2186-2213

11. Vanneste J. 2000: Mode conversion for Rossby waves over topography: comments on "Localized coupling between surface and bottom-intensified flow over topography", *J. Phys. Oceanogr.*, in press

NONOSCILLATORY ADVECTION SCHEMES WITH WELL-BEHAVED ADJOINTS

JOHN THUBURN
Department of Meteorology
University of Reading, UK
Email: swsthubn@met.rdg.ac.uk

THOMAS W. N. HAINE
Department of Earth and Planetary Sciences
Johns Hopkins University, USA
Email: Thomas.Haine@jhu.edu

1. Introduction

There are many applications in atmosphere and ocean science that require the calculation of sensitivities. Examples include variational data assimilation [7], parameter estimation [9], initialization of ensemble forecasts [1], and finding sensitive areas for the purpose of targeting observations [6]. Sensitivity calculations can often be accomplished efficiently using the adjoint of a tangent linear model, since the computational cost of a single integration of an adjoint is typically no more than a few times the cost of a single integration of its parent forward model.

Because of the linearity assumption behind the tangent linear model and its adjoint, a crucial factor limiting the applicability of an adjoint is nonlinearity in the parent forward model and the evolution it describes. For example, in variational data assimilation for numerical weather prediction the assimilation period is usually limited to about 1 day or less, because on longer timescales the flow no longer has a sufficiently linear dependence on its initial conditions. For similar reasons, adjoints have limited usefulness for quantifying climate sensitivities [3].

A particularly acute form of nonlinearity occurs when the parent model contains switches. For example, condensation in an atmospheric model might switch on when air becomes saturated. The most severe nonlinearity occurs when the model state at some final time depends discontinuously on the initial state; then sensitivities can be unbounded. A second, less severe but still problematic, kind of nonlinearity occurs when the derivative of the final state with respect to the initial state is discontinuous; then sensitivities can be discontinuous or ambiguous. We will show below that this second kind of nonlinearity occurs for typical nonoscillatory advection schemes. Consequently, the results of sensitivity calculations using such schemes depend on exactly how the calculation is carried out; and if it is carried out using an adjoint then the results depend on exactly how the adjoint is implemented. In this case the switching nonlinearity and the resulting ambiguity in sensitivities arise solely from the way the advection terms are discretized; the continuous advection equation contains no such switches.

2. Examples of ambiguous sensitivities

A simple test case is used here to illustrate the nature of the ambiguous sensitivities that can arise with nonoscillatory advection schemes. The test uses a one-dimensional periodic domain, discretized using 20 grid points. Initial conditions for a forward integration are specified as a delta function at the 5th grid point (fig. 1a). An advection scheme is then used to advect the profile left to right at a Courant number of 0.5 for 20 steps. Figure 2b shows the final state when the advection scheme is a third order upwind scheme (QUICKEST [4]) made nonoscillatory by including the Universal Limiter [5]. The set of states connecting fig. 1a and fig. 1b constitute the trajectory about which the scheme will be linearized in the adjoint calculation.

Let the notation q_i^n stand for the mixing ratio of the advected quantity at the ith grid point at the nth time step. We now ask what is the sensitivity $G_j^0 = \partial J/\partial q_j^0$ of a certain functional J of the final state to changes in the initial data q_j^0. For the test case discussed here we take J to be simply the value at the 15th grid point at the final time $J = q_{15}^{20}$. The sensitivity at the final time $G_j^{20} = \partial J/\partial q_j^{20}$ is therefore equal to 1 at grid point 15 and zero elsewhere (fig. 1c).

Two methods were used to compute the sensitivities to initial conditions. The first used an ensemble of forward integrations. For each grid point i the initial condition at point i is perturbed by an amount ε and the forward integration repeated. If the change in the final value at grid point 15 is Δ then the ith component of the sensitivity to initial conditions is $G_i^0 \approx \Delta/\varepsilon$. This estimate would be exact and independent of the value of ε if the advection scheme were linear. Since the scheme is in fact nonlinear we may hope that G_i^0 would exist and that $\Delta/\varepsilon \to G_i^0$ as $\varepsilon \to 0$. Therefore we used a small value $\varepsilon = 0.001$ but checked for convergence by rerunning the ensemble with $\varepsilon = -0.001$.

The results of these ensemble sensitivity calculations are shown in fig. 1e. The solid line is for $\varepsilon = 0.001$ while the dashed line is for $\varepsilon = -0.001$. The two curves are clearly different, indicating that the calculated sensitivities depend on the details of exactly how they are calculated. Further tests confirmed that the two curves do not converge to each other as $|\varepsilon|$ is reduced further.

The second method to compute sensitivities used the adjoint of the tangent linear model

$$G_j^{n-1} = \sum_i G_i^n \frac{\partial q_i^n}{\partial q_j^{n-1}}, \tag{1}$$

which is valid provided $\partial q_i^n / \partial q_j^{n-1}$ exists. It allows the sensitivities at earlier times to be computed from those at the final time. Figure 1d shows the sensitivity at the initial time computed in this way. The results are different from those of either of the two ensemble calculations. Thus, three different calculations give three different results for the sensitivities at the initial time. This is what we mean by saying that calculated sensitivities may be "ambiguous".

This ambiguity arises because $\partial q_i^n / \partial q_j^{n-1}$ does not depend continuously on the q^{n-1}'s and, strictly, is not defined at the points of discontinuity. The essence of the problem is captured by the following simple example. Virtually all nonoscillatory advection schemes involve the computation of the minimum (or maximum) of certain sets of numbers, or some equivalent calculation. Consider the function $f(x, y) = \min(x, y)$. Then for $x < y$

$$\left(\frac{\partial f}{\partial x}, \frac{\partial f}{\partial y}\right) = (1, 0) \tag{2}$$

while for $x > y$

$$\left(\frac{\partial f}{\partial x}, \frac{\partial f}{\partial y}\right) = (0, 1). \tag{3}$$

However, the partial derivatives are discontinuous at $x = y$. This explains why the two ensemble calculations could give such different results: when the situation analogous to $x = y$ occurs in the control integration (or sufficiently close to it) positive and negative values of ε lead to two different regimes where analogues of either (2) or (3) apply.

The example above also shows that when the situation analogous to $x = y$ occurs the partial derivatives, strictly, no longer exist so the adjoint itself, strictly, no longer exists. In order to perform the adjoint sensitivity calculation some arbitrary value must be assigned to the partial derivatives when they are undefined. Equations (2) and (3) give two possible choices. A third is

$$\left(\frac{\partial f}{\partial x}, \frac{\partial f}{\partial y}\right) = \left(\frac{1}{2}, \frac{1}{2}\right). \tag{4}$$

In fact the adjoint calculation shown in fig. 1d used the analogue of (4). Figure 2 shows the results of similar calculations using the analogues of (2) and (3). They are different from each other and different from fig. 1d, showing again that for this nonoscillatory advection scheme the results of sensitivity calculations depend on exactly how the calculation is implemented.

Repeating the sensitivity test case with other nonoscillatory advection schemes typical of those in widespread use showed that in all cases the calculated sensitivities depended on exactly how the calculation was implemented [8].

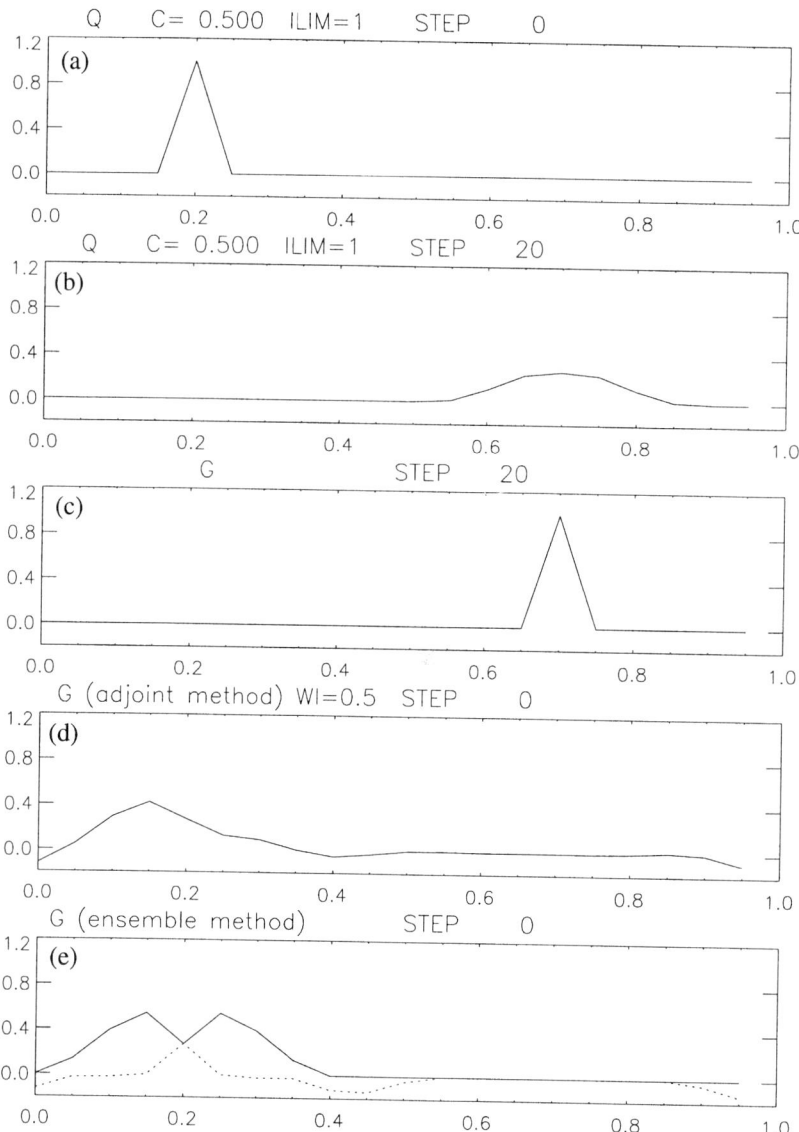

Figure 1. Results of sensitivity test case for the QUICKEST scheme with the Universal Limiter. (a) Initial state for control forward trajectory; (b) Final state for control forward trajectory; (c) Sensitivity at final time; (d) Sensitivity at initial time from an adjoint calculation; (e) Sensitivity at initial time from two ensembles of perturbed forward integrations: solid curve $\varepsilon = 0.001$, dashed curve $\varepsilon = -0.001$.

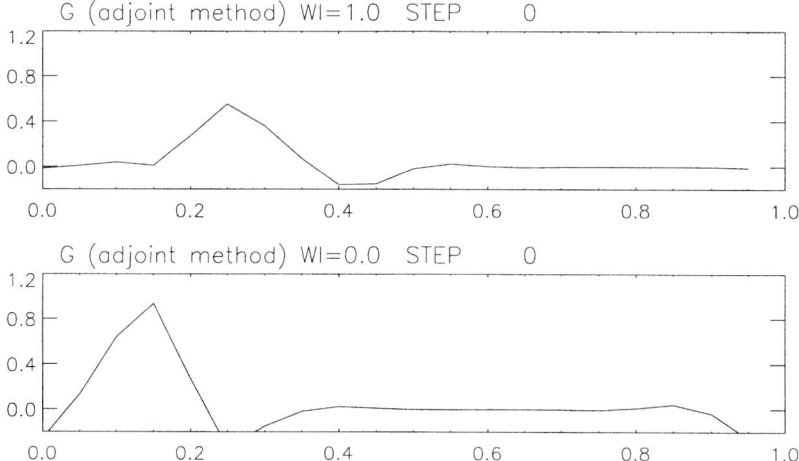

Figure 2. Results of sensitivity test case for the QUICKEST scheme with the Universal Limiter. Sensitivity at initial time from two alternative versions of the adjoint calculation.

3. Semilinear schemes have badly-behaved adjoints.

It would be desirable to have a nonoscillatory advection scheme whose adjoint was well-behaved in sensitivity calculations, and it might be wondered whether such a scheme could be constructed, without sacrificing other desirable properties, for example by smoothing the switches inherent in more conventional nonoscillatory schemes. Here we note a result implying that this goal cannot, in fact, be achieved.

Let $\mathbf{A}$ be the discrete advection operator that maps $\mathbf{q}^{n-1}$ into $\mathbf{q}^n$: $q_i^n = A_i(\mathbf{q}^{n-1})$. Then we will say that the scheme is semilinear if it satisfies

$$A_i(\alpha \mathbf{q} + \beta \mathbf{c}) = \alpha A_i(\mathbf{q}) + \beta \quad (5)$$

for all profiles $\mathbf{q}$ and where α and β are constants and $\mathbf{c}$ and is a constant unit profile $c_i = 1 \; \forall i$. Full linearity implies semilinearity, though the converse is not true. Semilinearity implies, among other things, that the results of an advection calculation are independent of the units that the advected quantity is expressed in, and that advected species whose mixing ratios are initially related by a linear functional relation retain that linear relation under advection. These properties are shared by the continuous advection equation (which is linear in the advected quantity), and are important for certain applications.

An advection scheme will have a well-defined adjoint, with none of the ambiguous sensitivities illustrated above, provided $\partial A_i / \partial q_j$ exists and is a continuous function of the data. It can be shown [8] that a scheme that is semilinear and for which $\partial A_i / \partial q_j$

exists and is continuous must actually be fully linear. Conversely, a scheme that is semilinear but not fully linear cannot have a well-behaved adjoint.

This result has important implications for nonoscillatory advection schemes because, by Godunov's theorem [2], a scheme that is nonoscillatory and better than first order accurate must necessarily be nonlinear. Such schemes may be semilinear, and many practical schemes are because of the desirability of the semilinearity property. However, because they are semilinear but not fully linear they must have badly-behaved adjoints.

4. Conclusion

It has been shown that practical nonoscillatory advection schemes are badly behaved in sensitivity calculations, leading to ambiguous results. There appears to be no way to modify such schemes, for example, by smoothing their inherent switches, so as to obtain good behaviour in sensitivity calculations without sacrificing other desirable properties such as semilinearity. Alternative approaches must therefore be sought, such as simplified versions of schemes for sensitivity calculations, or discretizing the adjoint of the continuous equations rather than using the adjoint of the discretized equations.

References

1. R. Gelaro, R. Buizza, T. N. Palmer and E. Klinker, Sensitivity analysis of forecast errors and the construction of optimal perturbations using singular vectors. *J. Atmos. Sci.*, **55**, 1012-1037 (1998).
2. S. K. Godunov, Finite difference method for numerical computation of discontinuous solutions of the equations of fluid dynamics (in Russian). *Mat. Sb.*, **47**, 271 (1959).
3. D. J. Lea, M. R. Allen, and T. W. N. Haine, Sensitivity analysis of the climate of a chaotic system. *Tellus, in press.* [[]]
4. B. P. Leonard, A stable and accurate convective modelling procedure based on quadratic upstream interpolation. *Comput. Methods Appl. Mech. Eng.*, **19**, 59-98 (1979).
5. B. P. Leonard, The ULTIMATE conservative difference scheme applied to unsteady one-dimensional advection. *Comput. Methods Appl. Mech. Eng.*, **88**, 17-74 (1991).
6. T. N. Palmer, R. Gelaro, J. Barkmeijer, and R. Buizza, Singular vectors, metrics, and adaptive observations. *J. Atmos. Sci.*, **55**, 633-653 (1998).
7. J. N. Thépuat and P. Courtier, Four-dimensional data assimilation using the adjoint of a multilevel primitive equation model. *Quart. J. Roy. Met. Soc.*, **117**, 1225-1254 (1991).
8. J. Thuburn and T. W. N. Haine, Adjoints of nonoscillatory advection schemes. *Submitted to J. Comput. Phys.* (2000).
9. D. Zupanski, A general weak constraint applicable to operational 4DVAR data assimilation systems. *Mon. Wea. Rev.*, **125**, 2274-2292 (1997).

A Statistical Equilibrium Model of Zonal Shears and Embedded Vortices in a Jovian Atmosphere

BRUCE TURKINGTON
Department of Mathematics and Statistics
University of Massachusetts
Amherst, MA 01003

Introduction. A prominent feature of two-dimensional and quasi-geostrophic turbulence is the formation of large-scale coherent structures among the small-scale fluctuations of the vorticity field. This separation-of-scales behavior is a consequence of the conservation of both energy and enstrophy by the dynamics, which results in a net flux of energy toward large scales and a net flux of enstrophy toward small scales. Many flows of this kind, whether free-decaying flows or weakly driven, can therefore be described approximately as coherent, deterministic structures on the large scales and disorganized, random motions on the small scales.

In the geophysical context, numerous examples of long-lived mean flows with these general characteristics are found in the Earth's oceans and atmosphere [11, 13]. Perhaps the most conspicuous examples, however, are observed in the active weather layer on Jupiter, where remarkably robust mean flows of this kind take the form of zonal jets and vortical spots [10, 4].

In this note, we build a model of a mid-latitude band in a Jovian atmosphere, using the methods of equilibrium statistical mechanics to connect the properties of the turbulent small scales of motion to the structure of the coherent flow on the large scale. For the underlying dynamics, we take barotropic, quasi-geostrophic flow on a β-plane with an effective zonal topography resulting from steady flow in a deep lower layer. This 1-1/2 layer model has been used for two decades to capture the essential features of the dynamics of the active upper layer [9]. For the statistical equilibrium theory, we adopt a modern methodology that has evolved over a half century from an insight of Onsager, who predicted that a system of many point vortices in a bounded domain will form a macroscopic vortex at high enough energy. The intervening years have produced the Joyce-Montgomery point-vortex theory, the Kraichnan energy-enstrophy theory and the Miller-Robert continuum vorticity theory. These theories are reviewed from the perspective of geophysical applications in [8], and are critiqued from a theoretical standpoint in [14]. Here, we formulate a simple, yet realistic, model of a specific physical problem and compare its predictions with recent observational data.

Our statistical equilibrium model incorporates the conservation of potential vorticity in a single-point probability distribution on the potential vorticity field, and imposes the conservation of total energy and circulation as constraints. In this formulation the model is defined by a Gibbs ensemble for which the generalized enstrophies (potential vorticity moments) are treated canonically and the energy and circulation invariants are treated microcanonically. Physical reasoning reminiscent of the dual cascade argument justifies this choice of ensemble, because the generalized enstrophies are sensitive the unresolved small-scale motions, while the energy and circulation depend solely on the large-scale flow structures [14].

By taking the continuum limit of the model, we obtain a variational principle that characterizes the equilibrium macrostate, which is the most probable, coarse-grained potential vorticity field. In this analysis we use an exponentially sharp form of the law of large numbers known in probability theory as a large deviation principle [5]. This key result allows us to deduce the macroscopic behavior of the turbulent system from limited information about its microscopic properties.

In the Jovian regime, the equilibrium macrostates are zonal jet patterns or anticyclonic, monopolar vortices embedded in such jets. All these distinguished steady flows are nonlinearly stable, as we show by constructing Lyapunov functionals for them [7]. Our stability argument extends a well-known construction of Arnold [1], which breaks down in many cases of interest including the zonal wind profile in the range of latitudes containing the Great Red Spot [4]. By closing this gap in the stability condition and by identifying these steady flows with most probable macrostates, our statistical equilibrium model offers an explanation of the formation and persistence of the zonal mean flows and the vortices that are created and destroyed within them.

Geostrophic Turbulence. For the microscopic dynamics that underlies the statistical equilibrium model we adopt the equations governing barotropic quasi-geostrophic turbulence [11]. These equations are equivalent to a single nonlinear advection equation for the potential vorticity

$$Q_t - Q_x \psi_y + Q_y \psi_x = 0, \qquad Q = \psi_{xx} + \psi_{yy} - \lambda^{-2}\psi + \beta y + h(y). \qquad (1)$$

The nondivergent velocity field v is is determined from the geostrophic streamfunction ψ by $v = (-\psi_y, \psi_x)$. We assume that the physical variables and fields are nondimensionalized, and we consider flow in the normalized zonal channel $\mathcal{X} = \{(x,y) : |x| < 2, |y| < 0.5\}$. On this domain we impose the boundary conditions, $\psi = 0$ on the walls $y = \pm 0.5$, and periodicity in x.

The model parameters λ and β are the nondimensionalized Rossby deformation radius and gradient of the Coriolis parameter $f(y)$, respectively. In terms of characteristic velocity and length scales, U and L, and the usual physical constants of rotating shallow water theory they are given by

$$\lambda = \frac{\sqrt{gH_0}}{f_0 L}, \qquad \beta = \frac{L^2}{U}\frac{2\Omega}{r_0}\cos\theta_0.$$

We take L to be the zonal channel width and U to be the r.m.s. velocity over the channel. The nondimensionalized height of bottom topography is $h(y)$, which is assumed to be zonal.

Equations (1) also govern the so-called 1-1/2 layer model, in which a shallow upper layer lies on a deep lower layer of denser fluid whose motion is unaffected by that in the upper layer. This is the model commonly used to describe the observed weather layer of the Jovian atmosphere [9, 10, 4]. In the applications to Jupiter, the lower layer flow is assumed to be steady, zonal and geostrophically balanced. The effective topography appearing in (1) is then given by $h(y) = -\lambda^{-2}\psi_2(y)$, where ψ_2 is the streamfunction for the flow in the lower layer, and a reduced gravity g' replaces g.

The conceptual basis of the statistical equilibrium theory is the notion that the microscopic dynamics (1) randomizes the scalar field Q while conserving the invariants of motion, which act as constraints. The goal of the model is then to characterize the typical macroscopic mean flows that persist on the large scales without resolving the small scales of motion of the evolving microstate. For this reason, we distinguish the the fine-grained, or microscopic, potential vorticity field, Q, from a corresponding macroscopic, coarse-grained potential vorticity field, say q. While the validity of the ergodic hypothesis underlying this approach is supported by direct numerical simulations, it is justified principally by the success of its predictions.

The conserved quantities associated with (1) are the total energy H and the total circulation C, which are given by,

$$H = \frac{1}{2}\int_{\mathcal{X}}[\psi_x^2 + \psi_y^2 + \lambda^{-2}\psi^2]\,dx\,, \qquad C = \int_{\mathcal{X}}[Q - \beta y - h(y)]\,dx\,, \qquad (2)$$

and the family of generalized enstrophies $A = \int_{\mathcal{X}} a(Q)\,dx$, where a is an arbitrary (moment) function. For simplicity, here we ignore the linear momentum that is also conserved in channel geometry. It is important to recognize that the "rugged" invariants H and C control the large scales of motion, while the "fragile" invariants A influence the structure of the turbulence on the small scales [14]. This fundamental insight determines the form of the statistical equilibrium ensemble that defines the model.

Gibbs Ensemble. The statistical equilibrium theory is derived by taking the continuum limit of a sequence of lattice models, defined on lattices $\mathcal{L}_n$ having n sites uniformly spaced in the domain $\mathcal{X}$. The phase space for the probabilistic lattice model is R^n, the lattice microstates $Q = (Q(s))$, $s \in \mathcal{L}_n$, being points in R^n. The small-scale structure of these random microstates is described by the probability distribution

$$\Pi_n(dQ) = \prod_{s \in \mathcal{L}_n} \rho(Q(s))\,dQ(s)\,. \qquad (3)$$

Here, $\rho(\sigma)$ is the single-point probability density of Q on $\mathcal{X}$, and σ denotes a real variable running over the range of Q. $\Pi_n(dQ)$ describes the statistical properties of the microstate Q before the conditioning due to the rugged invariants is imposed. Its form (3) is determined by the conservation of phase volume under a suitable discretized dynamics and by the conservation of the fragile invariants. In fact, $\Pi_n(dQ)$ is the canonical ensemble associated with the generalized enstrophies A_n under the identification $\rho(\sigma) = c e^{-a(\sigma)}$.

The statistical equilibrium model is defined by the conditional distribution, or microcanonical ensemble,

$$P_n^{E,\Gamma}(dQ) = \Pi_n \{ dQ \mid H_n(Q) = E, \ C_n(Q) = \Gamma \}, \tag{4}$$

at given values E and Γ of the energy and circulation. In the lattice model the functionals H_n and C_n in (4) are appropriate discretizations of the invariants H and C defined in (2). The vortex interactions governed by H_n are long-range, being determined essentially by the Green function $G(x, x')$ for the partial differential operator $\Delta - \lambda^{-2}$ on $\mathcal{X}$. This property gives this statistical equilibrium model the character of a "local mean-field theory."

Large Deviation Principle for the Coarse-grained Process. With the respect to this sequence of lattice models, we are interested in a double limit in which both the fine-graining index $n \to \infty$, and a corresponding coarse-graining index $r \to \infty$, with $n/r \to \infty$. In this continuum limit we examine the asymptotics of a certain coarse-graining of the random microstate Q, which we now define.

For macroscopic potential vorticity fields we use the square-integrable functions $q \in L^2(\mathcal{X})$. To establish a connection between the microscopic and macroscopic levels of description, we construct a "coarse-grained process" $\overline{Q}_{n,r}$ relative to a partition of the domain $\mathcal{X}$ into a uniform grid of r macrocells, each containing n/r sites of $\mathcal{L}_n$. Namely, we let $\overline{Q}_{n,r}$ be the average of the random microstate Q over each macrocell. Then, $\overline{Q}_{n,r}$ is an $L^2(\mathcal{X})$-valued stochastic process, and every realization of $\overline{Q}_{n,r}$ is a piecewise-constant function with respect to the partition of $\mathcal{X}$ into macrocells.

The probabilistic limiting behavior of the coarse-grained process $\overline{Q}_{n,r}$ is described in terms of the information functional

$$I(q) = \int_{\mathcal{X}} i(q(x)) \, dx, \qquad \text{where} \quad i(\sigma) = \sup_\tau [\sigma \tau - g(\tau)]. \tag{5}$$

Here, $g(\tau)$ is the cumulant generating function for the distribution $\rho(\sigma) d\sigma$, namely,

$$g(\tau) = \log \int_{-\infty}^{+\infty} \exp(\tau \sigma) \, \rho(\sigma) \, d\sigma, \qquad -\infty < \tau < +\infty. \tag{6}$$

The integrand $i(\sigma)$ in (5) is the convex conjugate function, or Legendre transform, of the convex function $g(\tau)$. The functional I, being determined explicitly by the given distribution $\Pi_n(dQ)$, quantifies the statistical properties of the random microstate Q.

With respect to the ensemble $P_n^{E,\Gamma}(dQ)$, the coarse-grained process $\overline{Q}_{n,r}$ satisfies the following large deviation principle in the continuum limit:

For any (regular) subset of macrostates, $B \subset L^2(\mathcal{X})$,

$$\lim_{r \to \infty} \lim_{n \to \infty} \frac{1}{n} \log P_n^{E,\Gamma} \{\overline{Q}_{n,r} \in B\}$$
$$= -\min\{I(q) : q \in B, H(q) = E, C(q) = \Gamma\} - S(E, \Gamma), \qquad (7)$$

where S is the microcanonical entropy

$$S(E, \Gamma) = -\min\{I(q) : H(q) = E, C(q) = \Gamma\} = \lim_{n \to \infty} \frac{1}{n} \log \Pi_n \{H_n = E, C_n = \Gamma\}.$$

This theorem is proved in [6, 7], where it is derived from Cramer's Theorem, the basic large deviation principle for sample means [5]. The form of I as an integral over the domain $\mathcal{X}$ is a consequence of the coarse-graining, in which sample means of the random variables $Q(s)$ are taken over sites s lying in each macrocell of the partition.

Among all admissible macrostates q on the microcanonical manifold $H = E, C = \Gamma$, the large deviation theorem characterizes the most probable macrostate $\bar{q} = \bar{q}(x, y; E, \Gamma)$ as the solution to the constrained minimization problem

$$\text{minimize} \quad I(q) \quad \text{subject to} \quad H(q) = E, \ C(q) = \Gamma. \qquad (8)$$

This maximum entropy macrostate $\bar{q}$ satisfies $I(\bar{q}) = -S(E, \Gamma)$. Any other admissible macrostate q has an exponentially small probability of being observed as a coarse-grained potential vorticity field, since (7) implies that in the continuum limit

$$P_n^{E,\Gamma} \{\overline{Q}_{n,r} \approx q\} \sim e^{-n[I(q) - I(\bar{q})]} \to 0.$$

The equilibrium macrostate $\bar{q}$ is thus the overwhelmingly most probable coarse-grained state compatible with the dynamical constraints, and hence it represents the long-lived, large-scale coherent structure of the turbulent system.

Stable, Steady Mean Flows. The predictive content of the model resides in the constrained variational principle (8) for the equilibrium states. The first-order conditions at a solution $\bar{q} = \bar{q}(E, \Gamma)$ are: $\delta(I + \theta H + \gamma C)(\bar{q}) = 0$, where $\bar{\psi}$ is the streamfunction corresponding to $\bar{q}$, and θ and γ are Lagrange multipliers corresponding to the constraints on H and C. These conditions reduce to the mean-field equation

$$\bar{q} = \Delta \bar{\psi} - \lambda^{-2} \bar{\psi} + \beta y + h(y) = g'(\theta \bar{\psi} - \gamma), \qquad (9)$$

since g' and i' are inverse functions. From (9) it is evident that every solution $\bar{q}$ defines a (deterministic) steady flow in which the mean potential vorticity depends on the mean streamfunction through the profile function g'. The branches of solutions to this nonlinear elliptic equation are parametrized by E and Γ.

The second-order conditions at an equilibrium state $\bar{q}$ are: $\delta^2(I + \theta H + \gamma C)(\bar{q}) \geq 0$, for all variations satisfying the linearization of the constraints at $\bar{q}$, namely, $\delta H(\bar{q}) = 0$, $\delta C(\bar{q}) = 0$. These second-order conditions are intimately related to the stability properties of the solutions. In fact, a Lyapunov functional for $\bar{q}$ can be constructed from the Lagrangian functional $I + \theta H + \gamma C$, using the method of the "augmented Lagrangian" [2]. Specifically, the functional

$$L(q) = I(q) + \theta H(q) + \gamma C(q) + c_E[H(q) - E]^2 + c_\Gamma[C(q) - \Gamma]^2, \qquad (10)$$

with sufficiently large constants c_E and c_Γ, can be shown to have a nondegenerate, unconstrained minimum at the equilibrium macrostate $\bar{q}$. From this property the nonlinear stability of the steady flow induced by $\bar{q}$ follows immediately [7].

It is noteworthy that the well-known nonlinear stability conditions of Arnold [1] are based on a Lyapunov functionals of the form (10), but with $c_E, c_\Gamma = 0$. For this reason, they are only sufficient conditions, which fail to cover some of the mean flows arising in the statistical equilibrium model as most probable states. Moreover, this gap in the classical Arnold stability condition often occurs for flows of great physical interest, including the zonal jets and embedded vortices observed in the Jovian atmosphere. By contrast, our refined stability argument based on the augmented Lyapunov functional (10) yields the stability of all equilibrium states.

Computed Solutions. In the 1-1/2 layer model of the Jovian atmosphere, we consider a channel $\mathcal{X}$ consisting of a zone and a belt. For an effective bottom topography h we take a sinusoid $h = B_2 \sin(2\pi y)$, which is positive in the anticyclonic, upwelling zone, and negative in the cyclonic, downwelling belt. For a single-point distribution ρ on the fine-grained potential vorticity, we choose a Gamma distribution with mean 0, variance 1, and skewness $2\epsilon < 0$. This choice fits the results of direct numerical simulations of balanced, rotating, shallow-water turbulence [12], in which distributions with anticyclonic skewness and exponential tails are typical. We set $\lambda = 0.2$, relying on the available estimates of deformation radius for Jupiter.

We compute branches of solutions $\bar{q}(x, y; E, \Gamma)$ to (8), using the numerical method developed in [15, 3]. The parameters $E, \Gamma, \beta, \epsilon$ are varied to obtain dimensionless solutions compatible with the observed scales $L \sim 10^7 m$ and $U \sim 30 m/s$. Another restriction on these parameters is the approximate relation $\lambda^{-2} \sim -\theta$, which is inferred from the analysis by Dowling of the observed zonal winds on Jupiter [4]. In particular, the equilibrium states have negative temperature, $\theta < 0$, throughout this Jovian regime.

As energy is increased holding the other parameters fixed, the computed solutions transition from zonal jets to embedded monopolar vortices. A representative sequence of three such mean flows is displayed below. In every case, even when the total circulation Γ is cyclonic, the coherent vortices are anticyclones, reflecting the anticyclonic skewness of ρ. When β is increased at fixed E and Γ, the vortices are absorbed into the zonal shears, following the Rhines effect [13, 3]. These shear flows typically have a triple-jet structure with one westward jet bounded by two eastward jets.

Through much of this parameter regime the entropy $S(E, \Gamma)$ is nonconcave with respect to E, indicating a breakdown of the equivalence of ensembles with respect to the energy constraint and of the Arnold stability criterion [7]. Nevertheless, these mean flows are most probable states for our statistical equilibrium ensemble and are nonlinearly stable by virtue of our refined criterion. Moreover, they exhibit the physical scales and qualitative features of the observed coherent structures in the Jovian weather layer.

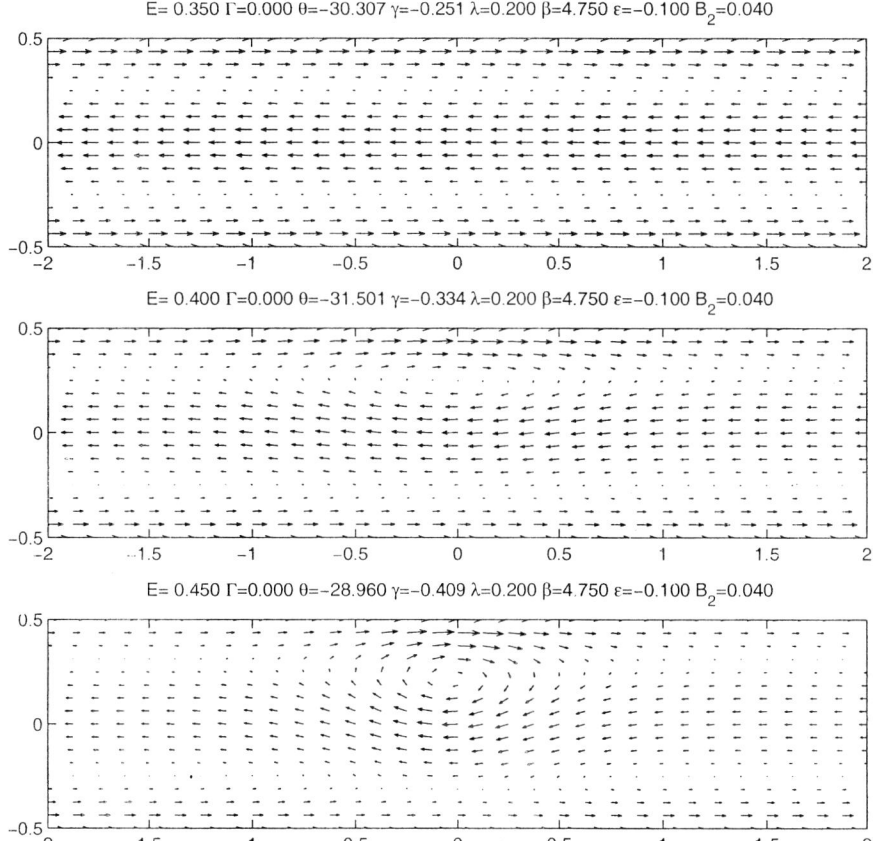

Acknowledgements. The author thanks Kyle Haven for generating the numerical solutions and graphics. This research was partially supported by grants from the Department of Energy (DE-FG02-99ER25376) and the National Science Foundation (NSF-DMS-9971204).

References

[1] V. Arnold. *Mathematical Methods in Classical Mechanics*. Springer-Verlag, New York, 1982.

[2] D. P. Bertsekas. *Constrained optimization and Lagrange multiplier methods*, Academic Press, New York, 1982.

[3] M. DiBattista, A. Majda and B. Turkington. Prototype geophysical vortex structures via large-scale statistical theory. *Geophys. Astrophys. Fluid Dyn.*, 89:235-283, 1998.

[4] T. E. Dowling. Dynamics of Jovian atmospheres. *Annual Rev. Fluid Mech.* 27:293-334, 1995.

[5] R. S. Ellis. *Entropy, Large Deviations, and Statistical Mechanics*. Springer-Verlag, 1985.

[6] R. S. Ellis, K. Haven, and B. Turkington. Large deviation principles and complete equivalence and nonequivalence results for pure and mixed ensembles. To appear in *J. Stat. Phys.*, 2000.

[7] R. S. Ellis, K. Haven, and B. Turkington. Nonequivalent statistical equilibrium ensembles and refined stability theorems for most probable flows. Preprint, 2000.

[8] G. Holloway. Eddies, waves, circulation and mixing: statistical geofluid mechanics. *Annual Rev. Fluid Mech.* 18:91-147, 1986.

[9] A. P. Ingersoll and P. G. Cuong. Numerical model of long-lived Jovian vortices. *J. Atmos. Sci.* 38:2067-2076, 1981.

[10] P. S. Marcus. Jupiter's Great Red Spot and other vortices. *Annual Rev. Astrophys.* 31:523-273, 1993.

[11] J. Pedlosky. *Geophysical Fluid Dynamics*. Springer, 1979.

[12] L. M. Polvani, J. C. McWilliams, M. A. Spall and R. Ford, The coherent structures of shallow-water turbulence: Deformation-radius effects, cyclone/anticyclone asymmetry and gravity-wave generation, *Chaos* 4(2): 177-186, 1994.

[13] P. Rhines. Geostrophic turbulence. *Annual Rev. Fluid Mech.* 11:404-441, 1979.

[14] B. Turkington. Statistical equilibrium measures and coherent states in two-dimensional turbulence. *Commun. Pure Appl. Math.* 52:781-809, 1999.

[15] B. Turkington and N. Whitaker, Statistical equilibrium computations of coherent structures in turbulent shear layers. *SIAM J. Sci. Comput.* 17:1414-1433, 1996.

THE IMPACT OF SMALL-SCALE TOPOGRAPHY ON LARGE-SCALE OCEAN DYNAMICS

J. VANNESTE
Department of Mathematics and Statistics,
University of Edinburgh, Edinburgh EH9 3JZ, UK

1. Introduction

Bottom topography influences the oceanic dynamics in a variety of ways, depending on its height, spatial scale, and on the flow regime. A physically relevant situation, which can be partly investigated analytically, is that of a topography whose scale is much smaller than the typical scale of the motion. Assuming periodic or random topography and small-amplitude flow, the standard technique of homogenization can be applied to derive averaged, or homogenized, evolution equations for the large-scale flow. In such equations, the usual (small-scale) topographic terms are replaced by non-trivial averaged terms which account for the large-scale effect of topography. The homogenization approach can thus be interpreted as providing an asymptotically consistent parameterization of topography in large-scale models.

In this paper, we describe the derivation of homogenized equations for the linearized barotropic quasi-geostrophic model (§2), following Vanneste (2000a); we discuss several limits for which simple analytical results are available (§3), including the cases of two-dimensional topography (Rhines and Bretherton, 1973) and sparse topography (Vanneste, 2000c; Benilov, 2000b); and we conclude by some remarks on extensions of the approach, in particular for the treatment of weakly nonlinear effects.

2. Homogenization of quasi-geostrophic equation

The linearized barotropic quasi-geostrophic equation can be written in the form

$$\partial_t \left(\nabla^2 \psi - \lambda^2 \psi\right) + \beta \partial_x \psi + \frac{f}{H}\partial(\psi, h) + r\nabla^2 \psi = 0, \qquad (2.1)$$

where ψ is the streamfunction, λ^{-1} the radius of deformation, $f + \beta y$ the Coriolis parameter, H the mean depth, h the topography height, and $\partial(\cdot, \cdot)$ the Jacobian operator. Dissipation mechanisms are parameterized by an Ekman friction with coefficient r; a viscous dissipation could also be included but is neglected here for simplicity.

The hypothesis of small-scale topography is made explicit by introducing a small parameter $\epsilon \ll 1$ and by writing the topography height as $h = h(\epsilon^{-1}\mathbf{x}) = h(\boldsymbol{\xi})$, where $\boldsymbol{\xi} := \epsilon^{-1}\mathbf{x}$ is a vector of fast spatial coordinates. The topography is assumed to be periodic or random, with a zero average: $\langle h \rangle = 0$. For simplicity we ignore a possible large-scale component of the topography as well as large-scale changes in its small-scale component; these are can be easily included.

Averaged equations are derived by introducing the multiple-scale expansion

$$\psi = \psi^{(0)}(\mathbf{x},t) + \epsilon \psi^{(1)}(\boldsymbol{\xi},\mathbf{x},t) + \cdots$$

into (2.1). The first non-trivial equation appears at $O(\epsilon^{-1})$, given by

$$\partial_t \nabla_\xi^2 \psi^{(1)} + \frac{f}{H}\partial_\xi(\psi^{(1)}, h) + r \nabla_\xi^2 \psi^{(1)} = \frac{f}{H} \nabla_\xi h \times \nabla \psi^{(0)}. \qquad (2.2)$$

Here, the subscript ξ indicates that the spatial derivatives of the corresponding operator are taken with respect to the fast spatial variables $\boldsymbol{\xi}$, and a projection on the vertical direction is implicit in the use of the cross product $\times$. Equation (2.2) can in principle be solved for a zero-average, periodic or random $\psi^{(1)}$; $\psi^{(1)}$ is obviously linear in $\nabla \psi^{(0)}$, so that we may write

$$\psi^{(1)}(\boldsymbol{\xi},\mathbf{x},t) = \frac{f}{H} \int_0^t \mathbf{w}(\boldsymbol{\xi}, t-\tau) \cdot \nabla \psi^{(0)}(\tau)\, \mathrm{d}\tau,$$

assuming $\psi^{(1)} = 0$ for $t < 0$. The vector $\mathbf{w}(\boldsymbol{\xi},t)$ satisfies the evolution equation

$$\partial_t \nabla_\xi^2 \mathbf{w} + \frac{f}{H}\partial_\xi(\mathbf{w}, h) + r \nabla_\xi^2 \mathbf{w} = \mathbf{z} \times \nabla_\xi h\, \delta(t), \qquad (2.3)$$

where $\mathbf{z}$ is a vertical unit vector. Since the coefficients in this equation depend explicitly on $\boldsymbol{\xi}$ only, it can be solved a priori for a given topography and, in the periodic case, in a single periodic domain. However, solving this so-called cell problem to find $\mathbf{w}$ generally requires numerical computations.

At order $O(1)$, an evolution equation for $\psi^{(2)}$ is found; its solvability requires that the term independent of $\psi^{(2)}$ has zero average. This condition provides the desired averaged evolution equation for the leading-order streamfunction $\psi^{(0)}$. Omitting the superscript (0) it is written as (see Vanneste (2000a) for details)

$$\partial_t \left(\nabla^2 \psi - \lambda^2 \psi\right) + \beta \partial_x \psi + r \nabla^2 \psi + \int_0^t \nabla \cdot \mathbf{r}_t(t-\tau) \cdot \nabla \psi(\tau)\, \mathrm{d}\tau = 0, \qquad (2.4)$$

where

$$\mathbf{r}_t = -\frac{f^2}{H^2}\langle (\mathbf{z} \times \nabla_\xi h) \otimes \mathbf{w}\rangle \qquad (2.5)$$

is a time-dependent but space-independent 2×2 tensor.

The last term in (2.4) encapsulates the averaged effect of topography on the large-scale quasi-geostrophic flow. It should be emphasized that this term drastically changes the nature of the quasi-geostrophic equation by introducing a dependence on the history of the flow. It is formally simple however: when a Laplace or Fourier transform in time is applied, it becomes equivalent to an anisotropic Ekman friction term, with a complex, frequency-dependent friction coefficient given by the transform of $\mathbf{r}_t(t)$.

The calculation of $\mathbf{r}_t$ requires the solution of (2.3). As already pointed out, this generally necessitates numerical means. Analytical progress is however possible in particular cases which are discussed in the next section.

3. Analytical results

3.1. ONE-DIMENSIONAL TOPOGRAPHY

If the topography depends only on one coordinate, $\eta = \epsilon^{-1}y$ say, (2.3) can be solved explicitly. We find that $\mathbf{w} = (w, 0)$, where w satisfies

$$\partial^2_{\eta\eta} w = -\partial_\eta h\, e^{-rt}\theta(t),$$

where $\theta(t)$ is the Heaviside function. Introducing this result into (2.5) shows that the only non-zero component of the tensor $\mathbf{r}_t$ is the component (1,1), given by $Ke^{-rt}\theta(t)$, with $K := f^2\langle h^2\rangle/H^2$. The averaged quasi-geostrophic equation is then conveniently reformulated as the system of two equations

$$\partial_t\left(\nabla^2\psi - \lambda^2\psi\right) + \beta\partial_x\psi + r\nabla^2\psi + K\partial_x\zeta = 0, \qquad (3.1)$$

$$\partial_t\zeta + r\zeta - \partial_x\psi = 0, \qquad (3.2)$$

by introducing the auxilliary variable ζ which is can be interpreted as a damped displacement accross the topography. This system, which is analogous to that describing the evolution of gravity waves in a two-dimensional Boussinesq fluid (for $\lambda = \beta = r = 0$), has been analyzed by several authors with $r = 0$ (Rhines and Bretherton, 1973; Reznik and Tsybaneva, 1999). (See also Benilov (2000a) for a study of instability in the presence of a mean flow.) As an illustration we consider the propagation of waves with spatio-temporal structure proportional to $\exp[i(kx + ly - \omega t)]$. Assuming $\beta = 0$, the frequency ω is readily found to be

$$\omega = \pm\left(\frac{Kk^2}{k^2 + l^2 + \lambda^2} - \delta^2 r^2\right)^{1/2} - ir(1-\delta), \quad \text{with} \quad \delta := \frac{\lambda^2}{2(k^2 + l^2 + \lambda^2)}.$$

When the Ekman friction is negligible, i.e. when $r \to 0$, and when $\lambda = 0$, this dispersion relation reduces to that obtained by Rhines and Bretherton (1973) and is equivalent to the gravity-wave dispersion relation, with $K^{1/2}$ the analogue of the Brünt–Väisälä frequency.

3.2. ISOLATED SEAMOUNTS

Analytical results for a two-dimensional topography can be derived when the topography consists of isolated features (i.e. features separated by distances asymptotically large compared to their size) which are distributed randomly. In this case,

(2.3) may be solved for a single feature in an infinite domain and an asymptotic approximation for $\mathbf{r}_t$ is obtained by ensemble averaging the result.

In Vanneste (2000c), an ensemble of cylindrical seamounts with random radii and heights was considered. It turns out that a single probability density function describing the seamounts enters the leading-order calculation of $\mathbf{r}_t$, namely $Q(H_t)$, the probablity for the seafloor height at any given point to be between H_t and $H_t + dH_t$. The tensor $\mathbf{r}_t(t)$ is best expressed in terms of its Laplace transform, $\hat{\mathbf{r}}_t(s)$, where s is the Laplace variable. It is found to be given by

$$\hat{\mathbf{r}}_t(s) = 2\alpha u \int_{-\infty}^{\infty} \frac{Q(H_t)}{1+(u/\omega_t)^2} \, dH_t \, \mathbf{I} + O(\alpha^2), \qquad (3.3)$$

where $u := s + r$, $\omega_t := fH_t/(2H)$ and $\mathbf{I}$ is the identity tensor. Here α is the area fraction of the ocean's bottom covered by seamounts, a small parameter. (See also Benilov (2000b) for similar results.)

As an example, let us consider the distribution

$$Q(H_t) = \frac{2a^3}{\pi(H_t^2 + a^2)^2},$$

with variance a^2, for the seamount heights. Noting that $\operatorname{Re}(u) > 0$ is imposed by the definition of the Laplace transform, the integration in (3.3) can be performed explicitly, leading to

$$\hat{\mathbf{r}}_t(s) = 2\alpha u \frac{\omega_a^2}{(\omega_a + u)^2} \mathbf{I} + O(\alpha^2), \qquad (3.4)$$

where $\omega_a = fa/(2H) > 0$. A dispersion relation for harmonic waves is easily derived from this expression. Taking the conservative limit $r \to 0$ (with $s = u = -i\omega$) for simplicity, we introduce (3.4) into (2.4) which we solve for ω perturbatively, using the expansion $\omega = \omega^{(0)} + \alpha \omega^{(1)} + O(\alpha^2)$. This yields $\omega^{(0)} = \omega_r := -\beta k/(k^2 + l^2 + \lambda^2)$, and

$$\omega^{(1)} = \frac{-2(k^2+l^2)}{(k^2+l^2+\lambda^2)} \frac{\omega_r \omega_a^2}{(\omega_r^2+\omega_a^2)^2} \left[(\omega_a^2 - \omega_r^2) + i\omega_a \omega_r\right].$$

The flat-bottom Rossby-wave frequency ω_r is of course recovered at leading order. The $O(\alpha)$ correction indicates that topography induces a decrease (in absolute value) of the Rossby-wave frequency, i.e. a slow down of their propagation, if $\omega_a > |\omega_r|$ and an increase otherwise. It also leads to a finite damping associated with the generation of a small-scale flow, even in the limit $r \to 0$. Vanneste (2000c) reached similar conclusions for a Gaussian distribution of seamount heights. Following Benilov (2000b), we emphasize that the use of (3.4) derived assuming $\operatorname{Re} u > 0$ is consistent, even though when $r = 0$ it leads to $\operatorname{Re} u = \operatorname{Im} \omega < 0$; this is because the Laplace inversion technique prescribes the use of an analytic continuation — precisely given by (3.4) — when $\operatorname{Re} u < 0$.

3.3. LONG-WAVE REGIME

If the scale of the motion is much larger than the radius of deformation λ^{-1}, and if the strength of the Ekman friction is sufficient, the time derivative in the auxilliary equation (2.3) can be neglected while the friction term remains. In this case, $\mathbf{w}(t) = \tilde{\mathbf{w}}\delta(t)$, where $\tilde{\mathbf{w}}$ is a time-independent vector that can be computed (numerically in general) by solving

$$\frac{f}{H}\partial_\xi(\tilde{\mathbf{w}}, h) + r\nabla_\xi^2 \tilde{\mathbf{w}} = \mathbf{z} \times \nabla_\xi h.$$

Similarly, the tensor $\mathbf{r}_t(\tau)$ is impulsive in τ, i.e. $\mathbf{r}_t = \tilde{\mathbf{r}}_t \delta(\tau)$, and the averaged equation (2.4) becomes

$$\partial_t \left(\nabla^2 \psi - \lambda^2 \psi\right) + \beta \partial_x \psi + r\nabla^2 \psi + \nabla \cdot \tilde{\mathbf{r}}_t \cdot \nabla \psi(\tau) = 0,$$

where

$$\tilde{\mathbf{r}}_t = -\frac{f^2}{H^2}\langle (\mathbf{z} \times \nabla_\xi h) \otimes \tilde{\mathbf{w}} \rangle.$$

Thus, the effect of the topography in this limit is simply the introduction of an additional (and possibly anisotropic) Ekman friction. Calculation of $\tilde{\mathbf{r}}_t$ in Vanneste (2000a) shows that for sufficiently large topographies this additional friction can dominate the original one, with coefficient r; in particular, in the limit of small r it may be shown that $\tilde{\mathbf{r}}_t \sim r^{1/2}$.

4. Concluding remarks

The homogenization technique discussed in this paper allows the consistent derivation of an averaged quasi-geostrophic equation which describes small-amplitude motion over small-scale topography. In this equation, the effect of topography is parameterized by a history-dependent term of a form ressembling that of an Ekman damping. The history dependence emphasizes the complex effect of topography which drastically modifies wave propagation. This is shown explicitly in simple cases for which complete closed-form analytical results are available. Note, however, that the averaged equation may be employed for much more general situations than those which can be treated entirely analytically: this requires the numerical solution of the partial-differential equation (2.3) for the evolution $\mathbf{w}$. The decay in time of $\mathbf{w}$ is expected to be exponential with a rate of the order of r, so that the solution over a moderate time interval is sufficient for practical purposes.

A limitation of the work reviewed here is the assumption of small-amplitude motion. It is easy to show that the linear averaged equations become invalid when $\psi = O(\epsilon)$. Rescaling ψ accordingly, equation (2.2) for $\psi^{(1)}$ is modified by the addition of the two terms:

$$\partial_\xi(\psi^{(1)}, \nabla_\xi^2 \psi^{(1)}) + \partial_{\xi x}(\psi^{(0)}, \nabla_\xi^2 \psi^{(1)}),$$

where $\partial_{\xi x}(\cdot,\cdot)$ denotes the Jacobian with respect to mixed fast and slow spatial variables $\boldsymbol{\xi}$ and $\mathbf{x}$. The first of these terms is nonlinear and thus makes the derivation of a closed form for $\psi^{(1)}$ impossible in general. When the topography is one dimensional, however, it vanishes, so that $\psi^{(1)}$ can be found as a (nonlinear) function of $\psi^{(0)}$, or more precisely, ζ (see §3.1). In fact, the derivation of the homogenized equation in this case closely follows the linear derivation leading to the system (3.1), the only difference being that the constant K is replaced by the function of ζ,

$$K(\zeta) = \frac{f^2}{H^2}\langle h(\eta+\zeta)h(\eta)\rangle$$

proportional to the correlation of the height field (Volosov, 1976b; Vanneste, 2000b). Thus there is no explicit advective nonlinearity in the averaged equations; the nonlinear effect is simply to modify the restoring force associated with topography which is quantified by K. Explicit solutions of the nonlinear averaged equations can be found in the form of nonlinear waves, solitary waves and fronts (Volosov, 1976a).

Acknowledgements

The author is grateful to G. M. Reznik for pointing out relevant references by Volosov (1976a,b).

References

Benilov, E. S.: 2000a, 'The stability of zonal jets in a rough-bottomed ocean on the barotropic beta plane'. *J. Phys. Oceanogr.* **30**, 733–742.
Benilov, E. S.: 2000b, 'Waves on the beta-plane over "sparse" topography'. *J. Fluid Mech.* In press.
Reznik, G. M. and T. B. Tsybaneva: 1999, 'Planetary waves in a stratified ocean of variable depth. Part 1. Two-layer model'. *J. Fluid Mech.* **388**, 115–145.
Rhines, P. B. and F. Bretherton: 1973, 'Topographic Rossby waves in a rough-bottom ocean'. *J. Fluid Mech.* **61**, 583–607.
Vanneste, J.: 2000a, 'Enhanced dissipation for quasi-geostrophic motion over small-scale topography'. *J. Fluid Mech.* **407**, 105–122.
Vanneste, J.: 2000b, 'Nonlinear dynamics over rough topography: barotropic and stratified quasi-geostrophic theory'. Preprint.
Vanneste, J.: 2000c, 'Rossby-wave frequency change induced by small-scale topography'. *J. Phys. Oceanogr.* **30**, 1820–1826.
Volosov, V. M.: 1976a, 'Contribution to the nonlinear theory of topographic Rossby waves'. *Oceanology* **16**, 427–431.
Volosov, V. M.: 1976b, 'Nonlinear topographic Rossby waves'. *Oceanology* **16**, 217–221.

NUMERICAL EXPERIMENTS ON INTRASEASONAL AND INTERANNUAL VARIATIONS OF THE TROPOSPHERE–STRATOSPHERE COUPLED SYSTEM

S. YODEN AND M. TAGUCHI
Department of Geophysics, Kyoto University
Kyoto, 606-8502, JAPAN

1. Introduction

Time variations of the atmospheric state have two periodic components known as a diurnal cycle and an annual cycle, which are periodic responses to the periodic variations of the external forcings due to the earth's rotation and revolution, respectively. Intraseasonal and interannual variations are defined as deviations from the periodic annual response; generally the intraseasonal variation means low-frequency variation with week-to-week or month-to-month time scales, while the interannual one means year-to-year variation. Some part of these variations is a response to the time variations of the external forcings or boundary conditions of the atmospheric system, while the rest is generated within the system. Well known forcings with interannual time scales are 11-year solar cycle, irregular and intermittent eruptions of volcanos and interannual variations of sea surface temperature, although the last one should be considered as internal variation of the coupled system of the atmosphere and oceans. Quasi-Biennial Oscillation(QBO) of the mean zonal wind in the equatorial stratosphere, which is basically caused by the interaction between the mean zonal wind and equatorial waves propagated from the troposphere, could be considered as variation of a lateral boundary of the mid-latitude stratosphere. On the other hand, external forcings are not so significant in the intraseasonal time scales; fundamentally intraseasonal variations could be considered as internal one which may exist even under constant external conditions. Such variation may be a linear periodic oscillation or nonlinear variation(periodic, quasi-periodic or chaotic) produced in the atmospheric system. Classification of such time variations was done systematically in Yoden(1997) with some simple low-order models.

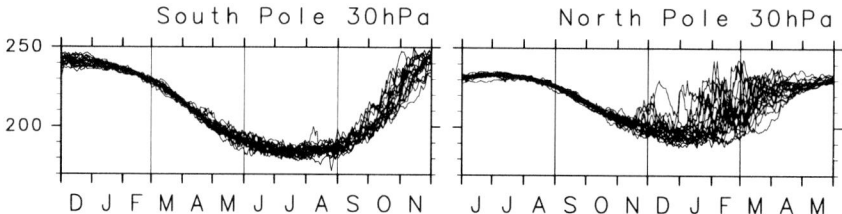

Figure 1. Seasonal variation of daily temperature at 30hPa at the South Pole(left) and the North Pole(right) drawn with NCEP/NCAR reanalysis data for 1979-1997 (Naito, private communication).

Figure 1 shows an example of the variations in the polar stratosphere; daily temperature at the South Pole has large intraseasonal and interannual variability in spring(SON in Fig.1), while large variability at the North Pole is seen in winter and early spring(DJFM). Large fluctuations at the North Pole are direct results of the occurrence of stratospheric sudden warming(SSW) event, which is associated with a rapid breakdown of the polar vortex from a cold, strong and undisturbed state to another warm, weak and disturbed one. The difference between the two hemispheres is quite noticeable. These intraseasonal and interannual variations are also seen in the monthly mean temperature data(e.g., Labitzke, 1982).

We have made some experiments with a hierarchy of numerical models in order to understand these variations in the polar stratosphere and its dynamical linkage to the troposphere. The importance of balanced attack with a hierarchy of numerical models was pointed out by Hoskins(1983) a long time ago, and his advocacy has influenced our experimentation. In this study, the main results are summarized and the usage of hierarchy of numerical models is discussed.

2. Understanding of stratospheric variations

Some simplified models on the interaction between the zonal mean zonal flow and planetary waves propagated from the troposphere have been developed to study the stratospheric variations including SSW events. Yoden(1987) discussed the intraseasonal and interannual variability of the northern hemisphere winter stratosphere by using a low-order model introduced by Holton and Mass(1976, hereafter referred to as HM).

The HM model is a highly-truncated spectral model in a middle-latitude β-channel. It consists of 81-variable ordinary differential equations with time after finite differencing in the vertical direction. Before taking the vertical difference, the dependent variables of the model are the zonal mean zonal flow and streamfunction for a single wave in the form of

$$\overline{u}(y, z, t) = U(z, t) \sin ly, \tag{1}$$
$$\psi'(x, y, z, t) = \text{Re}[\Psi(z, t)e^{ikx}]e^{z/2H} \sin ly. \tag{2}$$

Here the standard notation is used as in Andrews et al.(1987). These are governed by two partial differential equations of the zonal mean and the wave part of quasi-geostrophic potential vorticity equations, which are written schematically as,

$$\frac{\partial U}{\partial t} = \mathcal{F}(U, \Psi; \frac{dU_R}{dz}), \tag{3}$$

$$\frac{\partial \Psi}{\partial t} = \mathcal{G}(U, \Psi; h_B), \tag{4}$$

where important external parameters are shown explicitly; dU_R/dz is the mean zonal flow in radiative equilibrium which is used in the Newtonian heating/cooling radiation scheme to force the zonal mean zonal flow; and $h_B \equiv [f_0 \Psi/g]_{z=z_0}$ is the wave amplitude at the bottom boundary placed near the tropopause level. Wave-mean flow interactions are retained as a wave driving(drag) term in the mean equation (3) and the dependence of wave propagation on U in (4).

Yoden(1987) showed that multiple stable solutions of a cold/strong polar vortex and another warm/weak one may exist for the same external conditions in a finite range of h_B, and discussed a possible application of such concept for understanding the intraseasonal variability of the northern hemisphere winter stratosphere. Some theoretical models of SSW were reinterpreted with the same framework of the HM model. The essence of Matsuno's(1971) theory is impulsive initiation of a wave forcing in the troposphere; transient response to the increase of h_B for a short time interval may cause a rapid transition from the state of cold/strong polar vortex to the other warm/weak state. Another theory of SSW is the vacillation solution found by HM. They showed the periodic variation of the stratosphere that mimics repeated occurrence of SSW events with a period of 50-100 days may exist even for a time-constant h_B. The vacillation is a nonlinear internal variation of the mid-latitude stratospheric system with fixed external conditions.

These two theories made very opposite assumption on the dynamical linkage between the troposphere and the stratosphere. Matsuno(1971) assumed a "slave stratosphere"; the stratospheric variation is caused by the variation of its bottom boundary, i.e., the troposphere, without any stratospheric influence on the troposphere. On the other hand, HM assumed an "independent stratosphere" with a time-constant bottom boundary.

Seasonal and interannual variations of the stratospheric circulation have been studied theoretically with some "independent stratosphere" models. Yoden(1990) varied dU_R/dz in the HM model periodically with an annual component to investigate the response to the periodic forcing, but did not obtained any example of interannual variations(i.e., deviations from the periodic annual response). The periodic response(i.e., the seasonal variation) is qualitatively different depending on h_B, and the difference resembles that of the climatological seasonal march between the southern and the northern hemispheres; larger h_B corresponds to the northern hemisphere. By using similar dynamical conditions in a *hemispheric primitive-equation model*, Scott and Haynes(1998) found interannual variations even for a periodic annual forcing. The internal interannual variability arises because of the longer "memory" of the stratospheric flow at low latitudes. A given wind signal at low latitudes is less affected by radiative damping, of which time scale is much shorter than a year, due to the smaller Coriolis parameter. Internal variability due to the longer memory of low latitude winds may have a role in the interannual variability in the real stratosphere in addition to the variability in the external conditions such as the wave forcing in the troposphere and the low latitude winds influenced by the phase of QBO.

3. Numerical experiments to understand the troposphere-stratosphere coupled variations

The stratosphere only models described in the previous section, either "slave stratosphere" models or "independent stratosphere" ones, assume no downward influence from the stratosphere to the troposphere. However, some observational studies recently pointed out the coupled variability of the troposphere and the stratosphere with intraseasonal and interannual time scales(e.g., Thompson and Wallace, 1998; Hartmann et al., 2000). It is a deep signature of zonally-symmetric seesaw patterns of geopotential height, alternating between the polar region and mid-latitudes, from the surface to the lower stratosphere. They named the deep signature of the polar-vortex variation Arctic Oscillation(AO). Such annular variability is observed in all seasons in both hemispheres and the troposphere-stratosphere coupling is stronger in dynamically active season, namely winter in the northern hemisphere while spring in the southern hemisphere. Downward propagation of the AO signature from the stratosphere to the troposphere was also pointed out by Kuroda and Kodera(1999) and by Baldwin and Dunkerton(1999). Hartmann et al.(2000) argued its relation to SSW events and pointed out that the propagation route of planetary waves is very sensitive to the annular variations. These studies are indicative of the importance of two-way interactions between the troposphere and the stratosphere.

Recent progress in computing facilities enabled us to make some parameter sweep experiments, similar to those done with a low-order model over a decade ago, with more sophisticated three-dimensional models to understand the coupled variability of the troposphere and the stratosphere(Taguchi et al., 2000, Taguchi and Yoden, 2000). An atmospheric general circulation model(GCM) was used after some simplifications of physical processes; all the moist processes were taken out, the radiation code was replaced by a simple Newtonian heating/cooling scheme, Rayleigh friction was used at the surface, and sinusoidal surface topography was assumed in longitudinal direction with zonal wavenumber $m = 1$ or 2 component. It is a spectral primitive-equation model with 42 vertical levels from the surface to the mesopause and its horizontal resolution is given by a triangular

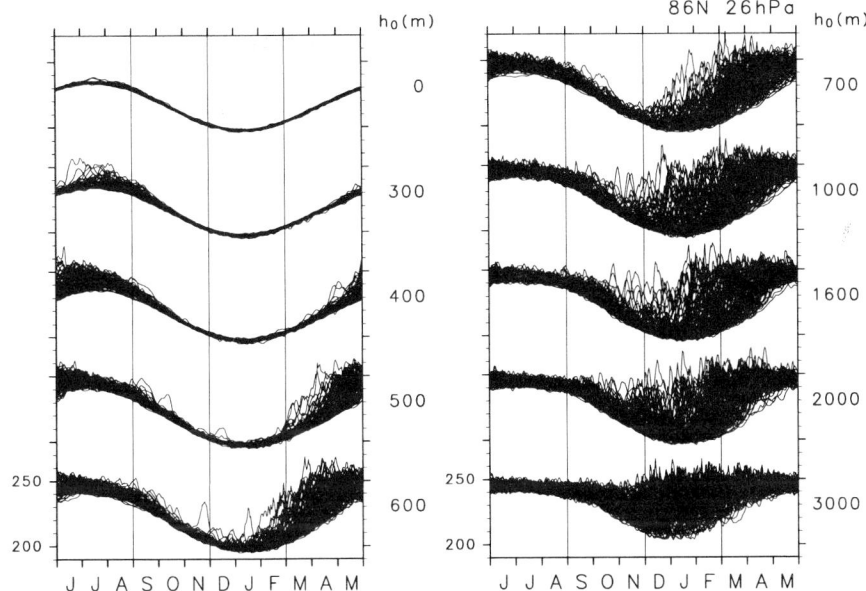

Figure 2. Seasonal variation of daily temperature at 2.6hPa at the Pole for 10 runs of 100-year integrations under a purely periodic annual forcing(Taguchi and Yoden, 2000). The topographic amplitude h_0 of zonal wavenumber 1 is changed from 0 m to 3000 m as the experimental parameter.

truncation at total wavenumber 21 in spherical harmonics. Thus the model has full dynamical process with $O(10^5)$ degrees of freedom.

Taguchi et al.(2000) made 110 runs of 1000-day integrations under a perpetual winter condition with different topographic amplitude of 0 m $\leq h_0 \leq$ 3000 m. Taguchi and Yoden(2000) also did 10 runs of 100-year integrations and 2 runs of 1000-year integrations under a purely periodic annual forcing. The topographic amplitude h_0 was chosen as the principal experimental parameter in order to examine the relative importance of forced planetary waves in the interactions with the zonal mean zonal flow and baroclinic disturbances.

Figure 2 is an example of the parameter sweep in the periodic annual-forcing experiment. If h_0 is equal to zero or a small value, the interannual variation of the polar stratosphere is very small, particularly in winter. The variation becomes large in spring for $h_0 = 400 \sim 500$m, which looks like the variation in the southern hemisphere as shown in Fig.1(left). If h_0 is increased further, the variation becomes large in winter. The results for $h_0 = 700$ or 1000 m look like the variation in the northern hemisphere. Note that the interannual variation in this experiment is caused only by the intraseasonal variations which have a random phase depending on each year. Details on the vertical dynamical link depending on h_0 and the role of forced planetary waves in the interactions are described in the two papers cited above.

4. Concluding remarks

A hierarchy of numerical models have been used to understand the intraseasonal and interannual variations of the troposphere-stratosphere coupled system. In the past, only simple low-order models with $O(10^{1\sim 2})$ variables could be used in parameter sweep experiments due to the limitation of computing facilities. We have to be careful with spurious results due to severe truncations, but they are still useful for conceptual description or illustration of the basic dynamics with limited components. Nowadays it becomes possible to use full dynamical models with $O(10^{4\sim 5})$ variables for parameter sweep experiments. Some idealization of physical processes helps us to understand the mechanism. We do not need to worry about the truncation effect. For quantitative arguments, more complex GCMs with $O(10^{4\sim 7})$ variables are necessary. But parameter sweep experiment with full GCMs is still limited.

Parameter sweep experiments are important to investigate nonlinear systems because of the limitation of linear interpolation in parameter space. Such experiments are useful for understanding the dynamical mechanism. If a system has stepwise transition to a state of complex behavior which is difficult to understand, more regular behavior before the transition might be easier to understand. From the understanding of such regular behavior we could speculate about the more complex behavior. An example of such parameter sweep experiment was shown for the variations of the troposphere-stratosphere coupled system. The coupling process is fundamentally two-way interaction in which roles of the mean zonal flow and planetary waves are important, and the generation of planetary waves is a highly nonlinear process in the range of realistic topographic amplitude.

Acknowledgments This work was supported in part by the Grant-in-Aid for the Research for the Future Program "Computational Science and Engineering" of the Japan Society for the Promotion of Science and by the Grant-in-Aid for Scientific Research of the Ministry of Education, Science, Sports and Culture of Japan.

References

Andrews, D.G., J.R. Holton and C.B. Leovy (1987) *Middle Atmosphere Dynamics*. Academic Press.
Baldwin, M.P. and T.J. Dunkerton (1999) *J. Geophys. Res.*, **27**, 30937–30946.
Hartmann, D.L., et al. (2000) *PNAS*, **97**, 1412–1417.
Holton, J.R. and C. Mass (1976) *J. Atmos. Sci.*, **33**, 2218–2225.
Hoskins, B.J. (1983) *Quart. J. Roy. Meteor. Soc.*, **109**, 1–21.
Kuroda, Y. and K. Kodera (1999) *J. Geophys. Res.*, **26**, 2375–2378.
Labitzke, K. (1982) *J. Meteor. Soc. Japan*, **60**, 124–139.
Matsono, T. (1971) *J. Atmos. Sci.*, **28**, 1479–1494.
Scott, R.K. and P.H. Haynes (1998) *Quart. J. Roy. Meteor. Soc.*, **124**, 2149–2173.
Taguchi, M., T. Yamaga and S. Yoden (2000) *J. Atmos. Sci.*, submitted.
Taguchi, M. and S. Yoden (2000) in preparation.
Thompson, D.W.J. and J.M. Wallace (1998) *Geophys. Res. Lett.*, **25**, 1297–1300.
Yoden, S. (1987) *J. Atmos. Sci.*, **44**, 1723–1733.
Yoden, S. (1990) *J. Atmos. Sci.*, **47**, 1845–1853.
Yoden, S. (1997) *Nonlinear Analysis, Theory, Methods & Applications*, **30**, 4607–4618.

List of Participants

Dr. Vladimir Alexeev
Danish Center for Earth System Science
Juliane Maries Vej 30
University of Copenhagen
DK-2100 Copenhagen 0
Denmark

Email: boba@dcess.ku.dk

Prof. Peter Bartello
Atmos. and Ocean. Sciences
McGill University
805 Sherbrooke West
Montreal, Quebec,
Canada H3A 2K6

bartello@zephyr.meteo.mcgill.ca

Dr. Eugene Benilov
Department of Mathematics & Statistics
University of Limerick
Limerick
Ireland

Eugene.Benilov@ul.ie

CDR. Christopher Butler, USN
Office of Naval Research International
Field Office – Europe
223 Old Marylebone Road
London, NW1 5TH
U.K.

cbutler@onreur.navy.mil

Dr. Carton Xavier
LPO – Ifremer
B.P. 70 – 29280 Plouzane
France

xcarton@ifremer.fr

Prof. Fendinand Baer
Department of Meteorology
University of Maryland
College Park
Maryland 20742
U.S.A

baer@atmos.umd.edu

Prof. J. Ray Bates
Dcess
Juliane Maries VEJ 30
DK-2100 Copenhagen 0
Denmark

jrb@gjy.ku.dk

Dr. Stephen Burk
Naval Research Laboratory
Marine Meteorology Division
Monterey, CA 93943-5502
U.S.A

burk@nrlmoy.navy.mil

Dr. Antonietta Capotondi
NOAA/CIRES
Climate Diagnostics Center
325 Broadway, Boulder,
CO 80303-3328
U.S.A.

mac@cdc.noaa.gov

Dr. Paola Cessi
Scripps Institution of
Oceanography
UCSD-0230
La Jolla, CA 92093-0230
U.S.A.

pcessi@ucsd.edu

Mr. Paul Choboter
Dept. of Mathematical Sciences
University of Alberta
Edmonton, Alberta
T6G 2G1, Canada

choboter@atlantic.math.ualberta.ca

Prof. Peter Chu
Dept of Oceanography
Naval Postgraduate School
Monterey
CA 93943
U.S.A.

chu@nps.navy.mil

Prof. H.C. Davies
Institute for Atmospheric Science, ETH
Honggerberg HPP
CH-8093 Zurich
Switzerland

davies@atmos.umnw.ethz.ch

Dr. Klara Finkele
Met Eireann
Glasnevin Hill
Dublin 9

Klara.Finkele@met.ie

Dr. Eugene Gath
Dept. of Mathematics & Statistics
University of Limerick
Limerick
Ireland

Eugene.Gath@ul.ie

Mr. Vivian Choy
Department of Mathematical Science
Loughborough University
Loughborough
Leicester
U.K. LE11 3TU

vchoy@neumann.maths.monash.edu.au

Dr. Simon Clarke
Dept. of Maths & Stats
Monash University
3800 Victoria
Australia

src@mail.maths.monash.edu.au

Prof. Dale Durran
Dept. of Atmospheric Sciences
University of Washington
Seattle, Washington,
U.S.A.

durrand@atmos.washington.edu

Dr. Michael Fox-Rabinovitz
ESSIC, CSS Bldg.
University of Maryland
College Park
MD 20742
U.S.A.

foxrab@atmos.umd.edu

Dr. Peter Gent
NCAR
Bou!der
Colorado
U.S.A.

gent@cgd.ucar.ed

Prof. Alexander Gluhovsky
Purdue University
1397 Civl Building
West Lafayette
IN47906
U.S.A.
aglu@purdue.edu

Mr. William I. Gustafson
Dept. of Land, Air and Water Resources
University of California, Davis
1 Shields Ave.
Davis, CA 95616-8627
U.S.A.

ivor@ucdavis.edu

Prof. George Haller
Division of Applied Mathematics,
Brown University
182 George Street
Providence, RI 02912
U.S.A

haller@cfm.brown.edu

Prof. Frank Hodnett
Dept. of Mathematics & Statistics
University of Limerick
Limerick
Ireland

frank.hodnett@ul.ie

Dr. Xin Huang
Woods Hole Oceanographic Institution
Woods Hole
MA02543
U.S.A.

rhuang@whoi.edu

Prof. Roger Grimshaw
Dept. Mathematical Sciences
Loughborough University
Loughborough
Leics, LE11 3TV
U.K.

R.H.S.Grimshaw@lboro.ac.uk

Prof. Michael Hayes
Dept. of Mathematical Physics
University College
Dublin

Michael.Hayes@ucd.ie

Mrs. Diane Hodnett
Dept. of Mathematics & Statistics
University of Limerick
Limerick
Ireland

dianehodnett@eircom.net

Dr. Anthony Hollingsworth
ECMWF
Shinfield Park
Reading, Berks, RG2 9AX
U.K.

a.hollingsworth@ecmwf.int

Dr. Chris W Hughes
Proudman Oceanographic Laboratory
Bidston Observatory
Bidston Hill
Prenton CH43 7RA
U.K.

cwh@pol.ac.uk

Dr. Nicholas Keeley
J.C.M.M.
Dept. of Meteorology
University of Reading
Reading, Berkshire RG6 4BB
U.K.

nkeeley@meto.gov.uk

Prof. Yoshifumi Kimura
Graduate School of Polymath
Nagoya University
Chikusa-ku
Nagoya 464-8602
Japan

kimura@math.nagoya-u.ac.jp

Prof. Piero Lionello
University of Lecce
Via per Arnesano
73100 Lecce
Italy

lionello@pd.infn.it

Dr. Ray McGrath
Met Eireann
Glasnevin Hill
Dublin 9
Ireland

Ray.McGrath@met.ie

Prof. Michael McIntyre
Centre for Atmospheric Science
Department of Applied Mathematics &
Theoretical Physics
Silver Street
Cambridge, CB3 9EW
U.K.

M.E.McIntyre@damtp.ac.uk

Dr. Nicholas Kevlahan
McMaster University
Hamilton, ON L8S 4K1
Canada

kevlahan@mcmaster.ca

Dr. Timothy La Row
Centre for Ocean Atmosphere Prediction
Studies, Florida State University
Suite 200, Johnson Bldg., Innovation Pk
Tallahassee, Florida 32306-2840,
U.S.A.

larow@coaps.fsu.edu

Dr. Peter Lynch
Met Eireann
Glasnevin Hill
Dublin 9
Ireland

Peter.Lynch@met.ie

Dr. John McGregor
CSIRO
PBI Aspendale
Victoria 3195
Australia

John.McGregor@dar.csiro.au

Mr. Raymond McNamara
Dept. of Mathematics & Statistics
University of Limerick
Limerick
Ireland

Raymond.McNamara@ul.ie

Mr. Aarne Mannik
Tartu Observatory
Toravere
Tartumaa 61601
Estonia

aarne@aai.ee

Dr. Alberto Maurizi
ISAO-CNR
Via Gobetti 101,
I-40129 Bologna
Italy

a.maurizi@isao.bo.cnr.it

Dr. Vasudeva Murthy
Tata Inst. of Fundamental Research
Indian Inst. of Science
Bangalore, 560 012
India

vasu@math.tifrbng.res.in

Prof. Basil Nicolaenko
Dept. Mathematics,
Arizona State University
Tempe, AZ 85287-1804
U.S.A.

byn@stokes.la.asu.edu

Prof. Penny Desmond
Southern Utah University
Physical Science Department
351 West Center Street
Cedar City, Utah 84720
U.S.A.

penny@suu.edu

Prof. Mankin Mak
University of Illinois at Urbana –
Champaign
105 S. Gregory
Urbana, IL 61801
U.S.A.

mak@atmos.uiuc.edu

Prof. H.K. Moffatt
Isaac Newton Institute for Mathematical
Sciences
20 Clarkson Road
Cambridge CB3 0EH
U.K.

hkm2@newton.cam.ac.uk

Mr. Nawri Nikolaj
Department of Meteorology
University of Maryland
11308 Cedar Lane
Beltsville MD 20705
U.S.A

nnawri@atmos.umd.edu

Dr. James J. O'Brien
Centre for Ocean-Atmospheric
Prediction
Studies, The Florida State University
Suite 200 RM Johnson Building
Tallahassee, FL 32306-2840
U.S.A.

obrien@coaps.fsu.edu

Prof. Joe Pedlosky
Clark 363 MS#21
Woods Hole Oceanographic Institution
Wood Hole
MA, 02543
U.S.A.

jpedlosky@whoi.edu

Mr. Mateusz Reszka
Department of Mathematical Sciences
University of Alberta
Edmonton, Alberta
T6G 2G1, Canada

mreszka@acubens.math.ualberta.ca

Dr. Lucrezia Ricciardulli
NOAA/CIRES/CDC
325 Broadway R/CDC1
Boulder, CO 80302
U.S.A.

lr@cdc.noaa.gov

Prof. Roger Samelson
College of Oceanic and Atmospheric
Sciences
Oregon State University
104 Ocean Admin Buildings
Corvallis, OR 97331-5503
U.S.A.

rsamelson@oce.orst.edu

Dr. Andreas Schiller
CSIRO
GPO Box 1538
Hobart
Tasmania
Australia

audreas.schiller@marine.csiro.au

Prof. Gary Shaffer
Danish Center for Earth System Science
University of Copenhagen
Juliane Maries Vej 30
2100 Copenhagen 0,
Denmark

gs@dcess.ku.du

Prof. G.M. Reznik
P.P. Shirshov Institute of
 Oceanology
Moscow
Russia

greznik@sio.rssi.ru

Prof. Rein Room
Tartu Observatory
Toravere 61602
Tartumaa
Estonia

room@aai.ee

Dr. Igor Sazonv
Dept. Applied Mathematics
University College Cork,
Ireland

i.sazonov@ucc.ie

Prof. Weiming Sha
Geophysical Institute
Graduate School of Sciences
Tohoku University
Aoba-ku, Sendai,
980-8578 Japan

sha@wind.geophys.tohoku.ac.jp

Mr Yuri Skrynnikov
Department of Mathematics and
Statistics
Monash University
Clayton VIC 3168
Australia

Yuri.Skrynnikov@maths.monash.edu.au

Prof. Dmitry Sonechkin
Hydro Meteo Center of Russia
Bolshoy Predtechensky Lane 9/13
Moscow 123242
Russia

dsonech@msku.mecom.ru

Dr. Georgi Sutyrin
Graduate School of Oceanography
University of Rhode Island
Bay Campus
Narragansett,
RI 02882
U.S.A.

gsutyrin@usa.net

Dr. Remi Tailleux
LMD-UPMC Paris 6
Case courrier 99
4 Place Jussieu
75252 Paris Cedex 05
France

tailleux@lmd.jussieu.fr

Prof. Bruce E. Turkington
Department of Mathematics & Statistics
University of Massachsette
Amherst MA01003
U.S.A.

turk@math.umass.edu

Prof. F-Y Wei
Chinese Academy of Meteorological
Sciences, 46 Baishiqiaolu
100081 Beijing
P.R. China

xieqiu@a-1.net.cn

Prof. David Straub
McGill University
805 Sherbrooke O
Montreal, PQ H2W 1S9
Canada

david@gumbo.meteo.mcgill.ca

Prof. Gordon Swaters
Department of Mathematical Sciences
University of Alberta
Edmonton, Alberta
T6G 2G1 Canada

gordon.swaters@ualberta.ca

Dr. John Thuburn
Department of Meteorology
University of Reading
PO Box 243, Earley Gate
Reading, RG6 6BB
U.K.

swsthubn@met.rdg.ac.uk

Dr. Jacques Vanneste
University of Edinburgh
Dept. of Mathematics and Statistics
King's Buildings, Mayfield Road
Edinburgh, EH15 1EY
U.K.

J.Vanneste@ed.ac.uk

Dr. Huijun Yang
University of South Florida
140 Seventh Ave. South
St. Petersburg, FL33701-5016
U.S.A.

yang@marine.usf.edu

Prof. M. Yaremchuk
International Pacific Research Center
University of Hawii
2525 Correa Road
Honolulu 96822
U.S.A.

maxy@soest.hawaii.edu

Prof. Vladimir Zeitlin
LMD, BP 99 University P. et M. Curie
4 Place Jussieu
75252 Paris Cedex 05,
France

zeitlin@lmd.ens.fr

Dr. Shigeo Yoden
Kyoto University
Department of Geophysics
Kyoto, 606-8502
Japan

yoden@kugi.kyoto-u.ac.jp

Mechanics

FLUID MECHANICS AND ITS APPLICATIONS
Series Editor: R. Moreau

Aims and Scope of the Series

The purpose of this series is to focus on subjects in which fluid mechanics plays a fundamental role. As well as the more traditional applications of aeronautics, hydraulics, heat and mass transfer etc., books will be published dealing with topics which are currently in a state of rapid development, such as turbulence, suspensions and multiphase fluids, super and hypersonic flows and numerical modelling techniques. It is a widely held view that it is the interdisciplinary subjects that will receive intense scientific attention, bringing them to the forefront of technological advancement. Fluids have the ability to transport matter and its properties as well as transmit force, therefore fluid mechanics is a subject that is particularly open to cross fertilisation with other sciences and disciplines of engineering. The subject of fluid mechanics will be highly relevant in domains such as chemical, metallurgical, biological and ecological engineering. This series is particularly open to such new multidisciplinary domains.

1. M. Lesieur: *Turbulence in Fluids*. 2nd rev. ed., 1990   ISBN 0-7923-0645-7
2. O. Métais and M. Lesieur (eds.): *Turbulence and Coherent Structures*. 1991
     ISBN 0-7923-0646-5
3. R. Moreau: *Magnetohydrodynamics*. 1990   ISBN 0-7923-0937-5
4. E. Coustols (ed.): *Turbulence Control by Passive Means*. 1990   ISBN 0-7923-1020-9
5. A.A. Borissov (ed.): *Dynamic Structure of Detonation in Gaseous and Dispersed Media*. 1991
     ISBN 0-7923-1340-2
6. K.-S. Choi (ed.): *Recent Developments in Turbulence Management*. 1991 ISBN 0-7923-1477-8
7. E.P. Evans and B. Coulbeck (eds.): *Pipeline Systems*. 1992   ISBN 0-7923-1668-1
8. B. Nau (ed.): *Fluid Sealing*. 1992   ISBN 0-7923-1669-X
9. T.K.S. Murthy (ed.): *Computational Methods in Hypersonic Aerodynamics*. 1992
     ISBN 0-7923-1673-8
10. R. King (ed.): *Fluid Mechanics of Mixing*. Modelling, Operations and Experimental Techniques. 1992   ISBN 0-7923-1720-3
11. Z. Han and X. Yin: *Shock Dynamics*. 1993   ISBN 0-7923-1746-7
12. L. Svarovsky and M.T. Thew (eds.): *Hydroclones*. Analysis and Applications. 1992
     ISBN 0-7923-1876-5
13. A. Lichtarowicz (ed.): *Jet Cutting Technology*. 1992   ISBN 0-7923-1979-6
14. F.T.M. Nieuwstadt (ed.): *Flow Visualization and Image Analysis*. 1993 ISBN 0-7923-1994-X
15. A.J. Saul (ed.): *Floods and Flood Management*. 1992   ISBN 0-7923-2078-6
16. D.E. Ashpis, T.B. Gatski and R. Hirsh (eds.): *Instabilities and Turbulence in Engineering Flows*. 1993   ISBN 0-7923-2161-8
17. R.S. Azad: *The Atmospheric Boundary Layer for Engineers*. 1993   ISBN 0-7923-2187-1
18. F.T.M. Nieuwstadt (ed.): *Advances in Turbulence IV*. 1993   ISBN 0-7923-2282-7
19. K.K. Prasad (ed.): *Further Developments in Turbulence Management*. 1993
     ISBN 0-7923-2291-6
20. Y.A. Tatarchenko: *Shaped Crystal Growth*. 1993   ISBN 0-7923-2419-6
21. J.P. Bonnet and M.N. Glauser (eds.): *Eddy Structure Identification in Free Turbulent Shear Flows*. 1993   ISBN 0-7923-2449-8
22. R.S. Srivastava: *Interaction of Shock Waves*. 1994   ISBN 0-7923-2920-1
23. J.R. Blake, J.M. Boulton-Stone and N.H. Thomas (eds.): *Bubble Dynamics and Interface Phenomena*. 1994   ISBN 0-7923-3008-0

Mechanics

FLUID MECHANICS AND ITS APPLICATIONS
Series Editor: R. Moreau

24. R. Benzi (ed.): *Advances in Turbulence V.* 1995   ISBN 0-7923-3032-3
25. B.I. Rabinovich, V.G. Lebedev and A.I. Mytarev: *Vortex Processes and Solid Body Dynamics.* The Dynamic Problems of Spacecrafts and Magnetic Levitation Systems. 1994
 ISBN 0-7923-3092-7
26. P.R. Voke, L. Kleiser and J.-P. Chollet (eds.): *Direct and Large-Eddy Simulation I.* Selected papers from the First ERCOFTAC Workshop on Direct and Large-Eddy Simulation. 1994
 ISBN 0-7923-3106-0
27. J.A. Sparenberg: *Hydrodynamic Propulsion and its Optimization.* Analytic Theory. 1995
 ISBN 0-7923-3201-6
28. J.F. Dijksman and G.D.C. Kuiken (eds.): *IUTAM Symposium on Numerical Simulation of Non-Isothermal Flow of Viscoelastic Liquids.* Proceedings of an IUTAM Symposium held in Kerkrade, The Netherlands. 1995   ISBN 0-7923-3262-8
29. B.M. Boubnov and G.S. Golitsyn: *Convection in Rotating Fluids.* 1995 ISBN 0-7923-3371-3
30. S.I. Green (ed.): *Fluid Vortices.* 1995   ISBN 0-7923-3376-4
31. S. Morioka and L. van Wijngaarden (eds.): *IUTAM Symposium on Waves in Liquid/Gas and Liquid/Vapour Two-Phase Systems.* 1995   ISBN 0-7923-3424-8
32. A. Gyr and H.-W. Bewersdorff: *Drag Reduction of Turbulent Flows by Additives.* 1995
 ISBN 0-7923-3485-X
33. Y.P. Golovachov: *Numerical Simulation of Viscous Shock Layer Flows.* 1995
 ISBN 0-7923-3626-7
34. J. Grue, B. Gjevik and J.E. Weber (eds.): *Waves and Nonlinear Processes in Hydrodynamics.* 1996   ISBN 0-7923-4031-0
35. P.W. Duck and P. Hall (eds.): *IUTAM Symposium on Nonlinear Instability and Transition in Three-Dimensional Boundary Layers.* 1996   ISBN 0-7923-4079-5
36. S. Gavrilakis, L. Machiels and P.A. Monkewitz (eds.): *Advances in Turbulence VI.* Proceedings of the 6th European Turbulence Conference. 1996   ISBN 0-7923-4132-5
37. K. Gersten (ed.): *IUTAM Symposium on Asymptotic Methods for Turbulent Shear Flows at High Reynolds Numbers.* Proceedings of the IUTAM Symposium held in Bochum, Germany. 1996   ISBN 0-7923-4138-4
38. J. Verhás: *Thermodynamics and Rheology.* 1997   ISBN 0-7923-4251-8
39. M. Champion and B. Deshaies (eds.): *IUTAM Symposium on Combustion in Supersonic Flows.* Proceedings of the IUTAM Symposium held in Poitiers, France. 1997   ISBN 0-7923-4313-1
40. M. Lesieur: *Turbulence in Fluids.* Third Revised and Enlarged Edition. 1997
 ISBN 0-7923-4415-4; Pb: 0-7923-4416-2
41. L. Fulachier, J.L. Lumley and F. Anselmet (eds.): *IUTAM Symposium on Variable Density Low-Speed Turbulent Flows.* Proceedings of the IUTAM Symposium held in Marseille, France. 1997
 ISBN 0-7923-4602-5
42. B.K. Shivamoggi: *Nonlinear Dynamics and Chaotic Phenomena.* An Introduction. 1997
 ISBN 0-7923-4772-2
43. H. Ramkissoon, *IUTAM Symposium on Lubricated Transport of Viscous Materials.* Proceedings of the IUTAM Symposium held in Tobago, West Indies. 1998   ISBN 0-7923-4897-4
44. E. Krause and K. Gersten, *IUTAM Symposium on Dynamics of Slender Vortices.* Proceedings of the IUTAM Symposium held in Aachen, Germany. 1998   ISBN 0-7923-5041-3
45. A. Biesheuvel and G.J.F. van Heyst (eds.): *In Fascination of Fluid Dynamics.* A Symposium in honour of Leen van Wijngaarden. 1998   ISBN 0-7923-5078-2

Mechanics

FLUID MECHANICS AND ITS APPLICATIONS
Series Editor: R. Moreau

46. U. Frisch (ed.): *Advances in Turbulence VII.* Proceedings of the Seventh European Turbulence Conference, held in Saint-Jean Cap Ferrat, 30 June–3 July 1998. 1998   ISBN 0-7923-5115-0
47. E.F. Toro and J.F. Clarke: *Numerical Methods for Wave Propagation.* Selected Contributions from the Workshop held in Manchester, UK. 1998   ISBN 0-7923-5125-8
48. A. Yoshizawa: *Hydrodynamic and Magnetohydrodynamic Turbulent Flows.* Modelling and Statistical Theory. 1998   ISBN 0-7923-5225-4
49. T.L. Geers (ed.): *IUTAM Symposium on Computational Methods for Unbounded Domains.* 1998   ISBN 0-7923-5266-1
50. Z. Zapryanov and S. Tabakova: *Dynamics of Bubbles, Drops and Rigid Particles.* 1999   ISBN 0-7923-5347-1
51. A. Alemany, Ph. Marty and J.P. Thibault (eds.): *Transfer Phenomena in Magnetohydrodynamic and Electroconducting Flows.* 1999   ISBN 0-7923-5532-6
52. J.N. Sørensen, E.J. Hopfinger and N. Aubry (eds.): *IUTAM Symposium on Simulation and Identification of Organized Structures in Flows.* 1999   ISBN 0-7923-5603-9
53. G.E.A. Meier and P.R. Viswanath (eds.): *IUTAM Symposium on Mechanics of Passive and Active Flow Control.* 1999   ISBN 0-7923-5928-3
54. D. Knight and L. Sakell (eds.): *Recent Advances in DNS and LES.* 1999 ISBN 0-7923-6004-4
55. P. Orlandi: *Fluid Flow Phenomena.* A Numerical Toolkit. 2000   ISBN 0-7923-6095-8
56. M. Stanislas, J. Kompenhans and J. Westerveel (eds.): *Particle Image Velocimetry.* Progress towards Industrial Application. 2000   ISBN 0-7923-6160-1
57. H.-C. Chang (ed.): *IUTAM Symposium on Nonlinear Waves in Multi-Phase Flow.* 2000   ISBN 0-7923-6454-6
58. R.M. Kerr and Y. Kimura (eds.): *IUTAM Symposium on Developments in Geophysical Turbulence* held at the National Center for Atmospheric Research, (Boulder, CO, June 16–19, 1998) 2000   ISBN 0-7923-6673-5
59. T. Kambe, T. Nakano and T. Miyauchi (eds.): *IUTAM Symposium on Geometry and Statistics of Turbulence* held at the Shonan International Village Center, Hayama (Kanagawa-ken, Japan November 2–5, 1999). 2001   ISBN 0-7923-6711-1
60. V.V. Aristov: *Direct Methods for Solving the Boltzmann Equation and Study of Nonequilibrium Flows.* 2001   ISBN 0-7923-6831-2

Kluwer Academic Publishers – Dordrecht / Boston / London

Mechanics

SOLID MECHANICS AND ITS APPLICATIONS
Series Editor: G.M.L. Gladwell

69. P. Pedersen and M.P. Bendsøe (eds.): *IUTAM Symposium on Synthesis in Bio Solid Mechanics.* Proceedings of the IUTAM Symposium held in Copenhagen, Denmark. 1999
ISBN 0-7923-5615-2
70. S.K. Agrawal and B.C. Fabien: *Optimization of Dynamic Systems.* 1999
ISBN 0-7923-5681-0
71. A. Carpinteri: *Nonlinear Crack Models for Nonmetallic Materials.* 1999
ISBN 0-7923-5750-7
72. F. Pfeifer (ed.): *IUTAM Symposium on Unilateral Multibody Contacts.* Proceedings of the IUTAM Symposium held in Munich, Germany. 1999   ISBN 0-7923-6030-3
73. E. Lavendelis and M. Zakrzhevsky (eds.): *IUTAM/IFToMM Symposium on Synthesis of Non-linear Dynamical Systems.* Proceedings of the IUTAM/IFToMM Symposium held in Riga, Latvia. 2000
ISBN 0-7923-6106-7
74. J.-P. Merlet: *Parallel Robots.* 2000   ISBN 0-7923-6308-6
75. J.T. Pindera: *Techniques of Tomographic Isodyne Stress Analysis.* 2000   ISBN 0-7923-6388-4
76. G.A. Maugin, R. Drouot and F. Sidoroff (eds.): *Continuum Thermomechanics.* The Art and Science of Modelling Material Behaviour. 2000
ISBN 0-7923-6407-4
77. N. Van Dao and E.J. Kreuzer (eds.): *IUTAM Symposium on Recent Developments in Non-linear Oscillations of Mechanical Systems.* 2000
ISBN 0-7923-6470-8
78. S.D. Akbarov and A.N. Guz: *Mechanics of Curved Composites.* 2000   ISBN 0-7923-6477-5
79. M.B. Rubin: *Cosserat Theories: Shells, Rods and Points.* 2000   ISBN 0-7923-6489-9
80. S. Pellegrino and S.D. Guest (eds.): *IUTAM-IASS Symposium on Deployable Structures: Theory and Applications.* Proceedings of the IUTAM-IASS Symposium held in Cambridge, U.K., 6–9 September 1998. 2000
ISBN 0-7923-6516-X
81. A.D. Rosato and D.L. Blackmore (eds.): *IUTAM Symposium on Segregation in Granular Flows.* Proceedings of the IUTAM Symposium held in Cape May, NJ, U.S.A., June 5–10, 1999. 2000
ISBN 0-7923-6547-X
82. A. Lagarde (ed.): *IUTAM Symposium on Advanced Optical Methods and Applications in Solid Mechanics.* Proceedings of the IUTAM Symposium held in Futuroscope, Poitiers, France, August 31–September 4, 1998. 2000
ISBN 0-7923-6604-2
83. D. Weichert and G. Maier (eds.): *Inelastic Analysis of Structures under Variable Loads.* Theory and Engineering Applications. 2000
ISBN 0-7923-6645-X
84. T.-J. Chuang and J.W. Rudnicki (eds.): *Multiscale Deformation and Fracture in Materials and Structures.* The James R. Rice 60th Anniversary Volume. 2001   ISBN 0-7923-6718-9
85. S. Narayanan and R.N. Iyengar (eds.): *IUTAM Symposium on Nonlinearity and Stochastic Structural Dynamics.* Proceedings of the IUTAM Symposium held in Madras, Chennai, India, 4–8 January 1999
ISBN 0-7923-6733-2
86. S. Murakami and N. Ohno (eds.): *IUTAM Symposium on Creep in Structures.* Proceedings of the IUTAM Symposium held in Nagoya, Japan, 3-7 April 2000. 2001   ISBN 0-7923-6737-5
87. W. Ehlers (ed.): *IUTAM Symposium on Theoretical and Numerical Methods in Continuum Mechanics of Porous Materials.* Proceedings of the IUTAM Symposium held at the University of Stuttgart, Germany, September 5-10, 1999. 2001
ISBN 0-7923-6766-9

Kluwer Academic Publishers – Dordrecht / Boston / London